Alternative Splicing in the Postgenomic Era

ADVANCES IN EXPERIMENTAL MEDICINE AND BIOLOGY

Recent Volumes in this Series

Volume 615
PROGRAMMED CELL DEATH IN CANCER PROGRESSION AND THERAPY
Edited by Roya Khosravi-Far and Eileen White

Volume 616
TRANSGENIC MICROALGAE AS GREEN CELL FACTORIES
Edited by Rosa León, Aurora Gaván, and Emilio Fernández

Volume 617
HORMONAL CARCINOGENESIS V
Edited by Jonathan J. Li

Volume 618
HYPOXIA AND THE CIRCULATION
Edited by Robert H. Roach, Peter Hackett, and Peter D. Wagner

Volume 619
INTERAGENCY, INTERNATIONAL SYMPOSIUM ON CYANOBACTERIAL HARMFUL ALGAL BLOOMS (ISOC-HAB)
Edited by H. Kenneth Hudnell

Volume 620
BIO-APPLICATIONS OF NANOPARTICLES
Edited by Warren C.W. Chan

Volume 621
AXON GROWTH AND GUIDANCE
Edited by Dominique Bagnard

Volume 622
OVARIAN CANCER
Edited by George Coukos, Andrew Berchuck, and Robert Ozols

Volume 623
ALTERNATIVE SPLICING IN THE POSTGENOMIC ERA
Edited by Benjamin J. Blencowe and Brenton R. Graveley

A Continuation Order Plan is available for this series. A continuation order will bring delivery of each new volume immediately upon publication. Volumes are billed only upon actual shipment. For further information please contact the publisher.

Alternative Splicing in the Postgenomic Era

Edited by

Benjamin J. Blencowe, Ph.D.

Banting and Best Department of Medical Research and Department of Molecular and Medical Genetics, Terrence Donnelly Center for Cellular and Biomedical Research, Univerity of Toronto, Toronto, Canada

Brenton R. Graveley, Ph.D.

Department of Genetics and Developmental Biology, University of Connecticut Health Center, Farmington, Connecticut, USA

Springer Science+Business Media, LLC

Landes Bioscience

Springer Science+Business Media, LLC
Landes Bioscience

Printed in the U.S.A.

Springer Science+Business Media, LLC, 233 Spring Street, New York, New York 10013, U.S.A.
http://www.springer.com

Please address all inquiries to the publishers:
Landes Bioscience, 1002 West Avenue, 2nd Floor, Austin, Texas 78701, U.S.A.
Phone: 512/ 637 5060; FAX: 512/ 637 6079
http://www.landesbioscience.com

Alternative Splicing in the Postgenomic Era, edited by Benjamin J. Blencowe and Brenton R. Graveley, Landes Bioscience / Springer Science+Business Media, LLC dual imprint / Springer series: Advances in Experimental Medicine and Biology

ISBN 978-1-4419-2651-7 e-ISBN 978-0-387-77374-2

Library of Congress Cataloging-in-Publication Data

A CIP catalog record of this book is available from the Library of Congress.

FOREWORD

Francis Crick formulated part of the central dogma as "information never flows from protein to nucleic acids." In fact, I am not aware of an exception to this proposal. However, the process of transferring information from DNA to protein became more biologically interesting with the discovery of alternative splicing. In this case, the gene sequence can be processed at the RNA stage to generate multiple proteins with different functions. An example is the alternative splicing of exons from the gene encoding CD44, a cell surface protein that influences cell migration and homing to specific sites. In resting cells, the CD44 gene primarily expresses a single splice isoform that reinforces the stationary state. In cells activated by growth factors, the gene is also expressed in a variety of other isoforms which promote cell migration and proliferation. Hence the interpretation of the DNA sequence of an organism frequently depends upon the environment and/or the stages of development of cells within an organism. In other words, the nature of information transferred from the gene sequence depends upon the cellular state, i.e., a gene is only defined in the context of the state of the cell.

The human genome sequence has been interpreted to encode some 20,000 genes, which is very similar to the number of genes encoded in the genome of a minute worm, *C. elegans*. However, because of the ambiguity in the definition of a gene caused by alternative splicing, this comparison ignores the differences in the genetic complexity of the two organisms. Over half of all genes in humans are alternatively spliced while this phenomenon is quite rare in *C. elegans*. Furthermore, many human genes produce tens and hundreds of different protein isoforms, more than current technology can easily monitor or catalogue. Thus, at the protein level, the complexity of human cells is much larger than that of the worm, allowing the obvious increase in biological functions.

This line of reasoning suggests that the increase in genetic complexity in humans is encoded in the factors which control processes such as alternative splicing and not in the number of genes with varying functions. In a fashion, this situation seems reasonable from an evolutionary perspective as changes in regulatory factors controlling alternative splicing could generate wide ranging variation in functional possibilities, thus providing a substrate for incremental evolution. Alternatively, it can be argued that this configuration would restrict evolutionary changes, as alterations in a regulatory factor controlling alternative splicing might disrupt the critical activities of many genes.

Humans, in an almost religious need, have sought to understand the biological basis of their higher order in the world. Myths have elevated humans to direct decedents of gods, and, on a physical level, scientists have long searched for the unique biological features which account for our singular abilities. Our most singular and unique feature is our brain, which has evolved very rapidly over the course of several million years. This evolution might not solely be the result of incremental selection acting on mutations in many different genes. Although not entirely clear, much of this complexity could be the consequence of mutations affecting the alternative splicing of many genes expressed in the brain. Several laboratories are currently investigating the factors controlling alternative splicing in neural cells. Recent progress in this area described in this book provides a basis for future work on this daunting problem.

The 30th anniversary of the discovery of RNA splicing will be celebrated in 2007. The past three decades have produced an exciting picture of the composition and functions of spliceosomal complexes. However, much remains to be determined about the specific roles of individual splicing factors, and the integration of information from atomic structures with functional studies is at an early stage. The control of alternative splicing culminates in the formation of spliceosome complexes on a specific set of splice sites, and thus understanding the nature of steps in this process is critical. It is anticipated that proteins that bind RNA in a sequence-specific fashion will be the major components in regulation of alternative splicing. A subset of these proteins share domains with repeats of alternating serine and arginine residues, and the state of phosphorylation of these domains is thought to be critical for their activities in splicing. Integrating the steps in formation of the spliceosomal complexes with the activities of sequence-specific RNA binding factors and their state of phosphorylation is necessary in order to understand the regulation of alternative splicing. Our current knowledge of the function of RNA binding proteins with roles in alternative splicing is presented in this book.

The integration of major steps in the synthesis of an mRNA is a new and profound theme in the study of gene expression. Of particular relevance to this book, RNA splicing has been shown to be coupled to the processes of transcription, cleavage and polyadenylation, transport to the cytoplasm and the efficiency of initiation of translation. This raises the possibility that factors influencing the nature of initiation and elongation of transcription by RNA polymerase II could influence the nature of alternative splicing. In turn, factors deposited on nuclear RNA through the action of the spliceosome signal the rapid decay of some mRNAs containing stop codons upstream of introns and can stimulate the rate of translation of mRNAs as they emerge into the cytoplasm. Cell biologists have long suspected that the spatial organization of the nucleus is related to the nature of genes expressed in different regions. It is possible that events controlling alternative splicing will someday be integrated into this holistic view of the cells.

The future of the study of alternative splicing lies in the integration of results from "systems-level" approaches with the current picture provided by biochemical studies. One objective of this integration is to elucidate the "splicing code" and "alternative splicing code." The former can be loosely defined as a set of parameters embodied in a computer program which accurately predicts the exon/intron structures of genes given the DNA sequence. Parameters such as splice site sequences and exon/intron enhancer and silencer sequences have already been proposed as the basis of the "splicing code." These parameters will also be important in the "alternative splicing code," together with additional variables and sequences that are important for the regulation of alternative splicing. These bioinformatic developments will be greatly aided by systems-level studies of alternative splicing. Using high throughput methods to measure the patterns of alternative splicing for many genes at a cellular level across a range of conditions or states should provide the necessary number of data points for the computational modeling of the regulation of RNA splicing governed by specific sequences and ultimately by changes in the activities of specific factors. A technical breakthrough that will permit further development of the two "codes," the availability of rapid and inexpensive DNA sequencing, is about to emerge. The ability to determine the sequences of 10 million DNA segments at modest cost in time and money will rapidly advance our understanding of alternative splicing. The resulting advances will probably stimulate the next book on this subject. Until then, enjoy the chapters in this exciting current volume.

Phillip A. Sharp
Massachusetts Iinstitute of Technology
Cambridge, Massachusetts, USA

PREFACE

Alternative splicing created a major stir when it was first discovered in transcripts from the immunoglobulin locus by the laboratories of David Baltimore and Leroy Hood in the early 1980s. The differential selection of splice sites to produce transcripts encoding the membrane bound and soluble forms of immunoglobulin revealed an elegant mechanism by which the structure and function of proteins can be diversified from a limited repertoire of genes. The increased employment of DNA sequencing methods yielded additional examples of alternative splicing in the ensuing years. In parallel, several laboratories began to explore the mechanisms by which the core splicing machinery can be regulated. In these formative years of the field, the extent of alternative splicing in cellular transcripts was entirely unknown.

Fast forward to the present. In the wake of major international efforts culminating in the complete sequences of many genomes, as well as the sequencing of large populations of transcripts for several species, it is now apparent that in some metazoan organisms, including ourselves, alternative splicing is the rule, not the exception. In particular, the finding that the number of protein coding genes in species from yeast to humans cannot account for the complexity of an organism brought alternative splicing into the limelight as a major mechanism underlying diversification of the genetic repertoire.

The combination of sequencing efforts and microarray profiling experiments have revealed that at least two thirds of human genes contain one or more alternative exons, yet the full extent of alternative splicing that occurs in the human genome, or any other genome, is not known. Databases have been assembled to catalog the multitude of splice isoforms that have already been detected in sequenced transcripts. Despite the prevalence of alternative splicing, the stark reality is that we understand the functional significance of only a minute fraction of the known splice variants. Moreover, although the mechanisms responsible for the regulation of alternative splicing have been studied in some depth for several genes, only a small number of factors that function to regulate alternative splicing in a cell type- or developmental stage-specific manner are known. Moreover, it is still unclear how most of the known factors regulate alternative splicing. Superimposed on these issues pertaining to the normal functions and regulation of alternative splicing is the knowledge that alterations in the splicing process are common in human diseases. This fact has prompted increased efforts towards understanding the relationship between mis-regulated alternative splicing and disease.

Given the considerable challenges ahead, we felt that it would be timely and appropriate to assemble the first book focused on the topic of alternative splicing. This book is intended to not only provide a broad perspective of recent progress in the field, but also to draw attention to some of the major questions that are currently under investigation, and also what we can anticipate learning from continued developments in the next several years of research. We are very pleased that many internationally-renowned researchers in the splicing community agreed to contribute to the book. The result is a compilation of complementary chapters that review a broad spectrum of important topics. The content of these chapters is summarized below.

An important goal in splicing research is to understand the role of nuclear organization in the control of alternative splicing. The nucleus is divided into several subnuclear compartments and splicing occurs in only a subset of these locations. Moreover, nuclear organization can change in response to the activity of factors that are required to transcribe and splice a particular pre-mRNAs. These and other topics relevant to the cell biology of alternative splicing are discussed in Chapter 1 by Carmo-Fonseca and Carvalho.

In Chapter 2, Matlin and Moore take us from the cellular perspective into the realm of molecular approaches used to characterize the spliceosome, the macromolecular machine that catalyzes intron removal. The spliceosome is a remarkably complex machine consisting of in excess of 200 proteins and five small nuclear RNA components. Chapter 2 initially provides the major "parts list" of the spliceosome and therefore serves as a useful reference for the chapters that follow. However, Matlin and Moore also describe in detail the remarkably dynamic transitions involved in the formation of active spliceosomes and also their subsequent disassembly. Knowledge of spliceosome assembly intermediates is of course critical for understanding alternative splicing, since many of the steps in the spliceosome assembly pathway are targets for regulation.

The availability of the genome sequences of multiple organisms has facilitated tremendous growth in the use of bioinformatics and genome-wide techniques to study alternative splicing. In Chapter 3, Xing and Lee discuss several bioinformatic and microarray approaches that have been used to elucidate the extent of alternative splicing, and the role of alternative splicing in modulating protein function, proteome complexity, and in shaping genome evolution. With regard to the later topic, comparative genomics has facilitated the identification of several previously unrecognized sequence elements that are under positive selection pressure and participate in alternative splicing.

Many of the genes that encode alternatively spliced pre-mRNAs produce a relatively small number of splice variants. In Chapter 4, Park and Graveley describe a few genes that encode pre-mRNAs that are alternatively spliced to an extraordinary extent. The most striking example is the *Drosphila Dscam* gene, which can generate over 38,000 alternatively spliced mRNA isoforms. The authors review the current state of knowledge regarding the mechanisms involved in the regulation of *Dscam* alternative splicing and also the complex alternative splicing of other genes.

In Chapter 5, Calarco et al review recent technologies, including microarray and sequenced-based approaches, that have enabled alternative splicing to be studied on a global scale. Their chapter also reviews some of the important insights that have been gained from these global approaches. Also provided in this chapter is an assessment of emerging sequencing and other non-microarray-based technologies and how these have the potential to further transform our knowledge of global regulatory features of alternative splicing.

In Chapter 6, Chasin critically reviews the various bioinformatic and experimental approaches that have been used to identify sequence motifs involved in the regulation of alternative splicing. The issues of how to assess and interpret the "embarrassment of riches", as pertaining to the extensive coverage of exon space by the recently discovered motifs, are also discussed in this chapter.

Proteins that bind to positive and negative-acting cis- regulatory motifs introduced in Chapter 6 are reviewed in Chapters 7 and 8. Among the most heavily studied proteins that function to control alternative splicing are members of the SR family and SR-related proteins. These proteins all contain one or more RS domains rich in alternating arginine and serine residues. Lin and Fu review this intriguing family of proteins in Chapter 7, focusing on the current state of knowledge regarding the mechanisms by which they act, as well as their biological roles based on recent studies using knockout mice.

Another major class of splicing regulators are the heterogeneous nuclear ribonucleoproteins (hnRNPs). Proteins belonging to this ubiquitous family were originally identified based on their association with heteronuclear RNA and were initially thought to function passively as RNA packaging proteins. However, many studies have shown clearly that hnRNPs have interesting, specific, and diverse functions in RNA metabolism, including alternative splicing. In Chapter 8, Martinez-Contreras et al review current knowledge of these proteins and how they are known to function in alternative splicing.

Alternative splicing is a tremendous mechanism for increasing protein diversity and is used extensively in the nervous and immune systems. In Chapter 9, Ule and Darnell discuss the neural-specific mammalian splicing regulators Nova-1 and Nova-2. These two proteins, which have been extensively studied in the Darnell lab, play a critical role in controlling the alternative splicing of a large repertoire of transcripts from genes that comprise a regulatory network that shapes neural activity. Ule and Darnell further review their recent work uncovering the rules by which clusters of binding sites for Nova proteins in exons and introns can dictate whether a target exon is skipped or included.

Over the past several years, it has become clear that alternative splicing does not occur separately from other steps in gene expression. Rather, it is closely coupled with several other cellular processes that can influence each other. Several chapters are devoted to this topic. First, in Chapter 10, Lynch describes how signal transduction pathways triggered at the surfaces of cells modulate the alternative splicing of specific pre-mRNAs. In Chapter 11, Kornblihtt describes several ways in which transcription can influence alternative splicing. This includes the composition of the promoter driving transcription and the rate of transcription elongation. Finally, in Chapter 12, Lareau et al describe how alternative splicing is coupled to nonsense-mediated decay (NMD). There are numerous examples of pre-mRNAs that are alternatively spliced such that one isoform is subject to degradation by the NMD pathway, while other isoforms are not, and regulated alternative splicing coupled to NMD is known to play an important role in the autoregulation of specific RNA binding proteins. While there are additional cellular processes that are coupled to alternative splicing, these last three chapters provide an excellent overview of how specific steps of gene regulation can be coordinated to achieve an important biological outcome.

The book is rounded off by a chapter from Orengo and Cooper that describes the role of alternative splicing in human disease. These authors highlight tauopathies, myotonic dystrophy, cancer, and Prader-Willi Syndrome as examples of diseases that are caused by disruptions in alternative splicing. Such splicing changes can arise by either cis-acting mutations or, more interestingly, by changes in the activities of trans-acting factors that modulate the splicing of a variety of pre-mRNAs. Future research will no doubt provide numerous additional examples in which human diseases can be caused by mutations that affect the alternative splicing of one or more pre-mRNAs.

In summary, we hope that this first book dedicated to alternative splicing will serve as a valuable resource for both experts and non-experts alike. Clearly, research in the burgeoning field of alternative splicing in the postgenomic era is poised to uncover new and important aspects of gene regulation, as well as functions of the proteome. We envision that we are on the verge of entering an era of exon-resolution biology, or "exonomics", in which every major biological study involving multi-exon genes will take into account the functions and properties of different splice variants produced from a gene. However, much work lies ahead in order to realize such a goal. For example, tools and approaches are desperately needed that will afford an understanding of the functions and regulation of the multitude of splice variants that are being identified by high-throughput methods at an ever increasing rate. Such an advance will be necessary before programs of gene regulation underlying fundamental biological processes associated with higher eukaryotes can be fully understood. In the meantime, we hope that the chapters in this volume will help stimulate further interest and appreciation of the major questions concerning the field of alternative splicing and how we might tackle these important questions in the years ahead.

We would like to thank Sidrah Ahmad for assistance with editing, as well as Cynthia Conomos and Celeste Carlton at Landes Bioscience for all of their time and effort in the production of this book.

Benjamin J. Blencowe
Brenton R. Graveley

ABOUT THE EDITORS...

BENJAMIN J. BLENCOWE is a Professor in the Banting and Best Department of Medical Research and Department of Molecular and Medical Genetics at the University of Toronto. He is also a member of the University of Toronto's Center for Cellular and Biomolecular Research, a recently constructed multidisciplinary research institute. Dr. Blencowe's research focuses on mechanisms underlying the regulation of pre-mRNA processing in mammalian cells. In recent years, he and his colleagues have developed and employed tools for studying alternative splicing regulation on a global scale. Dr. Blencowe received a BSc (hons) in Microbiology from Imperial College of Science and Technology, University of London in 1988. He received his PhD in Biochemistry in 1991 from the University of London, after working at the laboratory of Prof. Angus I. Lamond at the European Molecular Laboratory in Heidelberg. He was a Postdoctoral Fellow (1992-1996) and Research Fellow (1996-1998) in the laboratory of Prof. Phillip A. Sharp (Nobel Laureate) in the Center for Cancer Research, Massachusetts Institute of Technology, Cambridge, prior to moving to the University of Toronto in 1998. Dr. Blencowe is the recipient of several national and international research awards and is one of the most highly cited researchers is his field.

ABOUT THE EDITORS...

BRENTON R. GRAVELEY is currently an Associate Professor in the Department of Genetics and Developmental Biology at the University of Connecticut Health Center. His laboratory works on alternative splicing in *Drosophila* and human embryonic stem cells as well as the roles of small RNAs in controlling regeneration in planarians. In addition to his research activities, Dr. Graveley serves as the Director of the University Microarray Facility, is a member of the editorial board of *RNA*, and serves on study sections for the NIH. Dr. Graveley received his B.A. at the University of Colorado, Boulder in Molecular, Cellular, and Developmental Biology where he worked in the laboratory of David M. Prescott and studied genome rearrangements in the ciliate *Oxytricha nova*. His graduate studies were performed at the University of Vermont, Burlington in the laboratory of Gregory M. Gilmartin, where he studied 3' end formation in HIV-1. Dr. Graveley began studying alternative splicing as a Jane Coffin Childs postdoctoral fellow in the laboratory of Tom Maniatis in the Department of Molecular and Cellular Biology at Harvard University in Cambridge, Massachusetts, USA.

PARTICIPANTS

Benjamin J. Blencowe
Banting and Best Department
of Medical Research
and Department of Molecular
and Medical Genetics
Terrence Donnelly Center
for Cellular and Biomolecular
Research
University of Toronto
Toronto, Ontario
Canada

Steven E. Brenner
Department of Molecular
and Cell Biology
Biophysics Graduate Group
Department of Plant
and Microbial Biology
University of California, Berkeley
Berkeley, California
USA

Angela N. Brooks
Department of Molecular
and Cell Biology
University of California, Berkeley
Berkeley, California
USA

John A. Calarco
Banting and Best Department
of Medical Research
and Department of Molecular
and Medical Genetics
Terrence Donnelly Center
for Cellular and Biomolecular
Research
University of Toronto
Toronto, Ontario
Canada

Maria Carmo-Fonseca
Instituto de Medicina Molecular
Universidade de Lisboa
Lisboa
Portugal

Célia Carvalho
Instituto de Medicina Molecular
Universidade de Lisboa
Lisboa
Portugal

Benoit Chabot
Département de Microbiologie
et d'Infectiologie
Université de Sherbrooke
Sherbrooke, Québec
Canada

Lawrence A. Chasin
Department of Biological Sciences
Columbia University
New York, New York
USA

Philippe Cloutier
Département de Microbiologie
et d'Infectiologie
Université de Sherbrooke
Sherbrooke, Québec
Canada

Thomas A. Cooper
Departments of Pathology
and Molecular and Cellular Biology
Baylor College of Medicine
Houston, Texas
USA

Robert B. Darnell
Laboratory of Molecular
Neuro-Oncology
The Rockefeller University
New York, New York
USA

Jean-François Fisette
Département de Microbiologie
et d'Infectiologie
Université de Sherbrooke
Sherbrooke, Québec
Canada

Xiang-Dong Fu
Department of Cellular
and Molecular Medicine
University of California, San Diego
La Jolla, California
USA

Brenton R. Graveley
Department of Genetics
and Developmental Biology
University of Connecticut
Health Center
Farmington, Connecticut
USA

Joanna Y. Ip
Banting and Best Department
of Medical Research
and Department of Molecular
and Medical Genetics
Terrence Donnelly Center for Cellular
and Biomolecular Research
University of Toronto
Toronto, Ontario
Canada

Alberto R. Kornblihtt
Laboratorio de Fisiología
y Biología Molecular
Departamento de Fisiología,
Biología Molecular y Celular
IFIBYNE-CONICET
Universidad de Buenos Aires
Ciudad Universitaria
Buenos Aires
Argentina

Liana F. Lareau
Department of Molecular
and Cell Biology
University of California, Berkeley
Berkeley, California
USA

Christopher Lee
Molecular Biology Institute
Department of Chemistry
and Biochemistry
Institute for Genomics and Proteomics
Center for Computational Biology
University of California, Los Angeles
Los Angeles, California
USA

Shengrong Lin
Department of Cellular
and Molecular Medicine
University of California, San Diego
La Jolla, California
USA

Kristen W. Lynch
Department of Biochemistry
University of Texas
Southwestern Medical Center
Dallas, Texas
USA

Arianne J. Matlin
Howard Hughes Medical Institute
Department of Biochemistry
Brandeis University
Waltham, Massachusetts
USA

Rebeca Martinez-Contreras
Département de Microbiologie
et d'Infectiologie
Université de Sherbrooke
Sherbrooke, Québec
Canada

Qi Meng
Department of Plant
and Microbial Biology
University of California, Berkeley
Berkeley, California
USA

Melissa J. Moore
Howard Hughes Medical Institute
Department of Biochemistry
Brandeis University
Waltham, Massachusetts
USA

James P. Orengo
Departments of Pathology
and Molecular and Cellular Biology
Baylor College of Medicine
Houston, Texas
USA

Jung Woo Park
Department of Genetics
and Developmental Biology
University of Connecticut
Health Center
Farmington, Connecticut
USA

Timothée Revil
Département de Microbiologie
et d'Infectiologie
Université de Sherbrooke
Sherbrooke, Québec
Canada

Arneet L. Saltzman
Banting and Best Department
of Medical Research
and Department of Molecular
and Medical Genetics
Terrence Donnelly Center for Cellular
and Biomolecular Research
University of Toronto
Toronto, Ontario
Canada

Phillip A. Sharp
Center for Cancer Research
and Department of Biology
Massachusetts Institute of Technology
Cambridge, Massachusetts
USA

Lulzim Shkreta
Département de Microbiologie
et d'Infectiologie
Université de Sherbrooke
Sherbrooke, Québec
Canada

David A. W. Soergel
Biophysics Graduate Group
University of California, Berkeley
Berkeley, California
USA

Jernej Ule
MRC Laboratory of Molecular
Biology
Cambridge
England

Yi Xing
Department of Internal Medicine
Roy J. and Lucille A. Carver
College of Medicine
University of Iowa
Iowa City, Iowa
USA

CONTENTS

CHAPTER 1

Nuclear Organization and Splicing Control

Maria Carmo-Fonseca* and Célia Carvalho

Abstract

Although major splicing regulatory mechanisms rely on the presence of cis-acting sequence elements in the precursor messenger RNA (pre-mRNA) to which specific protein and RNA factors bind, splice choices are also influenced by transcription kinetics, promoter-dependent loading of RNA-binding proteins and nucleo-cytoplasmic distribution of splicing regulators. Within the highly crowded eukaryotic nucleus, molecular machines required for gene expression create specialized microenvironments that favor some interactions while repressing others. Genes located far apart in a chromosome or even in different chromosomes come together in the nucleus for coordinated transcription and splicing. Emerging tools to dissect gene expression pathways in living cells promise to provide more detailed insight as to how spatial confinement contributes to splicing control.

Splicing and the Nucleus

Much like the nucleus, splicing of precursor messenger RNA (pre-mRNA) by the spliceosome is a defining feature of eukaryotes. The origins of the nucleus and spliceosomal introns are interwoven, possibly because the spatial confinement provided by the nuclear envelope enabled the splicing mechanism to evolve.

Cells are compartments in which chemical reactions take place. Without mechanisms to restrain the building blocks of life from free diffusion in the primitive ocean, the biomolecules would never have reached sufficient concentrations to react with one another. Molecular caging provided by an outer membrane was thus one of the crucial events leading to the formation of the first cell. The need for further containment fulfilled by the development of intracellular membranes represents the hallmark of eukaryotes. The origin of cells containing a nucleus, the membrane-bounded organelle that defines eukaryotes, is considered one of the most successful outcomes of evolution. The nucleus has unique structural features, such as the nuclear envelope, the nuclear pore complex and linear chromosomes that enabled major functional innovations: nucleo-cytoplasmic transport, novel cell-cycle controls and mitosis, sexual recombination and novel patterns of RNA processing including pre-mRNA splicing.

The Origin of Eukaryotic Cells

In parallel with the advent of the nucleus, the eukaryotic cell acquired endomembranes (i.e., the endoplasmic reticulum, endosomal/lysosomal and secretory systems), mitochondria, chloroplasts, peroxisomes and a cytoskeleton. Whereas there is general agreement on the endosymbiotic origin of organelles such as mitochondria and chloroplasts, the origin of the nucleus and the endomembrane

*Corresponding Author: Maria Carmo-Fonseca—Instituto de Medicina Molecular, Faculdade de Medicina, Av. Prof. Egas Moniz, 1649-028 Lisboa, Portugal. Email: carmo.fonseca@fm.ul.pt

Alternative Splicing in the Postgenomic Era, edited by Benjamin J. Blencowe and Brenton R. Graveley.

systems remain mysterious.[1,2] Classically, endomembranes are thought to have evolved by invagination and inward separation of vesicles from the plasma membrane. Subsequent differentiation of primitive endomembranes into endoplasmic reticulum, Golgi and lysosomes must have depended on the evolution of valves to isolate different membrane compartments and this was achieved by selective coated vesicle budding. According to a prevailing theory, plasma membrane invaginations budded off forming vesicles that then fused, giving rise to the endoplasmic reticulum and nuclear envelope, which are continuous with each other.[3] An alternative model postulates that the nucleus is the result of symbiosis between an archaeon and a eubacterium, so that in a eukaryotic cell the nucleus is of archael origin but the cytoplasm is of bacterial origin.[4]

Importantly, the nuclear envelope is not composed of two distinct separate membranes, as in mitochondria. Instead, it is topologically a single membrane that folds forming sharp curves at sites where nuclear pore complexes (NPCs) are embedded. Examining the structure of the proteins that form the yeast NPC Nup84 subcomplex, Devos et al[5] discovered that these proteins share a common architecture with vesicle coat components. It was therefore proposed that akin to vesicle coats, which bend the plasma- and endo-membranes for budding, NPC proteins could bend the nuclear membrane, thereby forming the highly curved regions at the pore periphery.[5] Consistent with this model, mutational and deletion analysis showed that in the absence of the homologue complex in vertebrates nuclei formed with a closed nuclear envelope lacking NPCs.[6,7]

A possible selective advantage of the assembly of a nuclear envelope on the chromatin surface of early eukaryotes was to protect the DNA from shearing damage caused by the novel molecular motors associated with the cytoskeleton. In turn (or in parallel), formation of the nuclear envelope allowed for slower splicing in trans by the spliceosome to evolve from group II introns in genes that were transferred from the protomitochondrion to the nucleus.[8-10] Assembly of the nuclear envelope plugged with nuclear pore complexes that segregated the RNA and protein-synthesis machinery was instrumental in preventing ribosomes from translating nuclear messenger RNA before splicing. Shine-Dalgarno sequences may also have been lost to prevent nuclear messengers from binding directly to nascent ribosomes, while capping evolved instead. The nuclear envelope co-evolved with a novel mechanochemical division machinery (the mitotic apparatus) and novel temporal controls of the cell cycle that allowed the multiplication of origins of replication and other unique features of eukaryotic chromatin and chromosomes. Indirectly, this co-evolution allowed for massive increases in eukaryotic genome and nuclear size. The larger nuclei and the origin of sexual recombination together are thought to have favored the rapid intragenomic spread of the spliceosomal introns.[8] Supporting the view that spliceosomal introns were present in the last common ancestor of extant eukaryotes, both introns and spliceosomal components have been discovered in many species that could have diverged from other eukaryotes very early in evolution (for a review, see ref. 11).

In higher eukaryotes, the nucleus breaks down and reforms at each mitosis. Inspired by Stephen Jay Gould's idea that the development of an organism is a progression through ancestral life forms,[12] we propose that the assembly of a postmitotic nucleus in metazoan cells recapitulates some of the events that lead to the birth of the first nuclei approximately 1 billion years ago. Nuclear break down involves chromatin condensation with a global shut down of gene expression, disassembly of the nuclear lamina by phosphorylation and solubilization of the nuclear lamins and absorption of the nuclear membranes into the endoplasmic reticulum (for a review, see ref. 13). During anaphase or telophase, depending on the cell type, membrane vesicles start to associate with decondensing chromosomes. Progressive membrane fusion coordinated with de novo NPC assembly leads to a complete enclosure of chromosomes by a functional nuclear envelope.[14,15] This is followed by nuclear growth induced by complete chromosome decondensation and import of nuclear proteins via the pores. The gene expression machinery that became distributed throughout the cytoplasm during mitosis is recycled back into the nucleus, with transcription factors being imported prior to pre-mRNA splicing factors.[16] This succession of events leading to nuclear formation after mitosis strongly suggests that the nucleus shares a common evolutionary origin with the cytoplasmic endomembrane systems.

Compartmentalization of Splicing and Splicing Factors

Active Genes Are Nonrandomly Positioned in the Nucleus

Pre-mRNA is cotranscriptionally spliced, as first demonstrated by direct electron microscopic visualization of nascent transcripts on spread chromatin[17,18] and later further shown by biochemical analysis of nascent pre-mRNAs.[19-21] At the light microscopy level, methods were developed to simultaneously detect DNA, intronic RNA and spliced RNA by fluorescence in situ hybridization.[22,23] Using these tools, several studies showed colocalization of spliced mRNA and DNA from the corresponding gene, thereby firmly establishing that splicing occurs at the sites of transcription.[22-25]

Having demonstrated that active pre-mRNA splicing sites coincide with actively transcribed intron-containing genes, it becomes possible to identify where splicing takes place in the nucleus by mapping the localization of active genes. Inside the nucleus genes are nonrandomly positioned, exposing individual loci to distinct microenvironments. For example, several genes have been shown to relocalize to distinct sub-nuclear domains depending on their transcriptional status. (For recent reviews see refs. 26,27.) For example, in developing B and T-cells, genes destined for silencing are recruited to heterochromatic pericentromeric domains where transcription is repressed.[28-30] Similarly, coating one of the two X chromosomes in female embryos by Xist RNA induces the formation of a silencing domain that excludes RNA polymerase II and transcription factors.[31] The nuclear periphery is also generally regarded as a transcriptionally repressive compartment that preferentially harbors gene-poor chromosomes; specific genes were found to be associated with the periphery when inactive and located away from the periphery when active.[32-34] Despite being predominantly occupied by silenced heterochromatin, the nuclear periphery is infiltrated by channels of actively transcribed euchromatin that communicate with nuclear pore complexes. At least in yeast, transcription at the nuclear periphery apparently requires association of the gene with the nuclear pore complex,[35-39] thereby facilitating mRNA export as predicted in Blobel's "gene gating hypothesis".[40] Indeed, recent genetic studies in *Saccharomyces cerevisiae* reveal that proteins required for mRNA export are involved in anchoring active genes to nuclear pore complexes.[41,42] Thus, relocation of genes within the nucleus may be a result of, rather than requirement for, transcriptional activity, mRNA processing and transport.

Using electron microscopy and pulse-labeling of nascent RNA, Cook and colleagues found approximately 2400 sites of RNA polymerase II activity scattered throughout the nucleoplasm of mammalian cultured cells.[43,44] As the number of active genes and other transcription units vastly exceeds the number of foci detected, each site or "transcription factory" was suggested to contain several (up to 30) active polymerases and associated transcripts. In good agreement with this prediction, genes as far as 40 Mb apart in the chromosome were shown to colocalize to a shared factory when transcribed,[45] and even genes located on different chromosomes were found to have the ability to congregate.[46,47] This physical grouping of genomically distant loci in the nucleus is apparently mediated by proteins that bind to common regulatory sequence elements present in physiologically related genes. (For recent reviews see refs. 48,49.) Spatial proximity may therefore contribute to coordinating the expression of functionally related genes, but how do distant genetic elements find each other in the nucleus? We propose that pre-mRNA synthesis and processing contributes to the establishment and maintenance of an epigenetic state, in a manner similar to that of DNA replication. In face of the growing evidence for a tight coupling between the different steps of gene expression in the nucleus, we speculate that during mRNA biogenesis multi-role protein factors may impinge a persistent mark on the chromatin. If chromosome decondensation at the end of mitosis would follow a temporal order controlled by specific marks, then subsets of chromatin loops could form simultaneously and separately from other genomic regions allowing a rapid and preferential interaction with each other. Future studies analyzing the dynamic positioning of multiple loci in living cells combined with genetic and RNAi-mediated knock-out/knock-down approaches are likely to shed new light into the poorly understood link between nuclear organization and control of gene expression at the level of transcription and splicing.

The Spliceosome Cycle and Splicing Factor Compartments

The spliceosome is the multi-megadalton machine that catalyses pre-mRNA splicing (refer to chapter by Matlin and Moore). The building blocks of the spliceosome are uridine-rich small nuclear RNAs (U snRNAs) packaged as U snRNPs (small nuclear ribonucleoprotein particles) that function in concert with numerous nonsnRNP proteins. Spliceosomes form anew on nascent pre-mRNAs and disassemble after introns are excised and exons ligated. Thus, spliceosomal components in the nucleus can be either actively engaged in splicing or waiting for the next call to assemble a spliceosome. Recruitment of splicing snRNPs and nonsnRNP proteins to nascent transcripts has been visualized in several systems, including insect polytene chromosomes,[50-52] amphibian germinal vesicles[53] and mammalian cells.[23,54-56] In the salivary gland polytene cells from *Chironomus tentans*, 10-15% of the spliceosomal components were estimated to be bound to pre-mRNA at active gene loci at a given moment, while the vast majority was present in the nucleoplasm and apparently not engaged with pre-mRNA.[57] In amphibian germinal vesicles, the splicing snRNPs that did not associate with loops of the lampbrush chromosomes were detected concentrated in numerous nucleoplasmic granules called "snurposomes".[53] Mammalian cells also contain a surplus of spliceosome components (HeLa cells are thought to contain more than a million extremely stable U1 snRNP particles) and at any given time the majority of these molecules are not associated with nascent transcripts. When the mammalian cell nucleus is viewed with the electron microscope, spliceosomal components are detected in morphologically distinct structures termed interchromatin granule clusters (IGCs) and perichromatin fibrils (for a comprehensive review, see ref. 58). The perichromatin fibrils correspond to nascent transcripts and appear scattered throughout the nucleoplasm, excluding regions of condensed chromatin.[59] Perichromatin fibrils are often closely associated with the periphery of interchromatin granule clusters, making it impossible to distinguish the two structures within the speckled pattern that characterizes the distribution of splicing factors observed by fluorescence microscopy (Fig. 1). Consistent with the presence of nascent transcripts (i.e., perichromatin fibrils) being in close vicinity to IGCs, as detected in studies employing electron microscopy, pulse-labeled nascent RNA from numerous actively transcribed genes has been visualized by fluorescence microscopy to overlap the periphery of nuclear speckles.[60-62]

While snRNPs and splicing proteins detected on perichromatin fibrils most likely correspond to active spliceosomes, a large body of evidence indicates that the spliceosomal components localized in interchromatin granule clusters are primarily not involved in splicing (for a recent review, see ref. 63.) Indeed, interchromatin granule clusters do not contain detectable levels of either DNA or nascent RNA; moreover, spliceosomal components move away from the interchromatin granule clusters to sites of active transcription in the nucleus and accumulate in the clusters when transcription is inhibited (Fig. 2).

In the living cell, splicing factors are constantly roaming the nucleus[64,65] so that upon activation of a gene the spliceosome rapidly assembles on the nascent pre-mRNA.[66,67] Conversely, gene inactivation increases the pool of "reserve" splicing factors that accumulate within enlarged interchromatin granule clusters. Consequently, the organization of the speckled pattern is a reflection of the transcriptional and splicing activity of the cell.[67,68]

Proteome analysis of nuclear fractions highly enriched in IGCs revealed a 63% overlap with the composition of the spliceosome,[69,70] and the vast majority of identified splicing proteins were shown to localize in IGCs. But why are spliceosome components attracted to IGCs? The IGCs contain a metabolically stable, nuclear-restricted population of poly(A) RNA that has been suggested to play a structural scaffolding role.[63] However, this view is difficult to reconcile with the recent finding that nuclear poly(A) RNA is completely mobile, diffusing constantly between the speckles and the nucleoplasm.[71] An important clue to understand the biology of IGCs was the discovery that over-expression of Clk/STY, a protein kinase that phosphorylates SR splicing proteins, abolished the typical speckled immunolocalization of spliceosome components.[72] Later electron microscopy studies revealed that over-expression of Clk/STY led to a complete disassembly of IGCs concomitant with inhibition of splicing activity.[73] These results argue that IGCs lack a stable scaffold that is maintained when splicing factors are released. It is more likely that

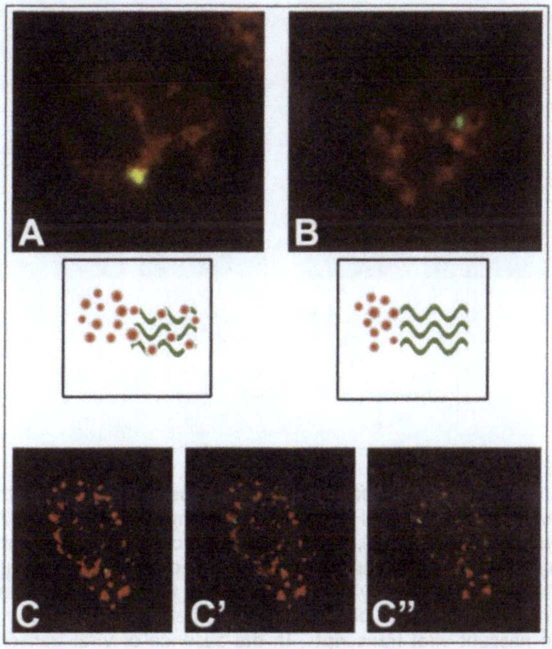

Figure 1. Transcribed genes localize in close proximity to nuclear speckles. A,B) Murine erythroleukemia (MEL) cells containing multiple copies of the human β-globin gene integrated in the genome as a tandem array were double-labeled for Sm proteins (red staining) and human β-globin RNA (green staining). The cell depicted in (A) expresses wild- type human β-globin RNA, while the cell shown in (B) expresses a mutant variant that is not spliced.[115] In both cells the nascent RNA is visualized as a green focus localized in the vicinity of a nuclear speckle. The schemes below the micrographs represent splicing factors in interchromatin granule clusters and in association with nascent transcripts. In the case of wild type transcripts (A), which are normally spliced, splicing factors assemble on the nascent transcripts; this results in concentration of Sm labeling at the RNA focus (the focus appears yellow due to superimposition of red and green staining). Note that at the light microscopic level, the concentration of splicing factors at the β-globin site of transcription is indistinguishable from a nuclear speckle. In the case of the mutant transcripts (B), which are not spliced,[115] splicing factors are not recruited to the site of transcription. C,C',C'') Human (HeLa) cells were double-labeled for splicing factor SC35 (red staining) and subtelomeric region 19p (green staining), a chromosomal region enriched in transcribed genes.[116] The panels depict three consecutive optical sections through the same cell; note that both alleles are in close proximity to nuclear speckles.

IGCs are formed by direct protein-protein interactions between splicing proteins; moreover, the data reported by Sacco-Bubulya and Spector 2002[73] strongly suggest that the assembly of IGCs is linked to the phosphorylation state of SR proteins. As phosphorylation of SR proteins also plays a critical role in splicing activity (refer to chapter by Lin and Fu), the question emerges as to whether similar protein-protein interactions are involved in IGC formation and spliceosome assembly.

Finally, what are the functions of IGCs? Do these structures represent simple aggregates of "reserve" spliceosome components waiting for a splicing opportunity, or are splicing proteins targeted to IGCs in order to become competent for splicing? Although pro and con arguments for each view have been extensively debated, a definitive answer to the question awaits further experimentation (for a review, see ref. 63).

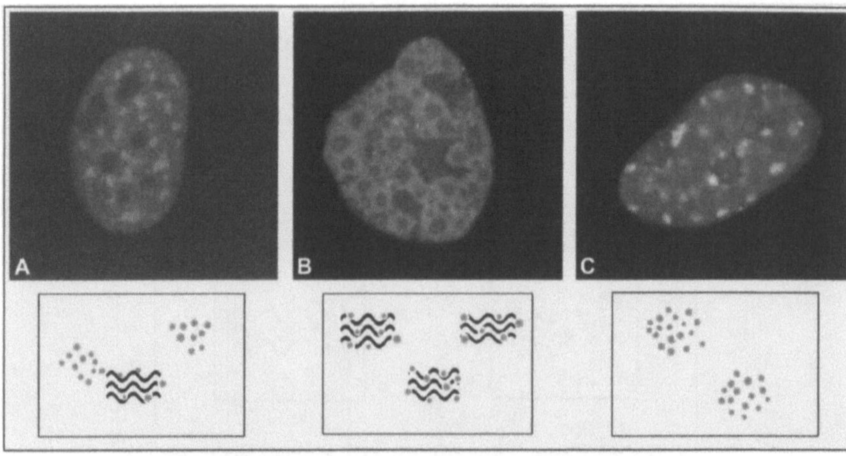

Figure 2. The distribution of splicing factors in the nucleus is a reflection of the transcriptional and splicing activity of the cell. Live HeLa cells expressing splicing factor U2AF[65] tagged with GFP were imaged. In control cells (A), which contain nascent transcripts scattered throughout the nucleoplasm, GFP-U2AF[65] accumulates in nuclear speckles. After infection with adenovirus (B), the nucleus is overloaded with viral transcripts and GFP-U2AF[65] no longer accumulates in speckles; the ring-like structures observed in these cells correspond to recruitment of the splicing factor to nascent viral transcripts. In the absence of viral transcripts (induced by treatment with a transcription inhibitor), GFP-U2AF[65] dissociates from the ring structures and accumulates predominantly in nuclear speckles (C).

Compartmentalization of snRNP Biogenesis

The major spliceosomal small nuclear ribonucleoprotein particles are the U1, U2, U5 and U4/U6 snRNPs. Each snRNP consists of one or two uridine-rich small nuclear RNAs (U1, U2, U5 and U4/U6 snRNAs) bound by a protein complex that comprises seven common Sm proteins and one or more proteins specific to each snRNP.[74] The Sm proteins B/B', D1, D2, D3, E, F and G are common to all spliceosomal snRNPs, except U6 and are arranged into a ring structure around a short (7 nt) highly conserved single-stranded uridine-rich sequence of the snRNA.[75-77] Biogenesis of splicing snRNPs in metazoa involves a complex sequence of reactions occurring at different locations within the cell. The snRNAs (except U6) are transcribed by RNA polymerase II and the initial precursors contain a 5' monomethylated m^7G cap and a short 3' extension. In the nucleus the nascent transcripts associate with a number of proteins and the resulting complex is rapidly exported to the cytoplasm. Export is dependent on interaction of the U snRNA cap structure with the nuclear cap-binding complex (CBC).[78,79] CBC is a heterodimeric complex composed of two subunits, CBP80 and CBP20, but neither subunit alone can interact with capped RNA; they must first heterodimerize. Phosphorylated adapter for RNA export (PHAX) then interacts with CBC bound to capped RNA, establishing a bridge between CBC on the one hand and RanGTP together with the export receptor CRM1/Xpo1 on the other.[80] Phosphorylated PHAX can bind to RanGTP/Xpo1 but does so more strongly in the presence of CBC-bound UsnRNA. As a consequence of this cooperative assembly of the complex neither PHAX nor CBC detectably leave the nucleus in the absence of RNA substrates.[80] In the cytoplasm, dissociation of the export complex is triggered by GTP hydrolysis mediated by RanBP1/2 and RanGAP,[81] binding of the Importin α/β heterodimer to CBC[82] and dephosphorylation of PHAX.[80] Once detached from export proteins, the snRNA is free to interact with Sm proteins. Although in vitro Sm cores assemble readily on uridine-rich RNAs, in cells this process involves the survival of motor neurons (SMN) complex.[83] Assembly of the Sm core is a prerequisite for removal of the snRNA 3' extension and hypermethylation of the 5' m^7G cap to $m^{2,2,7}G$ (m_3G or TMG).[74,84] The assembled Sm core and the modified cap then function as independent nuclear localization signals (NLS) for subsequent re-import into the nucleus. The m_3G cap is recognized by Snurportin1, an import adaptor that interacts with Importin β,[85] while the Sm-core mediated transport is linked to the nuclear import of SMN.[86]

After import to the nucleus the newly assembled snRNPs accumulate preferentially in Cajal Bodies (CBs) (Fig. 3)[87,88] and several lines of evidence indicate that snRNP maturation is completed at this location (for a recent review, see ref. 89). First, site-specific synthesis of 2'-*O*-methylated nucleotides and pseudouridines in the U1, U2, U4 and U5 snRNAs occurs in CBs[90] and is directed by guide RNAs that specifically localize to CBs.[91-93] Second, the association of the heterotrimeric

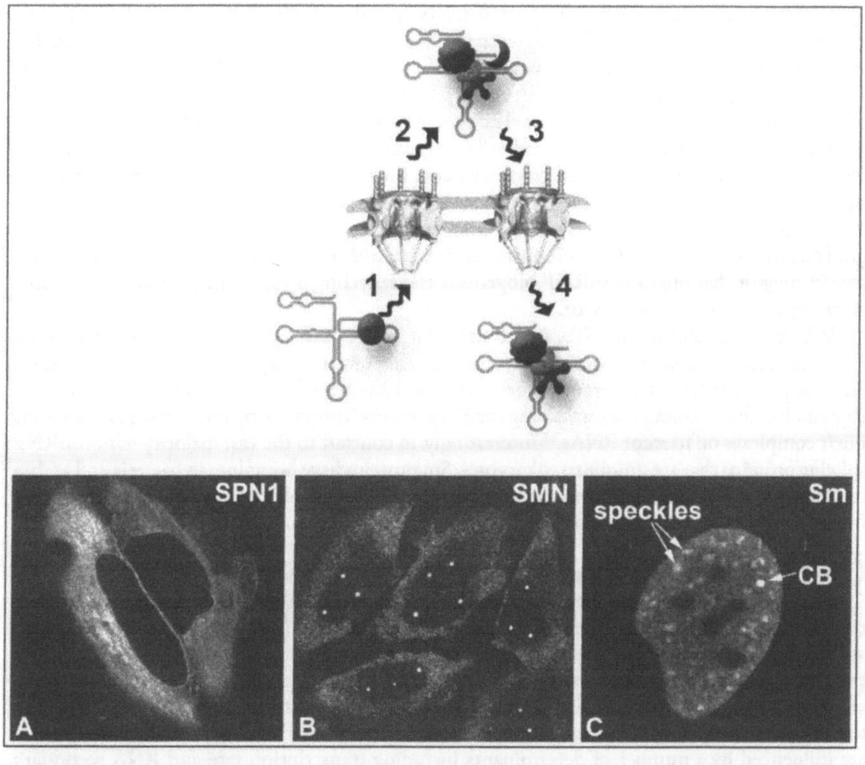

Figure 3. Compartmentalization of snRNP biogenesis. 1) The U1, U2, U4 and U5 snRNAs are synthesized as precursors in the nucleus; they contain a 5' monomethylated m^7G cap that binds CBC and export factors PHAX and CRM1/Xpo1 associated with RanGTP. 2) After translocation through the nuclear pore, the export complex dissociates and the Sm/SMN complex assembles. 3) The cap is hypermethylated and binds Snurportin1 (SPN1); together with the Sm/SMN complex, SPN1 drives nuclear import. 4) In the nucleus, snRNPs accumulate in Cajal bodies (CBs) to complete their maturation; later, snRNPs participate in spliceosome assembly and cycle through nuclear speckles. At steady state, SPN1 is predominantly detected in the cytoplasm, indicating that this protein dissociates rapidly from snRNPs after their nuclear import. SMN is also detected in the cytoplasm, but additionally concentrates in CBs, suggesting that its association with maturing snRNPs persists in the CBs. Sm proteins are predominantly detected throughout the nucleoplasm, excluding nucleoli and with additional concentration in CBs and nuclear speckles.

splicing factor SF3a with premature 15S U2 snRNP to form a functional 17S U2 snRNP particle occurs in CBs.[94] Additionally, CBs are likely to be sites where snRNPs are recycled. The U4, U5 and U6 snRNPs enter the spliceosome as a pre-assembled U4/U6.U5 tri-snRNP and after splicing the released individual particles must reassemble into a new U4/U6 di-snRNP and then into a U4/U6.U5 tri-snRNP, before entering the next splicing cycle. Consistent with the view that CBs may be the site where dismantled snRNPs are recycled into functional particles, it was shown that the U4/U6 di-snRNP recycling factor p110 is enriched in Cajal bodies and plays a role in targeting U6 snRNPs to CBs.[95] Furthermore, RNAi-mediated depletion of hPrp31 and hPrp6 (two proteins expected to be required for tri-snRNP assembly) caused an accumulation of U4/U6 di-snRNPs in Cajal bodies, suggesting that the formation of U4/U6.U5 tri-snRNPs occurs in these subnuclear structures.[96]

Direct visualization of newly imported snRNPs in living cells shows that the particles accumulate first in CBs and are later detected in the speckles, suggesting that the CB phase is a rate-limiting step in the functional maturation of spliceosomal snRNPs.[97,98] More recent studies further reported the disappearance of CBs upon depletion of either: SMN, which disrupted Sm core assembly,[99] PHAX, which blocked specifically the nuclear export of newly synthesized U snRNAs, or hTGS1, which impaired m^7G cap methylation.[100] In further agreement with the view that CBs are not predefined compartments into which snRNPs have to be targeted in order to complete maturation, cells from knock-out mice lacking the CB marker protein coilin are devoid of CBs yet viable.[101] In coilin knockout cells, Sm snRNAs and their modification guide RNAs colocalize in nucleoplasmic foci that are distinct from CBs.[90] Taken together, these results suggest that ongoing snRNP biogenesis and recycling is sufficient to create a dedicated microenvironment in the nucleus.

Why do spliceosomal snRNPs, which function in the nucleus, require a cytoplasmic phase for biogenesis? Because Sm proteins can form stable heptameric rings on their own with minimal RNA sequence specificity, it was proposed that the assembly of Sm cores on snRNAs in the cytoplasm has the advantage of preventing the Sm proteins from entering the nucleus and forming illicit complexes on nascent RNAs.[83] Interestingly, in contrast to the vast majority of nonsnRNP splicing proteins that are unique to eukaryotes, Sm proteins have homologs in bacteria and archea (reviewed in ref. 2). Could the cytoplasmic phase of snRNP assembly therefore be a direct descent from the primordial endosymbiont?

Compartmentalization of Splicing Factors and Splicing Regulation

The regulation of splicing involves both cis-acting sequence elements in the pre-mRNA and trans-acting splicing factors that associate with the pre-mRNA (refer to other chapters in this volume for detailed reviews). Many alternative splicing events involve a complex interplay between positive and negative regulators and it is often not possible to make a clear distinction between constitutive and alternative splicing elements and factors. Alternative splicing involves changes in the choice of sites for spliceosome assembly. Besides pre-mRNA binding proteins, such choices are influenced by a number of determinants including transcription rate and RNA secondary structure. In addition to tissue-specific splicing decisions, there is increasing evidence for splicing patterns being affected by both extra- and intracellular signaling to the splicing machinery (for a review, see ref. 102). One example is the control of splicing regulators by phosphorylation and sub-cellular localization, as first demonstrated by van der Houven et al.[103] Several splicing factors shuttle between the nucleus and the cytoplasm and some were shown to change their nucleo-cytoplasmic distribution in response to specific signals. For example, after stimulation of cells with osmotic shock or ultraviolet-C irradiation the hnRNP A1 protein was reported to become hyperphosphorylated and accumulated in the cytoplasm with a parallel change in alternative splicing of a reporter construct.[103] Similarly, an ultraviolet-C stress stimulus also reduced the nuclear concentration of splicing factor hSlu7 with a concomitant change in alternative splicing patterns of cellular genes.[104] Phosphorylation by signal transducers was further shown to cause cytoplasmic accumulation of hnRNP K,[105] Tra2β1[106] and PTB.[107] There is evidence suggesting that phosphorylation of a cargo protein modulates its binding affinity to importins/exportins and that the shuttling signals of several RNA-binding proteins contain potential consensus sites for protein kinases. Thus, it is likely that many proteins involved in RNA metabolism redistribute between the nucleus and the cytoplasm in response to signaling pathways (see ref. 107 and references therein). Clearly, relocalization of splicing factors from the nucleus to the cytoplasm can be concurrent with changes in alternative splicing, but the functional consequences of such changes in alternative splicing are not known.

Interestingly, the SRPK family of kinases specific for SR splicing factors localizes predominantly in the cytoplasm during interphase but translocates to the nucleus in late G2 cells.[108] As SR proteins are regulated by reversible phosphorylation and at least in vitro dephosphorylation is essential for splicing to take place, we propose that accumulation of SRPKs in the nucleus just before mitosis may contribute to the inhibition of splicing prior to nuclear break down. Conversely, at the end of mitosis, when the nuclear envelope reforms and transcriptional activity resumes, a precise temporal order for accumulation of splicing factors in the newly formed nucleus has been observed. The hnRNP C1/C2 proteins were detected first, followed by snRNPs and SR proteins and later on hnRNP A1.[16] This implies that nascent transcripts in a newly formed nucleus are transiently exposed to different relative concentrations of splicing activators (SR proteins) and repressors (hnRNPs). We speculate that this sequential order of splicing factor import creates a spatial-temporal gradient for regulated alternative splicing choices.

Perspectives

Recent advances in live cell imaging are dramatically changing our vision of the nucleus. For many years, beautiful electron- and light-microscopic pictures collected from fixed specimens fostered a view of a static and rigidly structured sub-nuclear organization.[109] Some of these traditional concepts are now being questioned by data supporting the dynamic movement of factors within the nucleus. Several independent studies confirm that most RNAs diffuse in the nucleus at rates of ~0.03-0.1 square μm per second, which is sufficient to translocate a fully mature messenger RNA to the cytoplasm within a few minutes after its generation anywhere inside the nucleus.[110] Because diffusion cannot be controlled, regulation of RNA trafficking is achieved by retention. For example, a quality control mechanism operates at the site of transcription that prevents release of incorrectly processed precursors (see ref. 110 and references therein). Retention at a nuclear microenvironment distinct from the site of transcription can also occur, ensuring the availability of a pool of mRNA that can be readily transported to the cytoplasm and translated into protein in response to an extracellular signal.[111]

Protein splicing factors are also thought to diffuse throughout the nucleus and associate stochastically with nascent transcripts.[65,67] Nevertheless, individual splicing proteins associate differentially with alternatively spliced transcripts.[112] According to the currently popular combinatorial model for alternative splicing regulation, which protein is recruited to each spliceosome depends on both the concentration of the protein in the cell and the regulatory elements present in the pre-mRNA. Whether retention/sequestration of splicing factors at specific nuclear microenvironments contributes to adjust their free pool in the nucleus, thereby restricting their availability at the sites of action and hence control alternative splicing is an appealing hypothesis that remains to be demonstrated.[110] In order to tackle this question it will be important to follow the dynamic behavior of a large number of splicing proteins and transcripts systematically. Such an endeavour will be greatly facilitated by emerging methodologies such as SILAC (stable-isotope labeling by amino acids in cell culture), a mass-spectroscopy-based technique that allows the monitoring of the fate of cellular proteins over time on a global scale.[113] In parallel, however, it is essential to be aware that gene expression is not synchronous in a population but rather varies stochastically from cell to cell (for a recent review, see ref. 114). Thus, a very important area of future work is the development of time-resolved global approaches performed on single cells.

Acknowledgements

We are thankful to our colleagues João Ferreira, Noélia Custódio, Margarida Gama-Carvalho and José Rino for critical discussions and for help in preparing the figures. Research in the authors' laboratory is supported by grants from Fundação para a Ciência e Tecnologia, Portugal and the European Commission.

References

1. Embley TM, Martin W. Eukaryotic evolution, changes and challenges. Nature 2006; 440(7084):623-30.
2. Kurland CG, Collins LJ, Penny D. Genomics and the irreducible nature of eukaryote cells. Science 2006; 312(5776):1011-4.
3. Cavalier-Smith T. The phagotrophic origin of eukaryotes and phylogenetic classification of Protozoa. Int J Syst Evol Microbiol 2002; 52(Pt 2):297-354.
4. Horiike T, Hamada K, Kanaya S et al. Origin of eukaryotic cell nuclei by symbiosis of Archaea in Bacteria is revealed by homology-hit analysis. Nat Cell Biol 2001; 3(2):210-4.
5. Devos D, Dokudovskaya S, Alber F et al. Components of coated vesicles and nuclear pore complexes share a common molecular architecture. PLoS Biol 2004; 2(12):e380.
6. Harel A, Orjalo AV, Vincent T et al. Removal of a single pore subcomplex results in vertebrate nuclei devoid of nuclear pores. Mol Cell 2003; 11(4):853-64.
7. Walther TC, Alves A, Pickersgill H et al. The conserved Nup107-160 complex is critical for nuclear pore complex assembly. Cell 2003; 113(2):195-206.
8. Cavalier-Smith T. Intron phylogeny: a new hypothesis. Trends Genet 1991; 7(5):145-8.
9. Cech TR. The generality of self-splicing RNA: relationship to nuclear mRNA splicing. Cell 1986; 44(2):207-10.
10. Martin W, Koonin EV. Introns and the origin of nucleus-cytosol compartmentalization. Nature 2006; 440(7080):41-5.
11. Roy SW, Gilbert W. The evolution of spliceosomal introns: patterns, puzzles and progress. Nat Rev Genet 2006; 7(3):211-21.
12. Gould SJ. Ontogeny and Phylogeny 1977.
13. Mattaj IW. Sorting out the nuclear envelope from the endoplasmic reticulum. Nat Rev Mol Cell Biol 2004; 5(1):65-9.
14. Antonin W, Mattaj IW. Nuclear pore complexes: round the bend? Nat Cell Biol 2005; 7(1):10-2.
15. D'Angelo MA anderson DJ, Richard E et al. Nuclear pores form de novo from both sides of the nuclear envelope. Science 2006; 312(5772):440-3.
16. Prasanth KV, Sacco-Bubulya PA, Prasanth SG et al. Sequential entry of components of the gene expression machinery into daughter nuclei. Mol Biol Cell 2003; 14(3):1043-57.
17. Beyer AL, Osheim YN. Splice site selection, rate of splicing and alternative splicing on nascent transcripts. Genes Dev 1988; 2(6):754-65.
18. Osheim YN, Miller OL Jr, Beyer AL. RNP particles at splice junction sequences on Drosophila chorion transcripts. Cell 1985; 43(1):143-51.
19. Bauren G, Wieslander L. Splicing of Balbiani ring 1 gene pre-mRNA occurs simultaneously with transcription. Cell 1994; 76(1):183-92.
20. LeMaire MF, Thummel CS. Splicing precedes polyadenylation during Drosophila E74A transcription. Mol Cell Biol 1990; 10(11):6059-63.
21. Wuarin J, Schibler U. Physical isolation of nascent RNA chains transcribed by RNA polymerase II: evidence for cotranscriptional splicing. Mol Cell Biol 1994; 14(11):7219-25.
22. Xing Y, Johnson CV, Dobner PR et al. Higher level organization of individual gene transcription and RNA splicing. Science 1993; 259(5099):1326-30.
23. Zhang G, Taneja KL, Singer RH et al. Localization of pre-mRNA splicing in mammalian nuclei. Nature 1994; 372(6508):809-12.
24. Custodio N, Carmo-Fonseca M, Geraghty F et al. Inefficient processing impairs release of RNA from the site of transcription. EMBO J 1999; 18(10):2855-66.
25. Zhang G, Zapp ML, Yan G et al. Localization of HIV-1 RNA in mammalian nuclei. J Cell Biol 1996; 135(1):9-18.
26. Gilbert N, Gilchrist S, Bickmore WA. Chromatin organization in the mammalian nucleus. Int Rev Cytol 2005; 242:283-336.
27. Taddei A, Hediger F, Neumann FR et al. The function of nuclear architecture: a genetic approach. Annu Rev Genet 2004; 38:305-45.
28. Brown KE, Baxter J, Graf D et al. Dynamic repositioning of genes in the nucleus of lymphocytes preparing for cell division. Mol Cell 1999; 3(2):207-17.

29. Brown KE, Guest SS, Smale ST et al. Association of transcriptionally silent genes with Ikaros complexes at centromeric heterochromatin. Cell 1997; 91(6):845-54.
30. Cobb BS, Morales-Alcelay S, Kleiger G et al. Targeting of Ikaros to pericentromeric heterochromatin by direct DNA binding. Genes Dev 2000; 14(17):2146-60.
31. Chaumeil J, Le Baccon P, Wutz A et al. A novel role for Xist RNA in the formation of a repressive nuclear compartment into which genes are recruited when silenced. Genes Dev 2006; 20(16):2223-37.
32. Kosak ST, Skok JA, Medina KL et al. Subnuclear compartmentalization of immunoglobulin loci during lymphocyte development. Science 2002; 296(5565):158-62.
33. Ragoczy T, Bender MA, Telling A et al. The locus control region is required for association of the murine beta-globin locus with engaged transcription factories during erythroid maturation. Genes Dev 2006; 20(11):1447-57.
34. Zink D, Amaral MD, Englmann A et al. Transcription-dependent spatial arrangements of CFTR and adjacent genes in human cell nuclei. J Cell Biol 2004; 166(6):815-25.
35. Brickner JH, Walter P. Gene recruitment of the activated INO1 locus to the nuclear membrane. PLoS Biol 2004; 2(11):e342.
36. Casolari JM, Brown CR, Komili S et al. Genome-wide localization of the nuclear transport machinery couples transcriptional status and nuclear organization. Cell 2004; 117(4):427-39.
37. Ishii K, Arib G, Lin C et al. Chromatin boundaries in budding yeast: the nuclear pore connection. Cell 2002; 109(5):551-62.
38. Rodriguez-Navarro S, Fischer T, Luo MJ et al. Sus,1 a functional component of the SAGA histone acetylase complex and the nuclear pore-associated mRNA export machinery. Cell 2004; 116(1):75-86.
39. Schmid M, Arib G, Laemmli C et al. Nup-PI: the nucleopore-promoter interaction of genes in yeast. Mol Cell 2006; 21(3):379-91.
40. Blobel G. Gene gating: a hypothesis. Proc Natl Acad Sci USA 1985; 82(24):8527-9.
41. Cabal GG, Genovesio A, Rodriguez-Navarro S et al. SAGA interacting factors confine sub-diffusion of transcribed genes to the nuclear envelope. Nature 2006; 441(7094):770-3.
42. Taddei A, Van Houwe G, Hediger F et al. Nuclear pore association confers optimal expression levels for an inducible yeast gene. Nature 2006; 441(7094):774-8.
43. Iborra FJ, Pombo A, Jackson DA et al. Active RNA polymerases are localized within discrete transcription "factories' in human nuclei. J Cell Sci 1996; 109(Pt 6):1427-36.
44. Jackson DA, Iborra FJ, Manders EM et al. Numbers and organization of RNA polymerases, nascent transcripts and transcription units in HeLa nuclei. Mol Biol Cell 1998; 9(6):1523-36.
45. Osborne CS, Chakalova L, Brown KE et al. Active genes dynamically colocalize to shared sites of ongoing transcription. Nat Genet 2004; 36(10):1065-71.
46. Ling JQ, Li T, Hu JF et al. CTCF mediates interchromosomal colocalization between Igf2/H19 and Wsb1/Nf.1 Science 2006; 312(5771):269-72.
47. Spilianakis CG, Lalioti MD, Town T et al. Interchromosomal associations between alternatively expressed loci. Nature 2005; 435(7042):637-45.
48. Dean A. On a chromosome far, far away: LCRs and gene expression. Trends Genet 2006; 22(1):38-45.
49. Fraser P. Transcriptional control thrown for a loop. Curr Opin Genet Dev 2006.
50. Bauren G, Jiang WQ, Bernholm K et al. Demonstration of a dynamic, transcription-dependent organization of pre-mRNA splicing factors in polytene nuclei. J Cell Biol 1996; 133(5):929-41.
51. Kiseleva E, Wurtz T, Visa N et al. Assembly and disassembly of spliceosomes along a specific pre-messenger RNP fiber. EMBO J 1994; 13(24):6052-61.
52. Sass H, Pederson T. Transcription-dependent localization of U1 and U2 small nuclear ribonucleoproteins at major sites of gene activity in polytene chromosomes. J Mol Biol 1984; 180(4):911-26.
53. Wu ZA, Murphy C, Callan HG et al. Small nuclear ribonucleoproteins and heterogeneous nuclear ribonucleoproteins in the amphibian germinal vesicle: loops, spheres and snurposomes. J Cell Biol 1991; 113(3):465-83.
54. Cmarko D, Verschure PJ, Martin TE et al. Ultrastructural analysis of transcription and splicing in the cell nucleus after bromo-UTP microinjection. Mol Biol Cell 1999; 10(1):211-23.
55. Neugebauer KM, Roth MB. Distribution of pre-mRNA splicing factors at sites of RNA polymerase II transcription. Genes Dev 1997; 11(9):1148-59.
56. Puvion E, Puvion-Dutilleul F. Ultrastructure of the nucleus in relation to transcription and splicing: roles of perichromatin fibrils and interchromatin granules. Exp Cell Res 1996; 229(2):217-25.
57. Wieslander L, Bauren G, Bernholm K et al. Processing of pre-mRNA in polytene nuclei of Chironomus tentans salivary gland cells. Exp Cell Res 1996; 229(2):240-6.
58. Spector DL. Macromolecular domains within the cell nucleus. Annu Rev Cell Biol 1993; 9:265-315.
59. Fakan S. Perichromatin fibrils are in situ forms of nascent transcripts. Trends Cell Biol 1994; 4(3):86-90.

60. Johnson C, Primorac D, McKinstry M et al. Tracking COL1A1 RNA in osteogenesis imperfecta. splice-defective transcripts initiate transport from the gene but are retained within the SC35 domain. J Cell Biol 2000; 150(3):417-32.
61. Shopland LS, Johnson CV, Byron M et al. Clustering of multiple specific genes and gene-rich R-bands around SC-35 domains: evidence for local euchromatic neighborhoods. J Cell Biol 2003; 162(6):981-90.
62. Wei X, Somanathan S, Samarabandu J et al. Three-dimensional visualization of transcription sites and their association with splicing factor-rich nuclear speckles. J Cell Biol 1999; 146(3):543-58.
63. Lamond AI, Spector DL. Nuclear speckles: a model for nuclear organelles. Nat Rev Mol Cell Biol 2003; 4(8):605-12.
64. Kruhlak MJ, Lever MA, Fischle W et al. Reduced mobility of the alternate splicing factor (ASF) through the nucleoplasm and steady state speckle compartments. J Cell Biol 2000; 150(1):41-51.
65. Phair RD, Misteli T. High mobility of proteins in the mammalian cell nucleus. Nature 2000; 404(6778):604-9.
66. Janicki SM, Tsukamoto T, Salghetti SE et al. From silencing to gene expression: real-time analysis in single cells. Cell 2004; 116(5):683-98.
67. Misteli T, Caceres JF, Spector DL. The dynamics of a pre-mRNA splicing factor in living cells. Nature 1997; 387(6632):523-7.
68. O'Keefe RT, Mayeda A, Sadowski CL et al. Disruption of pre-mRNA splicing in vivo results in reorganization of splicing factors. J Cell Biol 1994; 124(3):249-60.
69. Mintz PJ, Patterson SD, Neuwald AF et al. Purification and biochemical characterization of interchromatin granule clusters. EMBO J 1999; 18(15):4308-20.
70. Saitoh N, Spahr CS, Patterson SD et al. Proteomic analysis of interchromatin granule clusters. Mol Biol Cell 2004; 15(8):3876-90.
71. Politz JC, Tuft RA, Prasanth KV et al. Rapid, diffusional shuttling of poly(A) RNA between nuclear speckles and the nucleoplasm. Mol Biol Cell 2006; 17(3):1239-49.
72. Colwill K, Pawson T andrews B et al. The Clk/Sty protein kinase phosphorylates SR splicing factors and regulates their intranuclear distribution. EMBO J 1996; 15(2):265-75.
73. Sacco-Bubulya P, Spector DL. Disassembly of interchromatin granule clusters alters the coordination of transcription and pre-mRNA splicing. J Cell Biol 2002; 156(3):425-36.
74. Will CL, Luhrmann R. Spliceosomal UsnRNP biogenesis, structure and function. Curr Opin Cell Biol 2001; 13(3):290-301.
75. Achsel T, Stark H, Luhrmann R. The Sm domain is an ancient RNA-binding motif with oligo(U) specificity. Proc Natl Acad Sci U S A 2001; 98(7):3685-9.
76. Kambach C, Walke S, Young R et al. Crystal structures of two Sm protein complexes and their implications for the assembly of the spliceosomal snRNPs. Cell 1999; 96(3):375-87.
77. Stark H, Dube P, Luhrmann R et al. Arrangement of RNA and proteins in the spliceosomal U1 small nuclear ribonucleoprotein particle. Nature 2001; 409(6819):539-42.
78. Izaurralde E, Lewis J, Gamberi C et al. A cap-binding protein complex mediating U snRNA export. Nature 1995; 376(6542):709-12.
79. Izaurralde E, Lewis J, McGuigan C et al. A nuclear cap binding protein complex involved in pre-mRNA splicing. Cell 1994; 78(4):657-68.
80. Ohno M, Segref A, Bachi A et al. PHAX, a mediator of U snRNA nuclear export whose activity is regulated by phosphorylation. Cell 2000; 101(2):187-98.
81. Dahlberg JE, Lund E. Functions of the GTPase Ran in RNA export from the nucleus. Curr Opin Cell Biol 1998; 10(3):400-8.
82. Gorlich D, Henklein P, Laskey RA et al. A 41 amino acid motif in importin-alpha confers binding to importin-beta and hence transit into the nucleus. EMBO J 1996; 15(8):1810-7.
83. Yong J, Wan L, Dreyfuss G. Why do cells need an assembly machine for RNA-protein complexes? Trends Cell Biol 2004; 14(5):226-32.
84. Filipowicz W, Pogacic V. Biogenesis of small nucleolar ribonucleoproteins. Curr Opin Cell Biol 2002; 14(3):319-27.
85. Huber J, Cronshagen U, Kadokura M et al. Snurportin,1 an m3G-cap-specific nuclear import receptor with a novel domain structure. EMBO J 1998; 17(14):4114-26.
86. Narayanan U, Achsel T, Luhrmann R et al. Coupled in vitro import of U snRNPs and SMN, the spinal muscular atrophy protein. Mol Cell 2004; 16(2):223-34.
87. Carvalho T, Almeida F, Calapez A et al. The spinal muscular atrophy disease gene product, SMN: A link between snRNP biogenesis and the Cajal (coiled) body. J Cell Biol 1999; 147(4):715-28.
88. Sleeman JE, Lamond AI. Newly assembled snRNPs associate with coiled bodies before speckles, suggesting a nuclear snRNP maturation pathway. Curr Biol 1999; 9(19):1065-74.

89. Cioce M, Lamond AI. Cajal bodies: a long history of discovery. Annu Rev Cell Dev Biol 2005; 21:105-31.
90. Jady BE, Darzacq X, Tucker KE et al. Modification of Sm small nuclear RNAs occurs in the nucleoplasmic Cajal body following import from the cytoplasm. EMBO J 2003; 22(8):1878-88.
91. Darzacq X, Jady BE, Verheggen C et al. Cajal body-specific small nuclear RNAs: a novel class of 2'-O-methylation and pseudouridylation guide RNAs. EMBO J 2002; 21(11):2746-56.
92. Kiss AM, Jady BE, Darzacq X et al. A Cajal body-specific pseudouridylation guide RNA is composed of two box H/ACA snoRNA-like domains. Nucleic Acids Res 2002; 30(21):4643-9.
93. Kiss T. Small nucleolar RNA-guided posttranscriptional modification of cellular RNAs. EMBO J 2001; 20(14):3617-22.
94. Nesic D, Tanackovic G, Kramer A. A role for Cajal bodies in the final steps of U2 snRNP biogenesis. J Cell Sci 2004; 117(Pt 19):4423-33.
95. Stanek D, Rader SD, Klingauf M et al. Targeting of U4/U6 small nuclear RNP assembly factor SART3/p110 to Cajal bodies. J Cell Biol 2003; 160(4):505-16.
96. Schaffert N, Hossbach M, Heintzmann R et al. RNAi knockdown of hPrp31 leads to an accumulation of U4/U6 di-snRNPs in Cajal bodies. EMBO J 2004; 23(15):3000-9.
97. Sleeman JE, Ajuh P, Lamond AI. snRNP protein expression enhances the formation of Cajal bodies containing p80-coilin and SMN. J Cell Sci 2001; 114(Pt 24):4407-19.
98. Sleeman JE, Trinkle-Mulcahy L, Prescott AR et al. Cajal body proteins SMN and Coilin show differential dynamic behaviour in vivo. J Cell Sci 2003; 116(Pt 10):2039-50.
99. Shpargel KB, Matera AG. Gemin proteins are required for efficient assembly of Sm-class ribonucleoproteins. Proc Natl Acad Sci USA 2005; 102(48):17372-7.
100. Lemm I, Girard C, Kuhn AN et al. Ongoing U snRNP biogenesis is required for the integrity of Cajal bodies. Mol Biol Cell 2006; 17(7):3221-31.
101. Tucker KE, Berciano MT, Jacobs EY et al. Residual Cajal bodies in coilin knockout mice fail to recruit Sm snRNPs and SMN, the spinal muscular atrophy gene product. J Cell Biol 2001; 154(2):293-307.
102. Shin C, Manley JL. Cell signalling and the control of pre-mRNA splicing. Nat Rev Mol Cell Biol 2004; 5(9):727-38.
103. van der Houven van Oordt W, Diaz-Meco MT, Lozano J et al. The MKK(3/6)-p38-signaling cascade alters the subcellular distribution of hnRNP A1 and modulates alternative splicing regulation. J Cell Biol 2000; 149(2):307-16.
104. Shomron N, Alberstein M, Reznik M et al. Stress alters the subcellular distribution of hSlu7 and thus modulates alternative splicing. J Cell Sci 2005; 118(Pt 6):1151-9.
105. Habelhah H, Shah K, Huang L et al. ERK phosphorylation drives cytoplasmic accumulation of hnRNP-K and inhibition of mRNA translation. Nat Cell Biol 2001; 3(3):325-30.
106. Daoud R, Mies G, Smialowska A et al. Ischemia induces a translocation of the splicing factor tra2-beta 1 and changes alternative splicing patterns in the brain. J Neurosci 2002; 22(14):5889-99.
107. Xie J, Lee JA, Kress TL et al. Protein kinase A phosphorylation modulates transport of the polypyrimidine tract-binding protein. Proc Natl Acad Sci USA 2003; 100(15):8776-81.
108. Ding JH, Zhong XY, Hagopian JC et al. Regulated cellular partitioning of SR protein-specific kinases in mammalian cells. Mol Biol Cell 2006; 17(2):876-85.
109. Pederson T. Dynamics and genome-centricity of interchromatin domains in the nucleus. Nat Cell Biol 2002; 4(12):E287-91.
110. Gorski SA, Dundr M, Misteli T. The road much traveled: trafficking in the cell nucleus. Curr Opin Cell Biol 2006; 18:284-90.
111. Prasanth KV, Prasanth SG, Xuan Z et al. Regulating gene expression through RNA nuclear retention. Cell 2005; 123:249-63.
112. Mabon SA, Misteli T. Differential recruitment of pre-mRNA splicing factors to alternatively spliced transcripts in vivo. PLoS Biol. 2005; 3:e374.
113. Andersen JS, Lam YW, Leung AK et al. Nucleolar proteome dynamics. Nature 2005; 433:77-83.
114. Shav-Tal Y, Darzacq X, Singer RH. Gene expression within a dynamic nuclear landscape. EMBO J 2006; 25(15):3469-79.
115. Custódio N, Carvalho C, Condado I et al. In vivo recruitment of exon junction complex proteins to transcription sites in mammalian cell nuclei. RNA 2004; 10(4):622-33.
116. Riethman H, Ambrosini A, Castaneda C et al. Mapping and initial analysis of the human subtelomeric sequence assemblies. Genome Res 2004; 14:18-28.

CHAPTER 2

Spliceosome Assembly and Composition

Arianne J. Matlin and Melissa J. Moore*

Abstract

Cells control alternative splicing by modulating assembly of the pre-mRNA splicing machinery at competing splice sites. Therefore, a working knowledge of spliceosome assembly is essential for understanding how alternative splice site choices are achieved. In this chapter, we review spliceosome assembly with particular emphasis on the known steps and factors subject to regulation during alternative splice site selection in mammalian cells. We also review recent advances regarding similarities and differences between the in vivo and in vitro assembly pathways, as well as proofreading mechanisms contributing to the fidelity of splice site selection.

Introduction

Pre-mRNA splicing occurs within the context of a large macromolecular assemblage known as the spliceosome. Within this enormous (ca. 3 MDa) machine, intron excision occurs in two chemical steps: (1) cleavage at the 5' splice site (donor site), coupled to formation of a lariat structure in which the first nucleotide of the intron is linked via a 2'-5' phosphodiester bond to an intronic adenosine (the branch point) in the vicinity of the 3' splice site (acceptor site); and (2) joining (ligation) of the two exons, coupled to cleavage at the 3' splice site. The spliceosome is a highly dynamic complex containing five stable RNAs (small nuclear or snRNAs) and a host of stably bound and transiently interacting proteins (see below).

Assembly of the spliceosome on a nascent transcript requires recognition of several cis-acting RNA elements located within the intron: the 5' splice site GU, branch point A, polypyrimidine tract and 3' splice site AG. Whereas these elements are surrounded by highly invariant consensus sequences in yeast, in mammals they are generally situated within a more loosely conserved context.[1,2] This creates the potential for greater regulatory flexibility and combinatorial control in mammalian cells, where transcripts can be alternatively spliced. The degree to which splice sites conform to the consensus determines their intrinsic strength and consequently can impact the rate of spliceosome assembly. Correspondingly, the relative strengths of competing splice sites establish a default pathway for each potential splicing event. This default pathway can be overcome in response to physiological or developmental signals.[3] Combinatorial control by multiple trans-acting splicing regulators permits specific and differential recognition of short, degenerate signals and creates a situation in which variations in the concentration of a single factor can elicit a marked change in the splicing pattern.[4]

Composition and Structure

Recent estimates suggest that more than 300 different proteins are involved in mammalian pre-mRNA splicing.[5] Among these are core scaffolding proteins, RNA binding factors and proteins with domains rich in arginine/serine (RS) repeats subject to dynamic phosphorylation. Enzymatic

*Corresponding Author: Melissa J. Moore—Howard Hughes Medical Institute, Department of Biochemistry, Brandeis University, Waltham, MA 02454, USA.
Email: mmoore@brandeis.edu

Alternative Splicing in the Postgenomic Era, edited by Benjamin J. Blencowe and Brenton R. Graveley.

activities include numerous kinases and RNA-dependent ATPases, one GTPase and cis-trans prolyl isomerases. Structurally, the splicing machinery is organized as several more or less stable subunits (akin to ribosomal subunits) that come together during spliceosome assembly, aided by a plethora of additional transiently interacting factors. Key subunits are the U1, U2, U4/U6 and U5 snRNPs (small nuclear ribonucleoprotein complexes), each named for the U snRNA(s) they contain. Another stable subunit, the PRP19 complex, is entirely protein-based. In addition to its snRNA(s), each snRNP carries a set of common core factors (the Sm and Lsm ring proteins) plus 3 to 15 snRNP-specific proteins. Successive base pairing interactions among the snRNAs and between snRNAs and the intronic splice site consensus sequences are critical for the progression of spliceosome assembly and intron excision.

Remarkably, in addition to the major U2-dependent spliceosome, metazoan cells contain a minor spliceosome responsible for the removal of ca. 0.25% of all introns.[6] This so-called U12-dependent spliceosome contains an overlapping set of snRNAs (U11, U12, U4atac, U6atac and U5) and proteins, but also numerous unique factors. When U2 and U12-dependent introns reside in the same transcript, the two spliceosomes are able to coordinate exon definition (see below) and intron excision via their shared components, though introns containing a mixture of U2- and U12-dependent splice sites cannot be excised by either spliceosome.

Many of the individual factors involved in spliceosome assembly will be introduced where relevant in the sections below. However, two classes merit specific attention due to their numerous members. One class comprises members of the SR protein family (refer to chapter by Lin and Fu), which typically contain an N-terminal RRM (RNA recognition motif) and a C-terminal domain rich in RS dipeptides. This RS region is subject to extensive post-translational modification by phosphorylation[7] and can form specific interactions with either proteins or RNA. Various SR proteins function either as general splicing factors or as specific modulators of alternatively spliced exons.[8] The second family comprises the DExH/D box proteins.[9] These ATP-dependent RNA binding proteins can disrupt both RNA-RNA duplexes and RNA-protein interactions, or can act as sequence-independent RNA clamps to secure a section of RNA. Nearly every step in spliceosome assembly and disassembly requires the action of at least one DExH/D box protein, although for the most part, the exact roles of these proteins have yet to be determined.

Although the spliceosome is of similar size and complexity to the ribosome, our structural understanding of the splicing machinery is still in its infancy compared to current knowledge of the translation machinery. Whereas numerous X-ray structures of the individual ribosomal subunits and fully assembled ribosomes exist, only a limited set of low resolution cryo-EM structures have been determined for individual splicing complexes to date.[10] As will become clear below, the highly dynamic nature of spliceosome assembly, with numerous factors coming and going coupled to an extremely intricate choreography of structural contortions, will likely continue to challenge the limits of structural biology for years to come.

In Vitro Spliceosome Assembly

Studies of spliceosome assembly over many years have led to a prevailing view that the pathway proceeds in vitro through a series of distinct complexes and conformations leading to the catalytically active species. According to this model, each successive stage in the transcript-dependent association of snRNPs and other protein factors is a necessary precursor to the next.[11] Native gel electrophoresis, gel filtration and glycerol gradient centrifugation have all been successfully utilized to isolate discrete complexes (H, E, A, B and C) during spliceosome assembly.[12-17] These studies have been carried out most extensively in mammalian and *S. cerevisiae* extracts and more recently in extracts from *S. pombe*. The following discussion will focus mainly on mammalian systems in the context of alternative splicing, although some evidence from yeast will also be mentioned. Therefore, complexes will be described using mammalian nomenclature conventions.

In outline, spliceosome assembly (Fig. 1) begins with the ATP-independent formation of E (early) complex, in which U1 snRNP recognizes the 5' splice site while the protein factors SF1/mBBP and U2AF (Table 1) bind to the branch point and polypyrimidine tract/3' splice site AG,

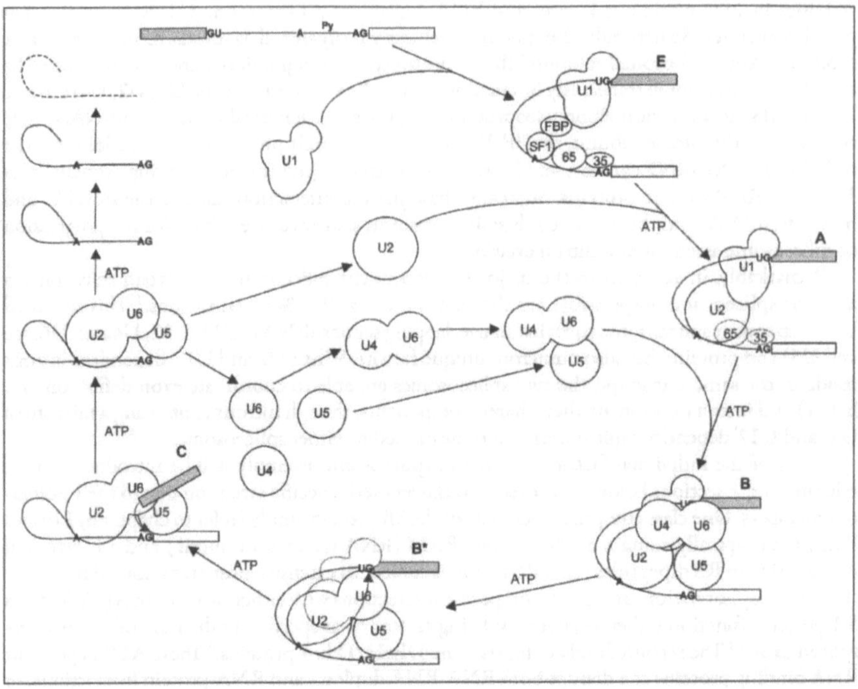

Figure 1. The spliceosome cycle. U1, U2 and U4/U6•U5 snRNPs are illustrated, in addition to selected proteins involved in E complex formation; other factors are also required (Table 1) but have been omitted for simplicity. The pre-mRNA substrate shown contains two exons (gray and white boxes) separated by an intron (black line). Abbreviations: Py, polypyrimidine tract; U1-U6, snRNPs; SF1, splicing factor 1/mBBP; 65, U2AF^{65}; 35, U2AF^{35}; FBP, formin binding protein (Modified from Moore et al 1993).

respectively. All subsequent steps are ATP-dependent. The prespliceosome, or A complex, forms upon stable interaction of U2 snRNP with the branch point region. This is followed by binding of the U4/U6•U5 tri-snRNP particle to the 5' splice site to produce B complex. Subsequent conformational changes destabilize U1 and U4 snRNP interactions to produce B* complex, which is poised to catalyze the first chemical step of splicing. A further set of rearrangements generates C complex, in which the second step of splicing occurs. Upon completion of the second step, additional structural reorganizations are required to release the spliced exons and disassemble the splicing machinery so that it can engage the next substrate.

The following sections provide more detailed discussions of each stage in the in vitro spliceosome assembly pathway, with particular focus on control points important for the regulation of alternative splicing.

H Complex

Another species relevant to the splicing process is H complex. Although not a necessary precursor to spliceosome assembly, H complex is important because it can compete with and regulate the canonical pathway. H complex formation requires neither ATP nor the presence of splice sites in the transcript. It is named after its heterogeneous nuclear ribonucleoprotein (hnRNP) constituents,[18] which comprise a diverse group of chromatin-associated factors containing a range of RNA binding domain types but no RS domains (refer to chapter by Martinez-Contreras

Table 1. Key factors involved in spliceosome assembly

Factor	Function
U1 snRNP	Mediates initial recognition of the 5′ splice site in E complex via base pairing with snRNA and contacts with U1C protein. Can be involved in exon definition interactions with SR proteins. Interaction with spliceosome weakened/displaced upon B complex assembly
U2 snRNP	Forms stable contact with branch point during A complex assembly. Includes SF3a and SF3b protein complexes that mediate additional contacts with pre-mRNA branch point region. U2 snRNA contributes to catalytic site of spliceosome along with U6 snRNA
U4/U6•U5 tri-snRNP	Tri-snRNP interacts stably with spliceosome during B complex formation U4: associates transiently with spliceosome; packages U6 in a catalytically inactive state U5: snRNA and protein components contact transcript and aligns exons for splicing reactions. Proteins contribute to active site as cofactors (see individual proteins) U6: snRNA contributes to catalytic site of spliceosome along with U2 snRNA
hnRNPs	Diverse RNA binding proteins with no RS domain that coat nascent transcripts and modulate (positively or negatively) nucleation of spliceosome assembly
SR proteins	Involved in constitutive and enhancer-dependent splicing. Mediate exon definition interactions via association with U1 snRNP and $U2AF^{35}$. RS domains form sequential contacts with branch point and 5′ splice site during spliceosome assembly
SF1/mBBP	Mediates initial recognition of the branch point in E complex
U2AF	Small subunit ($U2AF^{35}$) recognises the 3′ splice site and large subunit ($U2AF^{65}$) recognises the polypyrimidine tract in E complex, facilitating subsequent U2 snRNP recruitment. RS domain of $U2AF^{35}$ can participate in exon definition interactions, while RS domain of $U2AF^{65}$ contacts branch point directly
Formin binding proteins	Form a bridge between U1 snRNP and SF1/mBBP in E complex
UAP56	DExH/D box protein. Promotes stable U2 snRNP-branchpoint interaction via ATP-dependent weakening of $U2AF^{65}$ and/or SF1/mBBP interactions
Prp5	DExH/D box protein. Forms a bridge between U1 snRNP and U2 snRNP in A complex
Prp8	Integral component of U5 snRNP, involved in formation of contacts between the tri-snRNP and transcript. Also controls timing of 5′ splice site switch to activate B compex. Acts as a cofactor in RNA-catalysed splicing reactions and mediates proofreading of splice sites and branch point
Prp19	Component of a protein complex required for catalytic step I; enhances interactions of U5 and U6 snRNPs with spliceosome. Isy1 component may act together with U6 snRNA to stabilize a spliceosomal conformation that favours step I

continued on next page

Table 1. Continued

Factor	Function
U5-100K (Prp28)	DExH/D box protein, U5 snRNP component. ATPase activity required for displacement of U1 snRNP during 5′ splice site switch
U5-200K (Brr2)	DExH/D box protein, U5 snRNP component. ATPase activity required for destabilization of U4:U6 snRNA duplex during 5′ splice site switch
U5-116K (Snu114)	GTPase, U5 snRNP component. GTPase required for activation of Prp28 and Brr2 during 5′ splice site switch
Prp16	DExH/D box protein. Mediates ATP-dependent proofreading and conformational rearrangement between steps I and II of splicing
Prp17	Participates in ATP-dependent conformational rearrangements mediated by Prp16, following step I of splicing. Involved in 3′ splice site proofreading by Prp8
Slu7	Participates in ATP-independent interactions involving the U5 snRNA and Prp8 during step II of splicing. Potential cofactor for Prp22-mediated proofreading
Prp18	Participates in ATP-independent interactions involving the U5 snRNA and Prp8 during step II of splicing
Prp22	DExH/D box protein. Proofreads step II of splicing and catalyses ATP-dependent release of ligated mRNA product by disrupting contacts with Prp8
Prp43	Hydrolyzes ATP to release intron lariat following catalytic step II

et al). Many of these proteins act as sequence- and splice site-independent RNA packaging factors,[19] which are displaced as active splicing complexes form. Other hnRNPs exhibit sequence specificity and associate with pre-mRNAs in unique combinations and stoichiometries.[20,21] This latter set can affect early splice site recognition, both helping to ensure correct selection of constitutive splice sites over cryptic elements and modulating alternative splicing decisions.[22]

Specific hnRNPs interact, often cooperatively, with enhancer or silencer elements to promote or inhibit subsequent recruitment of the splicing machinery to a particular splice site (refer to chapter by Chasin).[23] For example, hnRNP A1 and the polypyrimidine tract binding protein (PTB or hnRNP I) can repress certain alternative splicing events either by directly blocking binding of the core splicing machinery to a transcript or by antagonizing positively-acting regulatory factors.[24-26] Well-characterized instances of repression by hnRNP A1 involve competition with SR family proteins that promote spliceosome assembly at particular sites.[27,28] In contrast, hnRNP H and the CELF family of hnRNPs can function as activators or repressors of different alternative splicing pathways depending on the context of their binding sites and, in the case of CELF proteins, on competition with PTB.[29-32]

E Complex

In competition with H complex is the mammalian E complex (Fig. 2). This competition represents a prime target for the regulation of alternative splicing.[33] Formation of E complex or the equivalent yeast commitment complex (CC) irreversibly commits a transcript to undergo splicing. Commitment is defined as the ability of the complex to be chased into a functional spliceosome even when challenged with excess competitor RNA. In *S. cerevisiae*, two distinct commitment complexes have been resolved by native gel electrophoresis:[34,35] CC1, which requires only the 5′ splice site, and CC2, which additionally involves branch site interactions.

In mammals, E complex assembly is initiated by U1 snRNP binding to the 5′ splice site.[15,34] This interaction is in part mediated by RNA base pairing between U1 snRNA and pre-mRNA sequences

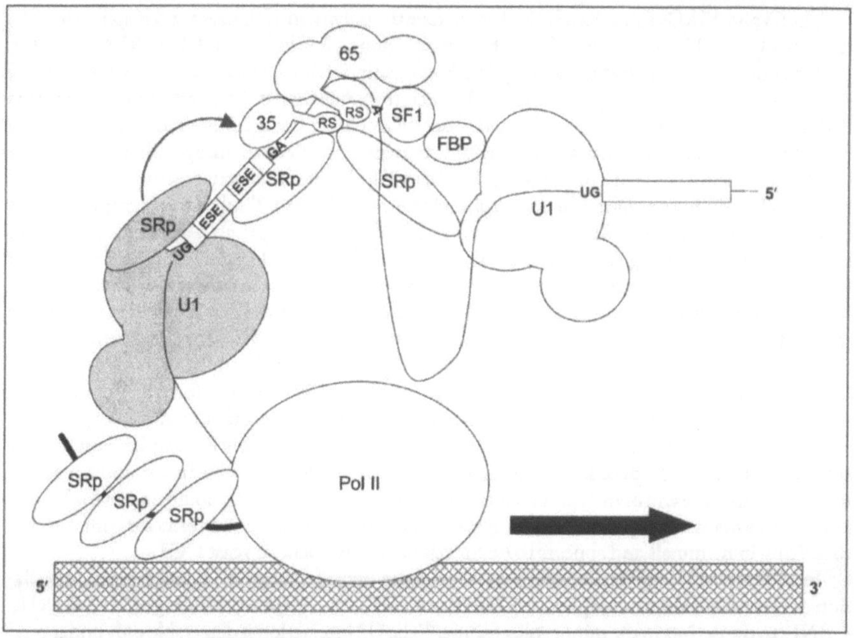

Figure 2. E complex assembled in its cotranscriptional context. The direction of Pol II-mediated transcription of the gene (hatched box) is indicated and SR proteins associated with the Pol II C-terminal domain (thick black line) are shown. Exon definition interactions between an upstream 3′ splice site and downstream 5′ splice site are illustrated by gray shading and the gray arrow. Abbreviations: U1, U1 snRNP; SRp, SR protein; ESE, exonic splicing enhancer; RS, arginine/serine-rich domain; 65, U2AF^{65}; 35, U2AF^{35}; SF1, splicing factor 1/mBBP; FBP, formin binding protein.

encompassing the conserved GU element. Recent studies have identified a set of low abundance human U1 snRNA variants that lack complementarity to the canonical GU dinucleotide, but could potentially interact with other previously identified natural 5' splice site variants.[36,37] In experiments using canonical 5' splice sites, extension of the complementarity between U1 snRNA and the transcript promotes E complex assembly in vitro and enhances the potency of an alternative splice site (although, conversely, it also inhibits later stages of spliceosome assembly by hindering U1 dissociation).[38,39] On the other hand, in *S. cerevisiae* at least, disruption of the base pairing neither abrogates U1-5' splice site association nor diminishes its specificity;[40] protein components of U1 snRNP, including the Sm proteins, also contribute to the recognition process and stabilize the complex.[41,42] The U1C protein in particular mediates sequence-specific contacts with the 5' splice site prior to the formation of RNA base pairing interactions.[38,43,44] Recent data support a model whereby the initial interaction of U1C with the 5' splice site prevents base pairing with U1 snRNA, but a subsequent U1C-mediated conformational change then promotes the canonical, stably base-paired arrangement.[45] Indeed, it has been suggested that the relative stability of RNA duplexes makes them undesirable for the earliest contacts between a transcript and the splicing machinery, but rather enables them to play a proofreading and stabilizing role following initial protein-mediated recognition.[43]

At the opposite end of the intron, initial 3' splice site recognition in mammals is accomplished by U2AF and SF1/mBBP.[46,47] These factors bind cooperatively to the RNA and their direct interaction with one another is regulated by SF1/mBBP phosphorylation.[48] They subsequently collaborate to promote stable association of U2 snRNP during A complex assembly.

The 65 and 35 kDa subunits of U2AF recognize the polypyrimidine tract and 3' splice site AG, respectively.[49,50] U2AF^{65} is a modular protein comprising an RS domain and three RRM domains. RRMs 1 and 2 display high affinity for pyrimidine-/uridine-rich sequences; recent crystallographic data indicate that their ability to recognize degenerate mammalian polypyrimidine tract elements derives from flexible side chains and bound water molecules.[51] The RS domain has been shown to form direct contacts with the branch point in E complex.[52,53] RNA structure probing data indicate that U2AF^{65} bends the transcript to bring the 5' and 3' splice sites into juxtaposition.[54] The importance of U2AF^{65} in the commitment to spliceosome assembly at a 3' splice site is emphasized by observations that its pre-mRNA interactions and/or function are prime targets for interference by alternative splicing regulators such as Sxl and PTB.[25,55] The smaller subunit, U2AF^{35}, contains two protein-protein interaction domains: one contacts U2AF^{65} and the other (an RS domain) may function in enhancer-dependent splicing.[56,57] Under certain conditions, it is required to stabilize the binding of U2AF^{65} to the sub-optimal polypyrimidine tract of AG-dependent introns.[49,58,59]

SF1/mBBP contains a zinc knuckle and a KH-QUA2 domain. In addition to contacting the branch point region via the KH-QUA2 domain,[60-62] SF1/mBBP may also form a bridge to U1C bound at the 5' splice site, through the formin binding proteins.[63] Recently, a mammalian precursor to E complex was resolved under conditions of U2AF depletion.[64] This E' complex contains U1 snRNP bound at the 5' splice site and SF1/mBBP at the branch point. Structure probing revealed that the 5' splice site and branch point regions were in close proximity in E' complex, but the branch point A was not itself required. This complex defines the minimal requirements for commitment to splicing in mammals and appears to be functionally equivalent to yeast CC1.

In addition to U1 snRNP, several lines of evidence suggest a weak, ATP-independent association of U2 snRNP with E complex. This is reminiscent of the concurrent binding of U11 and U12 snRNPs to transcripts in the minor spliceosome.[65] The U2 interaction is detectable only under mild conditions, by immunoprecipitation or MS2/MBP-mediated purification and does not require an intact branch point sequence.[66,67] Both U2 snRNA and some of its associated proteins have been shown to be necessary for E complex formation.[66-68] Nonetheless, a functional requirement for U2 snRNP in the commitment to splicing has not been demonstrated.

SR Proteins in E Complex

In mammals, the canonical splicing signals are highly degenerate and do not contain sufficient information to specify authentic splice sites. The affinity and specificity of the above recognition events for these degenerate sequences is greatly enhanced by an extensive network of cooperative interactions among the components of these complexes. However, the cotranscriptional nature of spliceosome assembly,[69] coupled with the fact that mammalian introns are long, suggests that introns are not the primary unit of recognition. Rather, in mammals, it is the exons that are initially recognized by the splicing machinery. This concept of exon definition involves cross-exon interactions between an upstream 3' splice site and a downstream 5' splice site.[70] Mutual stabilization of U2AF-3' splice site and downstream U1-5' splice site contacts can be mediated by members of the SR protein family that bind to exonic splicing enhancer sequences (ESEs). Excess SR proteins can even compensate for the absence of functional U1 snRNP in the commitment to splicing in HeLa extracts, although the accuracy of 5' splice site selection is compromised.[71,72] The involvement of SR proteins in committing transcripts to splicing provides opportunities for the regulation of alternative splice site selection, particularly through competition with hnRNPs.

Originally, the RS domains of SR proteins were thought to mediate only protein-protein interactions.[57,73] Their capacity to interact with the RS domains of U2AF^{35} and the 70 kDa protein of U1 snRNP (U1-70K) can enhance the recognition of weak splice sites.[74,75] More recently, however, several RS domains were shown to contact the pre-mRNA directly during spliceosome assembly.[52,76] Phosphorylation of RS domains regulates their activity[8] and may be required to reduce their non-specific affinity for RNA, hence enhancing RNA binding specificity.[77] It is important to note that the RNA- and protein-binding capacities of RS domains are not necessarily mutually exclusive and both could be targets for regulatory events involved in alternative splicing.

A Complex

The prespliceosome (A complex) is formed upon stable, ATP-dependent interaction of U2 snRNP with the pre-mRNA branchpoint (Fig. 3).[78-80] Formation of the short pre-mRNA-U2 snRNA intermolecular helix causes the branch point adenosine to protrude, facilitating subsequent nucleophilic attack on the 5' splice site.[81] The tight association of U2 snRNP with the pre-mRNA coincides with displacement of SF1/mBBP from the branch site.[61,80] U1 snRNP also becomes less tightly associated with the transcript than in E complex;[82] indeed, the presence of U1 is not absolutely required for U2 recruitment to the branch site.[83]

Base pairing of U2 snRNA to the branch point region is facilitated by U2AF[65] bound to the polypyrimidine tract. The U2AF[65] RS domain is thought to promote base pairing between U2 snRNA and the branch point by contacting the branch point and neutralizing the negatively charged phosphate backbone.[53] Subsequently, the RS domain of an ESE-bound SR protein appears to contact the branch site/U2 snRNA duplex, replacing the U2AF[65]-branch point interaction and stabilizing the prespliceosome.[52,76] This RNA-binding function of RS domains may compensate for the degeneracy of mammalian splicing signals relative to those of *S. cerevisiae*.[77]

The essential U2 snRNP-associated protein complexes SF3a and SF3b also contact RNA in A complex, interacting primarily upstream of the branch site.[84,85] The ATP dependence of U2 snRNP binding to the transcript derives from conformational changes required to accommodate SF3-RNA interactions.[86] In addition, the p14 subunit of SF3b directly contacts the branch point

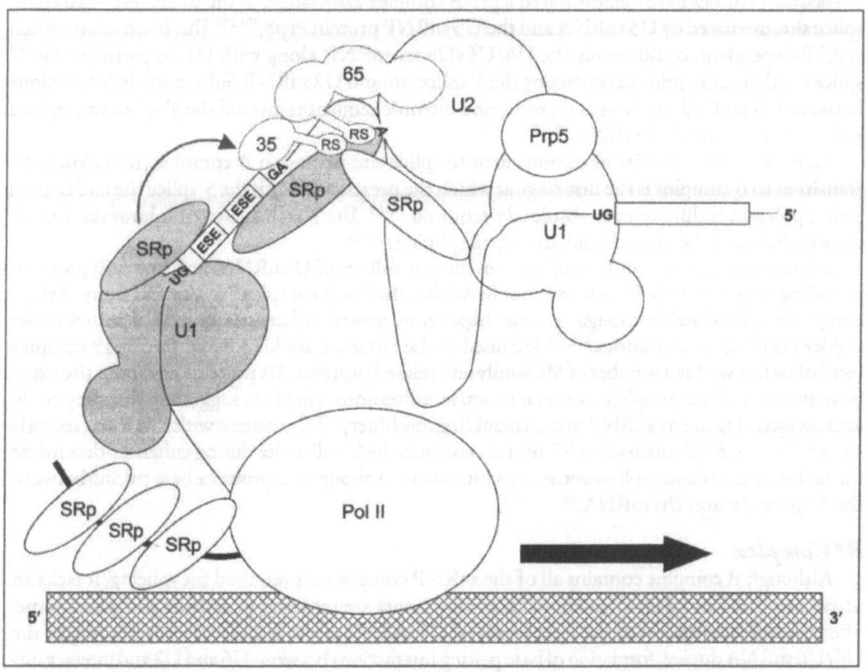

Figure 3. The prespliceosome (A complex) assembled in its cotranscriptional context. The direction of Pol II-mediated transcription of the gene (hatched box) is indicated and SR proteins associated with the Pol II C-terminal domain (thick black line) are shown. Exon definition interactions between an upstream 3' splice site and downstream 5' splice site are illustrated by gray shading and the gray arrow. Abbreviations: U1, U1 snRNP; U2, U2 snRNP; SRp, SR protein; ESE, exonic splicing enhancer; RS, arginine/serine-rich domain; 65, U2AF[65]; 35, U2AF[35].

adenosine,[87] an interaction that persists through subsequent spliceosomal complexes. p14 also contacts SF3b155, which forms a bridge to U2AF[65] during U2 snRNP recruitment, replacing the interaction of U2AF[65] with SF1/mBBP.[88,89]

Assembly of A complex requires the activity of two DExD/H-box ATPases: UAP56 and Prp5. UAP56 was identified as a U2AF[65]-interacting protein required for splicing and stable association of U2 snRNP with the branch point in vitro.[90] Studies of the *S. cerevisiae* ortholog Sub2 have suggested an essential function in ATP-dependent weakening of U2AF[65] and/or SF1/mBBP interactions to accommodate U2 binding to the transcript.[91] Prp5, in contrast, associates directly with both the U1 and U2 snRNPs and thus has been proposed to form a cross-intron or cross-exon bridge between the 5' splice site and the branch site in A complex.[92] Communication between U1 and U2 had previously been inferred from the observed stimulation of U2 binding by a 5' splice site or U1 snRNP.[93,94] Furthermore, the equivalent U11 and U12 components of the minor spliceosome are stably associated as a di-snRNP particle.[65,95] To promote A complex formation, Prp5 mediates an ATP-dependent conformational change in U2 that exposes the branch point-interacting region of the snRNA.[96-98]

B Complex

Initial in vitro analyses defined the progression from A to B complex as the ATP-dependent addition of U4, U5 and U6 snRNPs.[99,100] Multiple lines of evidence indicate that these factors exist as a preformed U4/U6•U5 tri-snRNP particle, in which the U4 and U6 snRNAs are extensively base paired[101,102] to prevent premature formation of the catalytic core of the spliceosome.[103] Subsequent studies have demonstrated a pre-B complex association of the tri-snRNP with the 5' splice site, mediated by U5 snRNA and the U5 snRNP protein Prp8.[104,105] This interaction, which is ATP-dependent, could enable the U4/U6•U5 tri-snRNP, along with U1, to participate in 5' splice site definition prior to contacting the 3' splice site and U2 snRNP. Subsequently, interactions between U5 and U6 snRNAs and exonic and intronic sequences around the 5' splice site replace earlier contacts with U1 snRNA.[106-108]

Consistent with reports of commitment to splice site pairing in A complex (see below), the transition to B complex is the first stage at which the presence of both the 5' splice site and branch point/polypyrimidine tract is absolutely required.[12,109] The 3' splice site AG, however, can be dispensable until the second catalytic step of splicing.[110,111]

Affinity-purified human B complex contains the full set of U snRNAs and over 110 proteins, including approximately 50 that have not been identified in A complex.[112] Whereas many of these components bind stably enough to resist heparin treatment, others can only be detected under milder purification conditions.[112,113] Included in the latter set are U2AF and the Prp19 complex (see below), as well as a number of SR family and related proteins. SR proteins have been shown to be required to chase prespliceosomes into active spliceosomes in vitro, suggesting that they could assist in escorting the tri-snRNP to the assembling machinery.[114] Consistent with this, it was recently shown that the RS domain of an SR protein contacts the 5' splice site during enhancer-dependent formation of the mature spliceosome. This interaction is thought to promote base pairing between the 5' splice site and U6 snRNA.[76]

B Complex*

Although B complex contains all of the snRNP components required for splicing, it lacks an active site. Substantial conformational rearrangements are required to activate the spliceosome. These include disruption of the interaction of U1 snRNP at the 5' splice site, unwinding of the U4/U6 snRNA duplex, formation of base pairing interactions between U6 and U2 and interactions between U6 snRNA and the 5' splice site to form B* complex.[103] These rearrangements contribute to the fidelity of 5' splice site recognition and reinforce contact with the branch point.

The 5' splice site switch requires the NTPase activities of three U5 snRNP components: Prp28, Brr2 and Snu114.[11,103,115] *S. cerevisiae* Prp28 (or its human ortholog U5-100K) displaces U1 snRNP, either by unwinding the U1 snRNA:5' splice site duplex or, more likely, by destabilizing the interaction between U1C and the pre-mRNA.[116,117] In contrast, the ATPase domain of yeast Brr2 (U5-200K in humans) is required to disrupt the U4:U6 snRNA duplex and destabilize the association of U4 snRNP with the spliceosome.[118,119] The activities of Prp28 and Brr2 appear to be inhibited by Prp8 until spliceosome assembly has occurred.[120] Snu114 (human U5-116K) further modulates ATPase activities of these proteins. It has been proposed that the association of the tri-snRNP with A complex triggers GTP hydrolysis by Snu114, altering its interaction with Prp8 in a manner that activates Prp28 and Brr2.[121-123] The conformational rearrangements resulting from this NTPase cascade are essential for the generation of the activated B* complex, which is poised to carry out the first chemical step of splicing.

B* complex formation also requires the Prp19 complex; stable integration of these proteins into the spliceosome is associated with remodelling of the U5 snRNP structure.[124,125] Depletion of Prp19 from HeLa nuclear extract blocks splicing after tri-snRNP addition but prior to the first transesterification reaction.[126] In yeast, the equivalent Prp19 complex (NTC) acts subsequent to U4 destabilization, enhancing the interaction of U5 and U6 with the activated spliceosome.[127,128] Also acting at this stage is Prp2, an essential yeast DExH/D box protein that dissociates from the spliceosome upon ATP hydrolysis.[129,130] Whereas the precise function of Prp2 in spliceosome activation is poorly understood, recent data suggest that its binding to B complex is specified by the associated factor Spp2.[131]

The activated B* complex catalyzes the first transesterification reaction to generate the free 5' exon and lariat-3' exon intermediates. The active site of the spliceosome is most likely composed of the U2 and U6 snRNAs[132-136] and the U5 snRNP protein Prp8.[137,138] Prp8 is believed to act primarily as a cofactor to position the U2 and U6 snRNAs in a conformation that forms an RNA-based catalytic core.[103,139,140] U6 snRNA is the most highly conserved RNA component of the spliceosome and contains two invariant motifs, ACAGAG and AGC; mutations in these sequences can block transesterification.[140,141] The ACAGAG element engages in base pairing interactions with the 5' splice site positioning it for the first catalytic step. Parallels can be drawn between the proposed active site structure of the spliceosome and the divalent cation-dependent ribozyme activities of the Group II self-splicing introns of eubacteria and eukaryotic organelles.[142-144] Indeed, in the presence of magnesium ions, isolated U2 and U6 snRNAs can catalyze a reaction resembling the first step of splicing.[145]

C Complex

Completion of the first catalytic step of splicing leads to formation of C complex, in which the second catalytic step takes place. Although the 3' splice site AG is not essential for C complex formation, it is required for efficient exon ligation.[5,146] Within the active site, the products of the first step of splicing must be realigned to displace the lariat and position the 5' exon for nucleophilic attack on the phosphodiester bond at the 3' splice site.[147,148]

Numerous contacts between the pre-mRNA, the U snRNAs and Prp8 are critical at this stage.[148,149] Within C complex, the 5' and 3' splice sites are held in close proximity by bridging interactions involving U2 and U6 snRNAs and also stem loop I of U5 snRNA, which interacts with both the 3' and 5' exons.[150,151] At this point, Prp8 contacts the branch point, 3' splice site, 5' splice site and the U5 and U6 snRNAs.[105,139,152] It has been proposed that Prp8 and U2 snRNA juxtapose U5 snRNA loop I with the catalytic core to position the 5' and 3' splice sites correctly.[153,154] Finally, the two terminal nucleotide residues of the intron engage in a critical non-Watson-Crick base pairing interaction.[155]

It has recently been proposed that a competitive equilibrium exists between the first and second step conformations of the spliceosome.[156,157] Changes in the stability of interactions between U6 snRNA and the 5' splice site are thought to modulate this equilibrium;[158] the duplex must be disrupted so that the branch structure can be removed from the first step catalytic centre in preparation for the second step.

The activities of several second step-specific proteins are required in C complex to mediate the related functions of remodeling and proofreading, ensuring correct 3' splice site recognition (see below). Conformational rearrangements leading to the second catalytic step are facilitated by the essential DExD/H box protein Prp16, which interacts transiently with the spliceosome and can unwind RNA duplexes in vitro.[159-161] Subsequent to ATP hydrolysis by Prp16, acting in concert with Prp17, the Slu7 and Prp18 proteins cooperate with U5 snRNA and Prp8 in an ATP-independent stage of the second step.[152,162-164] Also required at this stage is the DExH/D box protein Prp22, which is recruited to the 3' splice site, possibly via direct interaction with Slu7.[165,166]

Additionally, recent data indicate an essential role for dephosphorylation of U2 and U5 proteins by spliceosome-associated phosphatases (PP1 and PP2A) in the transition to the second step conformation.[167] The second transesterification reaction then takes place, ligating the exons and liberating the intron.

Product Release and snRNP Recycling

Upon completion of the second step of splicing, the multi-protein exon junction complex (EJC) is deposited at the boundary between the ligated exons, the RNA products are actively discharged from the spliceosome and the snRNPs and recycled for further rounds of catalysis.[168-170] Base pairing of U6 snRNA with the 5' splice site, U2 with the branch point and U5 with the exons is disrupted and the associations of U2, U4 and U6 are ultimately restored to their original configurations.[171]

Release of the ligated mRNA product requires the ATPase activity of Prp22, which disrupts mRNA contacts with Prp8.[172-174] Following dissociation of Slu7, Prp18 and Prp22, the excised intron lariat is liberated from the residual spliceosome by the ATPase Prp43, aided by Ntr1 and Ntr2.[175,176] The lariat is subsequently debranched and degraded.[177,178]

SnRNP recycling involves the re-association of U6 and U4 and then U5. Annealing of U4 and U6 snRNAs is catalyzed by the mammalian protein p110 (Prp24 in *S. cerevisiae*), which dissociates from U6 upon interaction of the U4/U6 di-snRNP with U5.[119,179-181] p110 is enriched in Cajal bodies which are thought to be the site of U4/U6 assembly.[182] In yeast, the NTC is also required for efficient recycling of U4/U6.[183]

Interactions at the Ends of Transcripts That Promote Spliceosome Assembly

Many alternative splicing pathways involve the selection of different promoters or ployadenylation signals at the 5' or 3' ends of a gene. The terminal exons of a transcript represent a special case of exon definition. Experiments involving uncapped transcripts, cap analog competitors and depletion of the cap-binding proteins indicate that interaction of U1 snRNP with the 5' splice site of the first exon is promoted by the 7-methyl-guanosine nuclear cap binding complex (CBC).[184] Similarly, polyadenylation signals stimulate recognition of the final 3' splice site: the C-terminal portion of poly(A) polymerase interacts specifically with U2AF[65] and enhances its interaction with the upstream intron, increasing the efficiency of splicing and coupling it to 3' end formation.[185]

The Holospliceosome Hypothesis

Some recent studies, combined with re-evaluation of longer-established observations, have challenged the traditional sequential view of spliceosome assembly. A 45S particle containing all five snRNPs, the Prp19 complex and additional associated splicing factors can be sedimented from yeast extracts,[186,187] raising the controversial possibility of concerted RNA binding by a pre-assembled complex. This penta-snRNP was prepared at low salt concentrations compatible with splicing and comparable to conditions under which the five snRNAs can be co-immunoprecipitated.[188-192] When supplemented with soluble splicing factors, the penta-snRNP could partially complement snRNA depleted extracts. Base pairing between the U4 and U6 snRNAs provided evidence that the penta-snRNP did not represent a fully mature spliceosome but rather a precursor. These findings are reconcilable with the previously-determined assembly pathway if the E, A and B complexes represent successive stabilizations, as opposed to recruitments, of factors bound to the RNA substrate. Interestingly, while free U2 snRNP was unable to exchange with U2 in the penta-snRNP, U1 exchange could occur. This suggests that a U2/U4/U6/U5 tetra-snRNP particle could potentially interact with a commitment/E complex containing U1 snRNP.[187] Similar tetra-snRNPs have been observed in HeLa nuclear extract,[95,193] although they have not been shown to be functional.

Further suggestions of penta-snRNP pre-assembly have arisen from additional data that apparently contradict the established chronology of events in spliceosome assembly. For example, mammalian E complexes purified using MBP-MS2 were found to contain functional U2 snRNP;[66] this stoichiometric association was not dependent on an intact branch site sequence and was proposed to represent a precursor to ATP-dependent stabilization to form A complex. Assembly of E complex was also shown to require specific modifications of U2 snRNA,[68] and U1-U2 snRNP interactions had previously been detected in HeLa nuclear extract.[194] Furthermore, an early ATP-dependent collaboration between U1 and U5 snRNPs in 5' splice site definition was observed in both HeLa and nematode cell extracts:[195] this was independent of U2 binding to the branch point and was found to involve Prp8 as a component of the U4/U6•U5 tri-snRNP.

Proponents of the holospliceosome hypothesis argue that the sequential spliceosome assembly model is based on the isolation or detection of distinct complexes in vitro under excessively stringent conditions that do not support splicing.[187] While the low ionic strength conditions used to detect the penta-snRNP might mimic the high protein concentrations in the nuclear environment, it has been countered that complex assembly on pre-transcribed RNA in nuclear extracts is a poor reflection of the cotranscriptional splicing process that occurs in vivo.[196] Furthermore, the relevance of the penta-snRNP to the more complex network of weak interactions that characterize mammalian splicing is unknown and it may be incompatible with exon definition and the processing of transcripts containing multiple exons and introns. It has also been suggested that the large excess of U1 and U2 over the other snRNPs in HeLa cells is likely to favor A complex assembly without obligate association of U4/U6•U5; indeed, prespliceosomes assembled in tri-snRNP-depleted extracts can be chased into spliced products by the addition of extract lacking U2.[197]

Insights from In Vivo Analysis of Spliceosome Assembly

In contrast to the studies described above, which have attempted to recapitulate spliceosome assembly in nuclear extracts, experiments have recently been carried out in *S. cerevisiae* to address the pathway by which spliceosomes form in vivo. These studies are critical because in vivo, the pre-mRNA substrate is intimately associated with both RNA polymerase and chromatin, and the appearance of the various splicing signals is temporally separated.[198,199] Thus, the in vivo situation is radically different from in vitro experiments using pre-transcribed naked pre-mRNA substrates.[200]

The in vivo studies utilized chromatin immunoprecipitation (ChIP) to analyse the interactions of selected proteins with nascent RNAs, which in turn are associated with genomic DNA. In summary, the results of these assays do not support the penta-snRNP model. Rather, the spatial pattern of U1 snRNP cross-linking to various genes is reproducibly distinct from that of U2 and U5. The peak of U1 binding occurs shortly after the 5' splice site is transcribed, while the peak in U2 and U5 snRNP binding does not occur until much later.[201] Depletion of U1 snRNA results in failure to recruit U2 and U5 to transcripts.[196] In combination with the inverse correlation between their cross-linking profiles, this suggests that successful recruitment of U2 and U5 requires U1 snRNP, although these snRNPs subsequently promote U1 snRNP dissociation. U5 snRNP association, in turn, is dependent on the presence of U2, and U2 depletion also leads to accumulation of stalled U1-containing commitment complexes. A lag in the cross-linking of U5 with respect to U2 could be resolved by ChIP only when a transcript containing a long second exon was analyzed.[202]

A number of caveats should be considered when interpreting ChIP data.[200] Importantly, the absence of a cross-link does not necessarily imply the absence of a particular interaction, since not all components or conformations of a complex are equally susceptible to detection by this method. Moreover, the relationship between time of association and distance along the gene is not necessarily straightforward. Additionally, the snRNP depletion experiments create an artificial situation; the observation that a U1/U2 snRNP-containing pre-spliceosome is able to assemble in the absence of U5 does not necessarily indicate that it does so under normal conditions. Nonetheless, the results of these ChIP assays are more consistent with the step-wise assembly pathway than the penta-snRNP model. In addition, these ChIP experiments have confirmed the cotranscriptional nature of splicing[199,202] and indicate that spliceosome assembly proceeds to completion before the entire gene has been transcribed.

Commitment to Particular Splice Sites

Ultimately, the key to alternative splicing decisions lies in understanding the splice site commitment process. This process occurs in two separable steps: (1) initial commitment of the transcript to the splicing pathway, followed by (2) committed pairing of a particular set of 5' and 3' splice sites. Commitment of a transcript to splicing in vitro occurs upon formation of E complex. In vivo, this event almost certainly occurs cotranscriptionally as splice sites emerge from the elongating polymerase. However, this early stage of commitment appears to be independent of splice site pairing, which occurs later in spliceosome assembly. In this section we discuss factors that influence this decision.

Trans-splicing is a process whereby one pre-mRNA substrate containing only a 5' splice can be spliced in trans to a second pre-mRNA substrate containing a 3' splice site. This reaction can proceed efficiently provided that the molecule containing the 3' splice site contains an SR protein binding site or a downstream 5' splice site.[203,204] Moreover, the 3' substrate in this reaction can be pre-assembled into A complex prior to exon ligation.[204] Thus, pairing of splice sites can be uncoupled from and occurs subsequent to, their initial recognition.

It appears that for pre-mRNAs that can be alternatively spliced, the splice sites become irreversibly paired in A complex.[205] This was demonstrated by an elegant kinetic trap assay using an RNA substrate containing a single 5' splice site upstream of a pair of competing 3' splice sites that could be selected alternatively depending on the activity of a synthetic splicing enhancer. Splice site pairing could be switched after stalling of spliceosome assembly at E complex, but not A complex, by adding a splicing activator that binds to the enhancer. These experiments suggest that splice site pairing is weak or dynamic prior to the ATP-dependent irreversible selection of a particular 3' element in A complex.[205]

Interestingly, for some introns, the 3' splice site AG dinucleotide that is used for exon ligation is not the same AG that is directly contacted by $U2AF^{35}$ early in spliceosome assembly.[206] Rather, the AG dinucleotide used for exon ligation is specified in an ATP-dependent step that follows lariat formation.[110] A linear search model has been proposed whereby the first AG downstream of the branch point is identified via a 5'→3' scanning mechanism.[207,208] Alternative models include selection according to optimal distance from the branch point or a combination of distance constraints and intrinsic splice site strength.[209,210]

An example from *Drosophila* neatly illustrates that commitment to the splicing reaction and splice site pairing can be uncoupled and demonstrates that alternative splicing can be regulated even at the latest stages of the splicing reaction. Autoregulation of *Sex-lethal (Sxl)* splicing requires an unusual 3' splice site arrangement containing two AG dinucleotides.[206] The distal AG and its associated polypyrimidine tract are essential for splicing and are initially recognized by both subunits of U2AF. However, after the first step of the splicing reaction, the proximal AG is preferentially selected for exon ligation. This splice site switch requires the spliceosomal protein SPF45, which commits and activates the proximal dinucleotide for splicing. SPF45 is antagonized by SXL protein after the first step of splicing, resulting in exon skipping. Interestingly, a similar mechanism was found to activate a cryptic 3' splice site in a human β-globin mutant associated with β-thalassaemia.[206]

Proofreading

Particularly in the context of alternative splicing, questions arise regarding the mechanisms that allow the spliceosome to accommodate flexibility and degeneracy in pre-mRNA sequence elements while maintaining the ability to recognize authentic splice sites accurately. The multiple sequential recognition events that occur at each of the pre-mRNA elements during spliceosome assembly help to ensure their correct identification. Furthermore, the spliceosomal DExD/H box ATPases can act as mediators of proofreading functions that ensure the fidelity of pre-mRNA splicing.[159,211] As discussed above, these proteins are required to facilitate rearrangements throughout spliceosome assembly and disassembly. The proofreading and conformational transitions are not separable functions, but represent two outcomes of the same activity, coupled to the irreversible step of ATP hydrolysis.[157,212] It is currently thought that the two catalytic states of the spliceosome are in a kinetic equilibrium, in which the balance between alternative conformations that promote substrate rejection or catalysis forms the basis of kinetic proofreading.[156,157]

Prp16 (Fig. 4A) was initially identified in a screen in *S. cerevisiae* cells displaying reduced fidelity of splicing.[159] A correlation was observed between impaired Prp16 ATPase activity and enhanced splicing efficiency of branch point mutant transcripts: an increased rate of transition from the first to the second catalytic step conformation was able to out compete the discard pathway for mutant lariat intermediates from the active site.[85,213] Similarly, U6 snRNA and Prp8 contribute to the accuracy of splice site selection by mediating opposite effects on the efficiencies of the two transesterification reactions.[152,214-216] U6 snRNA is involved in branch point and 3' splice site recognition, while Prp8 is additionally required for the fidelity of 5' splice site recognition. It has been proposed that Prp8 stabilizes an RNA tertiary structure that juxtaposes U6 snRNA with the two ends of the intron.[217,218] Genetic evidence from yeast studies suggests that phosphorylation of an unknown substrate by Sky1, a member of the SR protein kinase family, contributes to accurate AG recognition by Prp8 and Prp17.[219] The Isy1 component of the yeast Prp19 complex was recently found to suppress a Prp16 mutant that exhibits reduced fidelity of branch point selection.[220]

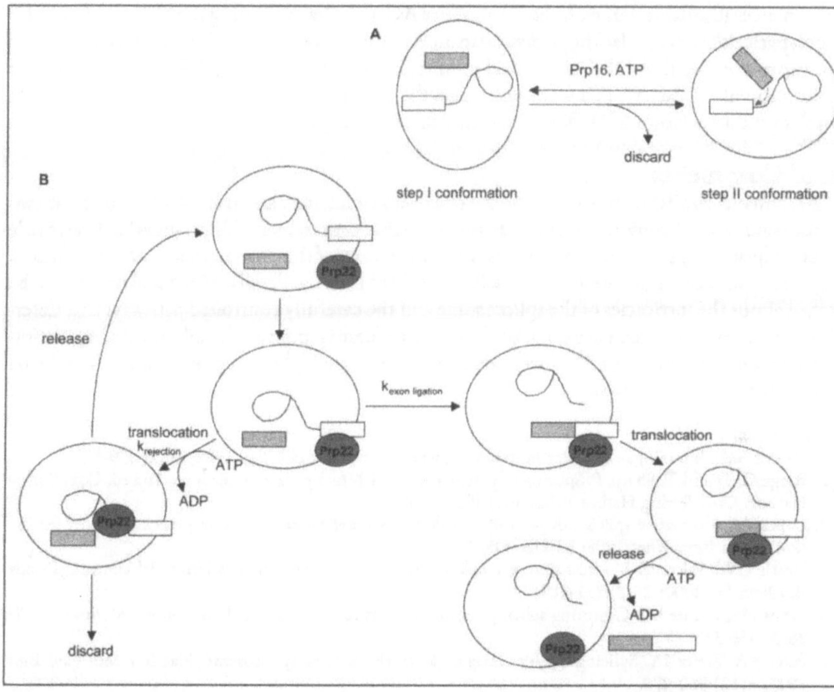

Figure 4. Spliceosomal proofreading mechanisms. A) Prp16-mediated proofreading. In concert with Isy1, Prp8 and U6 snRNA, Prp16 modulates the equilibrium between distinct first and second catalytic step conformations of the spliceosome. Mutant lariat intermediates are discarded from the active site, but mechanistic details are poorly understood at present. B) Prp22-mediated proofreading. (Adapted from Mayas et al 2006.) Competition is proposed to occur between exon ligation and ATP-dependent rejection of defective step I splicing products. For a wild type substrate, exon ligation is followed by Prp22-mediated mRNA dissociation ($k_{exon\ ligation} > k_{rejection}$); for a mutant substrate with slower step II kinetics, ATP hydrolysis by Prp22 results in rejection of the aberrant intermediate ($k_{exon\ ligation} < k_{rejection}$). Substrate rejection may be followed by irreversible discard of the intermediates or by return to a previous conformation

Deletion of Isy1 in a Prp16 wild type background caused a decrease in the accuracy of 3' splice site selection; these data suggest that Prp16 is released prematurely from the spliceosome in the absence of Isy1, allowing the second step of splicing to proceed on mutant substrates. Mutations in Isy1 and U6 snRNA display synthetic lethality, providing further evidence that Isy1 and U6 snRNA collaborate to enhance the fidelity of 3' splice site selection. The normal function of Isy1 could therefore be to act in concert with U6 snRNA to stabilize a spliceosomal conformation that favors the first step of splicing.

Subsequent to Prp16 action, yeast Prp22 mediates proofreading during the second step of splicing, further enhancing fidelity (Fig. 4B).[221] Prp22 appears to increase the accuracy of splice site selection either by rejecting defective first step splicing products or by promoting exon ligation in an ATP-dependent manner.[157] The mechanism of substrate rejection remains unclear, but Slu7, which interacts with Prp22[166] and is required for correct positional selection of 3' splice sites,[162] may act as a cofactor for Prp22.

Finally, the chromatin-associated protein DEK appears to promote correct 3' splice site selection by U2AF.[212,222] In the absence of DEK, U2AF cannot distinguish between AG and CG dinucleotides at the 3' splice site. It appears that phosphorylated DEK interacts with U2AF[35] and enhances its interaction with the 3' splice site AG, thereby preventing the binding of U2AF[65] to polypyrimidine tracts lacking a downstream AG dinucleotide. It will be of great interest to investigate the idea that DEK is involved in kinetic proofreading by modulating the activity of a spliceosomal ATPase.[222] This work also raises the possibility of a phosphorylation-dependent switch in the association of DEK with chromatin or splicing factors.[212]

Final Comments

2007 marks the 30th anniversary of the discovery of introns, and it has also been about two decades since the spliceosome was first identified. In that time, remarkable progress has been made in our understanding of the components and inner workings of this amazing machine. Nonetheless, particularly in relation to alternative splicing, the devil is in the details. Much still remains to be learned about the intricacies of the spliceosome and the carefully controlled pathways that determine the function of its components. With so many potential points of combinatorial regulation in spliceosome assembly, it may emerge that each alternative splicing event is determined by its own unique set of interactions.

References

1. Mount SM. A catalogue of splice junction sequences. Nucleic Acids Res 1982; 10(2):459-472.
2. Burge C, Tuschl T, Sharp, P. Splicing of precursors to mRNAs by the spliceosomes. 2nd ed. Cold Spring Harbor: Cold Spring Harbor Laboratory Press 1999.
3. Lopez AJ. Alternative splicing of pre-mRNA: developmental consequences and mechanisms of regulation. Annu Rev Genet 1998; 32:279-305.
4. Smith CW, Valcarcel J. Alternative pre-mRNA splicing: the logic of combinatorial control. Trends Biochem Sci 2000; 25(8):381-388.
5. Jurica MS, Moore MJ. Capturing splicing complexes to study structure and mechanism. Methods 2002; 28(3):336-345.
6. Patel AA, Steitz JA. Splicing double: insights from the second spliceosome. Nat Rev Mol Cell Biol 2003; 4(12):960-970.
7. Huang Y, Steitz JA. SRprises along a messenger's journey. Mol Cell 2005; 17(5):613-615.
8. Graveley BR. Sorting out the complexity of SR protein functions. RNA 2000; 6(9):1197-1211.
9. Rocak S, Linder P. DEAD-box proteins: the driving forces behind RNA metabolism. Nat Rev Mol Cell Biol 2004; 5(3):232-241.
10. Stark H, Luhrmann R. Cryo-electron microscopy of spliceosomal components. Annu Rev Biophys Biomol Struct 2006; 35:435-457.
11. Brow DA. Allosteric cascade of spliceosome activation. Annu Rev Genet 2002; 36:333-360.
12. Konarska MM, Sharp PA. Electrophoretic separation of complexes involved in the splicing of precursors to mRNAs. Cell 1986; 46(6):845-855.
13. Pikielny CW, Rymond BC, Rosbash M. Electrophoresis of ribonucleoproteins reveals an ordered assembly pathway of yeast splicing complexes. Nature 1986; 324(6095):341-345.
14. Grabowski PJ, Sharp PA. Affinity chromatography of splicing complexes: U2, U5 and U4 + U6 small nuclear ribonucleoprotein particles in the spliceosome. Science 1986; 233(4770):1294-1299.

15. Michaud S, Reed R. An ATP-independent complex commits pre-mRNA to the mammalian spliceosome assembly pathway. Genes Dev 1991; 5(12B):2534-2546.
16. Jamison SF, Crow A, Garcia-Blanco MA. The spliceosome assembly pathway in mammalian extracts. Mol Cell Biol 1992; 12(10):4279-4287.
17. Huang T, Vilardell J, Query CC. Prespliceosome formation in S.pombe requires a stable complex of SF1-U2AF(59)-U2AF(23). EMBO J 2002; 21(20):5516-5526.
18. McAfee JG, Huang, M, Soltaninassab, S et al. The packaging of pre-mRNA. Oxford: Oxford University Press 1997.
19. Thomas JO, Glowacka SK, Szer W. Structure of complexes between a major protein of heterogeneous nuclear ribonucleoprotein particles and polyribonucleotides. J Mol Biol 1983; 171(4):439-455.
20. Bennett M, Pinol-Roma S, Staknis D et al. Differential binding of heterogeneous nuclear ribonucleoproteins to mRNA precursors prior to spliceosome assembly in vitro. Mol Cell Biol 1992; 12(7):3165-3175.
21. Dreyfuss G. Structure and function of nuclear and cytoplasmic ribonucleoprotein particles. Annu Rev Cell Biol 1986; 2:459-498.
22. Pozzoli U, Sironi M. Silencers regulate both constitutive and alternative splicing events in mammals. Cell Mol Life Sci 2005; 62(14):1579-1604.
23. Black DL. Mechanisms of alternative pre-messenger RNA splicing. Annu Rev Biochem 2003; 72:291-336.
24. Charlet BN, Logan P, Singh G et al. Dynamic antagonism between ETR-3 and PTB regulates cell type-specific alternative splicing. Mol Cell 2002; 9(3):649-658.
25. Sharma S, Falick AM, Black DL. Polypyrimidine tract binding protein blocks the 5' splice site-dependent assembly of U2AF and the prespliceosomal E complex. Mol Cell 2005; 19(4):485-496.
26. Villemaire J, Dion I, Elela SA et al. Reprogramming alternative pre-messenger RNA splicing through the use of protein-binding antisense oligonucleotides. J Biol Chem 2003; 278(50):50031-50039.
27. Eperon IC, Makarova OV, Mayeda A, et al. Selection of alternative 5' splice sites: role of U1 snRNP and models for the antagonistic effects of SF2/ASF and hnRNP A1. Mol Cell Biol 2000; 20(22):8303-8318.
28. Mayeda A, Krainer AR. Regulation of alternative pre-mRNA splicing by hnRNP A1 and splicing factor SF2. Cell 24 1992; 68(2):365-375.
29. Chen CD, Kobayashi R, Helfman DM. Binding of hnRNP H to an exonic splicing silencer is involved in the regulation of alternative splicing of the rat beta-tropomyosin gene. Genes Dev 1999; 13(5):593-606.
30. Chou MY, Rooke N, Turck CW et al. hnRNP H is a component of a splicing enhancer complex that activates a c-src alternative exon in neuronal cells. Mol Cell Biol 1999; 19(1):69-77.
31. Gromak N, Matlin AJ, Cooper TA et al. Antagonistic regulation of alpha-actinin alternative splicing by CELF proteins and polypyrimidine tract binding protein. RNA 2003; 9(4):443-456.
32. Savkur RS, Philips AV, Cooper TA. Aberrant regulation of insulin receptor alternative splicing is associated with insulin resistance in myotonic dystrophy. Nat Genet 2001; 29(1):40-47.
33. Reed R. Initial splice-site recognition and pairing during pre-mRNA splicing. Curr Opin Genet Dev 1996; 6(2):215-220.
34. Seraphin B, Rosbash M. Identification of functional U1 snRNA-pre-mRNA complexes committed to spliceosome assembly and splicing. Cell 1989; 59(2):349-358.
35. Seraphin B, Rosbash M. The yeast branchpoint sequence is not required for the formation of a stable U1 snRNA-pre-mRNA complex and is recognized in the absence of U2 snRNA. EMBO J 1991; 10(5):1209-1216.
36. Burset M, Seledtsov IA, Solovyev VV. Analysis of canonical and noncanonical splice sites in mammalian genomes. Nucleic Acids Res 2000; 28(21):4364-4375.
37. Kyriakopoulou C, Larsson P, Liu L et al. U1-like snRNAs lacking complementarity to canonical 5' splice sites. RNA. 2006; 12(9):1603-1611.
38. Lund M, Kjems J. Defining a 5' splice site by functional selection in the presence and absence of U1 snRNA 5' end. RNA 2002; 8(2):166-179.
39. Rossi F, Forne T, Antoine E et al. Involvement of U1 small nuclear ribonucleoproteins (snRNP) in 5' splice site-U1 snRNP interaction. J Biol Chem 1996; 271(39):23985-23991.
40. Du H, Rosbash M. Yeast U1 snRNP-pre-mRNA complex formation without U1snRNA-pre-mRNA base pairing. RNA 2001; 7(1):133-142.
41. Zhang D, Abovich N, Rosbash M. A biochemical function for the Sm complex. Mol Cell 2001; 7(2):319-329.
42. Zhang D, Rosbash M. Identification of eight proteins that cross-link to pre-mRNA in the yeast commitment complex. Genes Dev 1999; 13(5):581-592.
43. Du H, Rosbash M. The U1 snRNP protein U1C recognizes the 5' splice site in the absence of base pairing. Nature 2002; 419(6902):86-90.
44. Heinrichs V, Bach M, Winkelmann G et al. U1-specific protein C needed for efficient complex formation of U1 snRNP with a 5' splice site. Science 1990; 247(4938):69-72.

45. Du H, Tardiff DF, Moore MJ et al. Effects of the U1C L13 mutation and temperature regulation of yeast commitment complex formation. Proc Natl Acad Sci USA 2004; 101(41):14841-14846.
46. Berglund JA, Abovich N, Rosbash M. A cooperative interaction between U2AF[65] and mBBP/SF1 facilitates branchpoint region recognition. Genes Dev 1998; 12(6):858-867.
47. Kramer A, Utans U. Three protein factors (SF1, SF3 and U2AF) function in pre-splicing complex formation in addition to snRNPs. EMBO J 1991; 10(6):1503-1509.
48. Wang X, Bruderer S, Rafi Z et al. Phosphorylation of splicing factor SF1 on Ser20 by cGMP-dependent protein kinase regulates spliceosome assembly. EMBO J 1999; 18(16):4549-4559.
49. Wu S, Romfo CM, Nilsen TW et al. Functional recognition of the 3' splice site AG by the splicing factor U2AF[35]. Nature 1999; 402(6763):832-835.
50. Zamore PD, Patton JG, Green MR. Cloning and domain structure of the mammalian splicing factor U2AF. Nature 1992; 355(6361):609-614.
51. Sickmier EA, Frato KE, Shen H et al. Structural basis for polypyrimidine tract recognition by the essential pre-mRNA splicing factor U2AF[65]. Mol Cell 2006; 23(1):49-59.
52. Shen H, Kan JL, Green MR. Arginine-serine-rich domains bound at splicing enhancers contact the branchpoint to promote prespliceosome assembly. Mol Cell 2004; 13(3):367-376.
53. Valcarcel J, Gaur RK, Singh R et al. Interaction of U2AF[65] RS region with pre-mRNA branch point and promotion of base pairing with U2 snRNA [corrected]. Science 1996; 273(5282):1706-1709.
54. Kent OA, Reayi A, Foong L et al. Structuring of the 3' splice site by U2AF[65]. J Biol Chem 2003; 278(50):50572-50577.
55. Valcarcel J, Singh R, Zamore PD et al. The protein Sex-lethal antagonizes the splicing factor U2AF to regulate alternative splicing of transformer pre-mRNA. Nature 1993; 362(6416):171-175.
56. Zhang M, Zamore PD, Carmo-Fonseca M et al. Cloning and intracellular localization of the U2 small nuclear ribonucleoprotein auxiliary factor small subunit. Proc Natl Acad Sci USA 1992; 89(18):8769-8773.
57. Zuo P, Maniatis T. The splicing factor U2AF[35] mediates critical protein-protein interactions in constitutive and enhancer-dependent splicing. Genes Dev 1996; 10(11):1356-1368.
58. Merendino L, Guth S, Bilbao D et al. Inhibition of msl-2 splicing by Sex-lethal reveals interaction between U2AF[35] and the 3' splice site AG. Nature 1999; 402(6763):838-841.
59. Zorio DA, Blumenthal T. Both subunits of U2AF recognize the 3' splice site in Caenorhabditis elegans. Nature 1999; 402(6763):835-838.
60. Berglund JA, Chua K, Abovich N et al. The splicing factor BBP interacts specifically with the pre-mRNA branchpoint sequence UACUAAC. Cell 1997; 89(5):781-787.
61. Liu Z, Luyten I, Bottomley MJ et al. Structural basis for recognition of the intron branch site RNA by splicing factor 1. Science 2001; 294(5544):1098-1102.
62. Peled-Zehavi H, Berglund JA, Rosbash M et al. Recognition of RNA branch point sequences by the KH domain of splicing factor 1 (mammalian branch point binding protein) in a splicing factor complex. Mol Cell Biol 2001; 21(15):5232-5241.
63. Bedford MT, Reed R, Leder P. WW domain-mediated interactions reveal a spliceosome-associated protein that binds a third class of proline-rich motif: the proline glycine and methionine-rich motif. Proc Natl Acad Sci USA 1998; 95(18):10602-10607.
64. Kent OA, Ritchie DB, Macmillan AM. Characterization of a U2AF-independent commitment complex (E') in the mammalian spliceosome assembly pathway. Mol Cell Biol 2005; 25(1):233-240.
65. Frilander MJ, Steitz JA. Initial recognition of U12-dependent introns requires both U11/5' splice-site and U12/branchpoint interactions. Genes Dev 1999; 13(7):851-863.
66. Das R, Zhou Z, Reed R. Functional association of U2 snRNP with the ATP-independent spliceosomal complex E. Mol Cell 2000; 5(5):779-787.
67. Hong W, Bennett M, Xiao Y et al. Association of U2 snRNP with the spliceosomal complex E. Nucleic Acids Res 1997; 25(2):354-361.
68. Donmez G, Hartmuth K, Luhrmann R. Modified nucleotides at the 5' end of human U2 snRNA are required for spliceosomal E-complex formation. RNA 2004; 10(12):1925-1933.
69. Listerman I, Sapra AK, Neugebauer KM. Cotranscriptional coupling of splicing factor recruitment and precursor messenger RNA splicing in mammalian cells. Nat Struct Mol Biol 2006; 13(9):815-822.
70. Berget SM. Exon recognition in vertebrate splicing. J Biol Chem 1995; 270(6):2411-2414.
71. Crispino JD, Blencowe BJ, Sharp PA. Complementation by SR proteins of pre-mRNA splicing reactions depleted of U1 snRNP. Science 1994; 265(5180):1866-1869.
72. Tarn WY, Steitz JA. SR proteins can compensate for the loss of U1 snRNP functions in vitro. Genes Dev 1994; 8(22):2704-2717.
73. Kohtz JD, Jamison SF, Will CL et al. Protein-protein interactions and 5'-splice-site recognition in mammalian mRNA precursors. Nature 1994; 368(6467):119-124.
74. Wu JY, Maniatis T. Specific interactions between proteins implicated in splice site selection and regulated alternative splicing. Cell 1993; 75(6):1061-1070.

75. Zhu J, Krainer AR. Pre-mRNA splicing in the absence of an SR protein RS domain. Genes Dev 2000; 14(24):3166-3178.
76. Shen H, Green MR. A pathway of sequential arginine-serine-rich domain-splicing signal interactions during mammalian spliceosome assembly. Mol Cell 2004; 16(3):363-373.
77. Shen H, Green MR. RS domains contact splicing signals and promote splicing by a common mechanism in yeast through humans. Genes Dev 2006; 20(13):1755-1765.
78. Chiara MD, Gozani O, Bennett M et al. Identification of proteins that interact with exon sequences, splice sites and the branchpoint sequence during each stage of spliceosome assembly. Mol Cell Biol 1996; 16(7):3317-3326.
79. MacMillan AM, Query CC, Allerson CR et al. Dynamic association of proteins with the pre-mRNA branch region. Genes Dev 1994; 8(24):3008-3020.
80. Rutz B, Seraphin B. Transient interaction of BBP/ScSF1 and Mud2 with the splicing machinery affects the kinetics of spliceosome assembly. RNA 1999; 5(6):819-831.
81. Query CC, Moore MJ, Sharp PA. Branch nucleophile selection in pre-mRNA splicing: evidence for the bulged duplex model. Genes Dev 1994; 8(5):587-597.
82. Michaud S, Reed R. A functional association between the 5' and 3' splice site is established in the earliest prespliceosome complex(E) in mammals. Genes Dev 1993; 7(6):1008-1020.
83. Query CC, McCaw PS, Sharp PA. A minimal spliceosomal complex A recognizes the branch site and polypyrimidine tract. Mol Cell Biol 1997; 17(5):2944-2953.
84. Gozani O, Feld R, Reed R. Evidence that sequence-independent binding of highly conserved U2 snRNP proteins upstream of the branch site is required for assembly of spliceosomal complex A. Genes Dev 1996; 10(2):233-243.
85. Kramer A, Ferfoglia F, Huang CJ et al. Structure-function analysis of the U2 snRNP-associated splicing factor SF3a. Biochem Soc Trans 2005; 33(Pt 3):439-442.
86. Newnham CM, Query CC. The ATP requirement for U2 snRNP addition is linked to the pre-mRNA region 5' to the branch site. RNA 2001; 7(9):1298-1309.
87. Will CL, Schneider C, MacMillan AM et al. A novel U2 and U11/U12 snRNP protein that associates with the pre-mRNA branch site. EMBO J 2001; 20(16):4536-4546.
88. Gozani O, Potashkin J, Reed R. A potential role for U2AF-SAP 155 interactions in recruiting U2 snRNP to the branch site. Mol Cell Biol 1998; 18(8):4752-4760.
89. Spadaccini R, Reidt U, Dybkov O et al. Biochemical and NMR analyses of an SF3b155-p14-U2AF-RNA interaction network involved in branch point definition during pre-mRNA splicing. RNA 2006; 12(3):410-425.
90. Fleckner J, Zhang M, Valcarcel J et al. $U2AF^{65}$ recruits a novel human DEAD box protein required for the U2 snRNP-branchpoint interaction. Genes Dev 1997; 11(14):1864-1872.
91. Kistler AL, Guthrie C. Deletion of MUD2, the yeast homolog of $U2AF^{65}$, can bypass the requirement for sub2, an essential spliceosomal ATPase. Genes Dev 2001; 15(1):42-49.
92. Xu YZ, Newnham CM, Kameoka S et al. Prp5 bridges U1 and U2 snRNPs and enables stable U2 snRNP association with intron RNA. EMBO J 2004; 23(2):376-385.
93. Barabino SM, Blencowe BJ, Ryder U et al. Targeted snRNP depletion reveals an additional role for mammalian U1 snRNP in spliceosome assembly. Cell 1990; 63(2):293-302.
94. Robberson BL, Cote GJ, Berget SM. Exon definition may facilitate splice site selection in RNAs with multiple exons. Mol Cell Biol 1990; 10(1):84-94.
95. Wassarman DA, Steitz JA. Interactions of small nuclear RNAs with precursor messenger RNA during in vitro splicing. Science 1992; 257(5078):1918-1925.
96. O'Day CL, Dalbadie-McFarland G, Abelson J. The Saccharomyces cerevisiae Prp5 protein has RNA-dependent ATPase activity with specificity for U2 small nuclear RNA. J Biol Chem 1996; 271(52):33261-33267.
97. Ruby SW, Chang TH, Abelson J. Four yeast spliceosomal proteins (PRP5, PRP9, PRP11 and PRP21) interact to promote U2 snRNP binding to pre-mRNA. Genes Dev 1993; 7(10):1909-1925.
98. Will CL, Urlaub H, Achsel T et al. Characterization of novel SF3b and 17S U2 snRNP proteins, including a human Prp5p homologue and an SF3b DEAD-box protein. EMBO J 2002; 21(18):4978-4988.
99. Black DL, Chabot B, Steitz JA. U2 as well as U1 small nuclear ribonucleoproteins are involved in pre-messenger RNA splicing. Cell 1985; 42(3):737-750.
100. Cheng SC, Abelson J. Spliceosome assembly in yeast. Genes Dev 1987; 1(9):1014-1027.
101. Lamond AI, Konarska MM, Grabowski PJ et al. Spliceosome assembly involves the binding and release of U4 small nuclear ribonucleoprotein. Proc Natl Acad Sci USA 1988; 85(2):411-415.
102. Will CL, Luhrmann R. Protein functions in pre-mRNA splicing. Curr Opin Cell Biol 1997; 9(3):320-328.
103. Turner IA, Norman CM, Churcher MJ et al. Roles of the U5 snRNP in spliceosome dynamics and catalysis. Biochem Soc Trans 2004; 32(Pt 6):928-931.

104. Newman AJ, Teigelkamp S, Beggs JD. snRNA interactions at 5' and 3' splice sites monitored by photoactivated crosslinking in yeast spliceosomes. RNA 1995; 1(9):968-980.
105. Teigelkamp S, Newman AJ, Beggs JD. Extensive interactions of PRP8 protein with the 5' and 3' splice sites during splicing suggest a role in stabilization of exon alignment by U5 snRNA. EMBO J 1995; 14(11):2602-2612.
106. Sawa H, Abelson J. Evidence for a base-pairing interaction between U6 small nuclear RNA and 5' splice site during the splicing reaction in yeast. Proc Natl Acad Sci USA 1992; 89(23):11269-11273.
107. Sawa H, Shimura Y. Association of U6 snRNA with the 5'-splice site region of pre-mRNA in the spliceosome. Genes Dev 1992; 6(2):244-254.
108. Wyatt JR, Sontheimer EJ, Steitz JA. Site-specific cross-linking of mammalian U5 snRNP to the 5' splice site before the first step of pre-mRNA splicing. Genes Dev 1992; 6(12B):2542-2553.
109. Lamond AI, Konarska MM, Sharp PA. A mutational analysis of spliceosome assembly: evidence for splice site collaboration during spliceosome formation. Genes Dev 1987; 1(6):532-543.
110. Anderson K, Moore MJ. Bimolecular exon ligation by the human spliceosome bypasses early 3' splice site AG recognition and requires NTP hydrolysis. RNA 2000; 6(1):16-25.
111. Rymond BC, Torrey DD, Rosbash M. A novel role for the 3' region of introns in pre-mRNA splicing of Saccharomyces cerevisiae. Genes Dev 1987; 1(3):238-246.
112. Deckert J, Hartmuth K, Boehringer D et al. Protein composition and electron microscopy structure of affinity-purified human spliceosomal B complexes isolated under physiological conditions. Mol Cell Biol 2006; 26(14):5528-5543.
113. Boehringer D, Makarov EM, Sander B et al. Three-dimensional structure of a pre-catalytic human spliceosomal complex B. Nat Struct Mol Biol 2004; 11(5):463-468.
114. Roscigno RF, Garcia-Blanco MA. SR proteins escort the U4/U6.U5 tri-snRNP to the spliceosome. RNA 1995; 1(7):692-706.
115. Will CL, Luhrmann R. Molecular biology. RNP remodeling with DExH/D boxes. Science. 2001; 291(5510):1916-1917.
116. Chen JY, Stands L, Staley JP et al. Specific alterations of U1-C protein or U1 small nuclear RNA can eliminate the requirement of Prp28p, an essential DEAD box splicing factor. Mol Cell 2001; 7(1):227-232.
117. Staley JP, Guthrie C. An RNA switch at the 5' splice site requires ATP and the DEAD box protein Prp28p. Mol Cell 1999; 3(1):55-64.
118. Laggerbauer B, Achsel T, Luhrmann R. The human U5-200kD DEXH-box protein unwinds U4/U6 RNA duplices in vitro. Proc Natl Acad Sci USA 1998; 95(8):4188-4192.
119. Raghunathan PL, Guthrie C. A spliceosomal recycling factor that reanneals U4 and U6 small nuclear ribonucleoprotein particles. Science 1998; 279(5352):857-860.
120. Kuhn AN, Reichl EM, Brow DA. Distinct domains of splicing factor Prp8 mediate different aspects of spliceosome activation. Proc Natl Acad Sci USA 2002; 99(14):9145-9149.
121. Bartels C, Klatt C, Luhrmann R et al. The ribosomal translocase homologue Snu114p is involved in unwinding U4/U6 RNA during activation of the spliceosome. EMBO Rep 2002; 3(9):875-880.
122. Brenner TJ, Guthrie C. Genetic analysis reveals a role for the C terminus of the Saccharomyces cerevisiae GTPase Snu114 during spliceosome activation. Genetics 2005; 170(3):1063-1080.
123. Brenner TJ, Guthrie C. Assembly of Snu114 into U5 snRNP requires Prp8 and a functional GTPase domain. RNA 2006; 12(5):862-871.
124. Makarov EM, Makarova OV, Urlaub H et al. Small nuclear ribonucleoprotein remodeling during catalytic activation of the spliceosome. Science 2002; 298(5601):2205-2208.
125. Tarn WY, Lee KR, Cheng SC. Yeast precursor mRNA processing protein PRP19 associates with the spliceosome concomitant with or just after dissociation of U4 small nuclear RNA. Proc Natl Acad Sci USA 1993; 90(22):10821-10825.
126. Makarova OV, Makarov EM, Urlaub H et al. A subset of human 35S U5 proteins, including Prp,19 function prior to catalytic step 1 of splicing. EMBO J 2004; 23(12):2381-2391.
127. Chan SP, Cheng SC. The Prp19-associated complex is required for specifying interactions of U5 and U6 with pre-mRNA during spliceosome activation. J Biol Chem 2005; 280(35):31190-31199.
128. Chan SP, Kao DI, Tsai WY et al. The Prp19p-associated complex in spliceosome activation. Science 2003; 302(5643):279-282.
129. Kim SH, Lin RJ. Spliceosome activation by PRP2 ATPase prior to the first transesterification reaction of pre-mRNA splicing. Mol Cell Biol 1996; 16(12):6810-6819.
130. King DS, Beggs JD. Interactions of PRP2 protein with pre-mRNA splicing complexes in Saccharomyces cerevisiae. Nucleic Acids Res 1990; 18(22):6559-6564.
131. Silverman EJ, Maeda A, Wei J et al. Interaction between a G-patch protein and a spliceosomal DEXD/H-box ATPase that is critical for splicing. Mol Cell Biol 2004; 24(23):10101-10110.
132. Butcher SE, Brow DA. Towards understanding the catalytic core structure of the spliceosome. Biochem Soc Trans 2005; 33(Pt 3):447-449.

133. Madhani HD, Guthrie C. A novel base-pairing interaction between U2 and U6 snRNAs suggests a mechanism for the catalytic activation of the spliceosome. Cell 1992; 71(5):803-817.
134. Madhani HD, Guthrie C. Randomization-selection analysis of snRNAs in vivo: evidence for a tertiary interaction in the spliceosome. Genes Dev 1994; 8(9):1071-1086.
135. Sun JS, Manley JL. A novel U2-U6 snRNA structure is necessary for mammalian mRNA splicing. Genes Dev 1995; 9(7):843-854.
136. Villa T, Pleiss JA, Guthrie C. Spliceosomal snRNAs: Mg(2+)-dependent chemistry at the catalytic core? Cell 2002; 109(2):149-152.
137. Ben-Yehuda S, Russell CS, Dix I et al. Extensive genetic interactions between PRP8 and PRP17/CDC40, two yeast genes involved in pre-mRNA splicing and cell cycle progression. Genetics 2000; 154(1):61-71.
138. Collins CA, Guthrie C. The question remains: is the spliceosome a ribozyme? Nat Struct Biol 2000; 7(10):850-854.
139. Grainger RJ, Beggs JD. Prp8 protein: at the heart of the spliceosome. RNA 2005; 11(5):533-557.
140. Valadkhan S. snRNAs as the catalysts of pre-mRNA splicing. Curr Opin Chem Biol 2005; 9(6):603-608.
141. Hilliker AK, Staley JP. Multiple functions for the invariant AGC triad of U6 snRNA. RNA 2004; 10(6):921-928.
142. Michel F, Ferat JL. Structure and activities of group II introns. Annu Rev Biochem 1995; 64:435-461.
143. Moore MJ, Sharp PA. Evidence for two active sites in the spliceosome provided by stereochemistry of pre-mRNA splicing. Nature 1993; 365(6444):364-368.
144. Sashital DG, Cornilescu G, McManus CJ et al. U2-U6 RNA folding reveals a group II intron-like domain and a four-helix junction. Nat Struct Mol Biol 2004; 11(12):1237-1242.
145. Valadkhan S, Manley JL. Splicing-related catalysis by protein-free snRNAs. Nature 2001; 413(6857):701-707.
146. Gozani O, Patton JG, Reed R. A novel set of spliceosome-associated proteins and the essential splicing factor PSF bind stably to pre-mRNA prior to catalytic step II of the splicing reaction. EMBO J 1994; 13(14):3356-3367.
147. Rhode BM, Hartmuth K, Westhof E et al. Proximity of conserved U6 and U2 snRNA elements to the 5' splice site region in activated spliceosomes. EMBO J 2006; 25(11):2475-2486.
148. Umen JG, Guthrie C. The second catalytic step of pre-mRNA splicing. RNA 1995; 1(9):869-885.
149. Reed RaP, L. Spliceosome assembly. Oxford: Oxford University Press 1997.
150. Newman AJ, Norman C. U5 snRNA interacts with exon sequences at 5' and 3' splice sites. Cell 1992; 68(4):743-754.
151. O'Keefe RT, Newman AJ. Functional analysis of the U5 snRNA loop 1 in the second catalytic step of yeast pre-mRNA splicing. EMBO J 1998; 17(2):565-574.
152. Umen JG, Guthrie C. Prp16p, Slu7p and Prp8p interact with the 3' splice site in two distinct stages during the second catalytic step of pre-mRNA splicing. RNA 1995; 1(6):584-597.
153. Newman AJ. The role of U5 snRNP in pre-mRNA splicing. EMBO J 1997; 16(19):5797-5800.
154. McGrail JC, Tatum EM, O'Keefe R T. Mutation in the U2 snRNA influences exon interactions of U5 snRNA loop 1 during pre-mRNA splicing. EMBO J 2006; 25(16):3813-3822.
155. Parker R, Siliciano PG. Evidence for an essential nonWatson-Crick interaction between the first and last nucleotides of a nuclear pre-mRNA intron. Nature 1993; 361(6413):660-662.
156. Konarska MM, Query CC. Insights into the mechanisms of splicing: more lessons from the ribosome. Genes Dev 2005; 19(19):2255-2260.
157. Query CC, Konarska MM. Splicing fidelity revisited. Nat Struct Mol Biol 2006; 13(6):472-474.
158. Konarska MM, Vilardell J, Query CC. Repositioning of the reaction intermediate within the catalytic center of the spliceosome. Mol Cell 2006; 21(4):543-553.
159. Burgess S, Couto JR, Guthrie C. A putative ATP binding protein influences the fidelity of branchpoint recognition in yeast splicing. Cell 1990; 60(5):705-717.
160. Schwer B, Guthrie C. PRP16 is an RNA-dependent ATPase that interacts transiently with the spliceosome. Nature 1991; 349(6309):494-499.
161. Wang Y, Wagner JD, Guthrie C. The DEAH-box splicing factor Prp16 unwinds RNA duplexes in vitro. Curr Biol 1998; 8(8):441-451.
162. Chua K, Reed R. The RNA splicing factor hSlu7 is required for correct 3' splice-site choice. Nature 1999; 402(6758):207-210.
163. James SA, Turner W, Schwer B. How Slu7 and Prp18 cooperate in the second step of yeast pre-mRNA splicing. RNA 2002; 8(8):1068-1077.
164. Zhou Z, Reed R. Human homologs of yeast prp16 and prp17 reveal conservation of the mechanism for catalytic step II of pre-mRNA splicing. EMBO J 1998; 17(7):2095-2106.
165. McPheeters DS, Schwer B, Muhlenkamp P. Interaction of the yeast DExH-box RNA helicase prp22p with the 3' splice site during the second step of nuclear pre-mRNA splicing. Nucleic Acids Res 2000; 28(6):1313-1321.

166. van Nues RW, Beggs JD. Functional contacts with a range of splicing proteins suggest a central role for Brr2p in the dynamic control of the order of events in spliceosomes of Saccharomyces cerevisiae. Genetics 2001; 157(4):1451-1467.
167. Shi Y, Reddy B, Manley JL. PP1/PP2A Phosphatases Are Required for the Second Step of Pre-mRNA Splicing and Target Specific snRNP Proteins. Mol Cell 2006; 23(6):819-829.
168. Lejeune F, Maquat LE. Mechanistic links between nonsense-mediated mRNA decay and pre-mRNA splicing in mammalian cells. Curr Opin Cell Biol 2005; 17(3):309-315.
169. Moore MJ, Query, CC, Sharp, PA. Splicing of precursors to mRNA by the spliceosome. 1st ed. Cold Spring Harbor: Cold Spring Harbor Laboratory Press 1993.
170. Tange TO, Nott A, Moore MJ. The ever-increasing complexities of the exon junction complex. Curr Opin Cell Biol 2004; 16(3):279-284.
171. Staley JP, Guthrie C. Mechanical devices of the spliceosome: motors, clocks, springs and things. Cell 1998; 92(3):315-326.
172. Schneider S, Campodonico E, Schwer B. Motifs IV and V in the DEAH box splicing factor Prp22 are important for RNA unwinding and helicase-defective Prp22 mutants are suppressed by Prp8. J Biol Chem 2004; 279(10):8617-8626.
173. Schwer B, Gross CH. Prp22, a DExH-box RNA helicase, plays two distinct roles in yeast pre-mRNA splicing. EMBO J 1998; 17(7):2086-2094.
174. Wagner JD, Jankowsky E, Company M et al. The DEAH-box protein PRP22 is an ATPase that mediates ATP-dependent mRNA release from the spliceosome and unwinds RNA duplexes. EMBO J 1998; 17(10):2926-2937.
175. Boon KL, Auchynnikava T, Edwalds-Gilbert G et al. Yeast ntr1/spp382 mediates prp43 function in postspliceosomes. Mol Cell Biol 2006; 26(16):6016-6023.
176. Tsai RT, Fu RH, Yeh FL et al. Spliceosome disassembly catalyzed by Prp43 and its associated components Ntr1 and Ntr2. Genes Dev 15 2005; 19(24):2991-3003.
177. Martin A, Schneider S, Schwer B. Prp43 is an essential RNA-dependent ATPase required for release of lariat-intron from the spliceosome. J Biol Chem 2002; 277(20):17743-17750.
178. Tanaka N, Schwer B. Mutations in PRP43 that uncouple RNA-dependent NTPase activity and pre-mRNA splicing function. Biochemistry 2006; 45(20):6510-6521.
179. Bell M, Schreiner S, Damianov A et al. p110, a novel human U6 snRNP protein and U4/U6 snRNP recycling factor. EMBO J 2002; 21(11):2724-2735.
180. Ghetti A, Company M, Abelson J. Specificity of Prp24 binding to RNA: a role for Prp24 in the dynamic interaction of U4 and U6 snRNAs. RNA 1995; 1(2):132-145.
181. Medenbach J, Schreiner S, Liu S et al. Human U4/U6 snRNP recycling factor p:110 mutational analysis reveals the function of the tetratricopeptide repeat domain in recycling. Mol Cell Biol 2004; 24(17):7392-7401.
182. Stanek D, Rader SD, Klingauf M et al. Targeting of U4/U6 small nuclear RNP assembly factor SART3/p110 to Cajal bodies. J Cell Biol 2003; 160(4):505-516.
183. Chen CH, Kao DI, Chan SP et al. Functional links between the Prp19-associated complex, U4/U6 biogenesis and spliceosome recycling. RNA 2006; 12(5):765-774.
184. Lewis JD, Izaurralde E. The role of the cap structure in RNA processing and nuclear export. Eur J Biochem 1997; 247(2):461-469.
185. Vagner S, Vagner C, Mattaj IW. The carboxyl terminus of vertebrate poly(A) polymerase interacts with U2AF[65] to couple 3'-end processing and splicing. Genes Dev 2000; 14(4):403-413.
186. Nilsen TW. The spliceosome: no assembly required? Mol Cell 2002; 9(1):8-9.
187. Stevens SW, Ryan DE, Ge HY et al. Composition and functional characterization of the yeast spliceosomal penta-snRNP. Mol Cell 2002; 9(1):31-44.
188. Abovich N, Legrain P, Rosbash M. The yeast PRP6 gene encodes a U4/U6 small nuclear ribonucleoprotein particle (snRNP) protein and the PRP9 gene encodes a protein required for U2 snRNP binding. Mol Cell Biol 1990; 10(12):6417-6425.
189. Arenas JE, Abelson JN. The Saccharomyces cerevisiae PRP21 gene product is an integral component of the prespliceosome. Proc Natl Acad Sci USA 1993; 90(14):6771-6775.
190. Banroques J, Abelson JN. PRP4: a protein of the yeast U4/U6 small nuclear ribonucleoprotein particle. Mol Cell Biol. 1989; 9(9):3710-3719.
191. Horowitz DS, Abelson J. A U5 small nuclear ribonucleoprotein particle protein involved only in the second step of pre-mRNA splicing in Saccharomyces cerevisiae. Mol Cell Biol 1993; 13(5):2959-2970.
192. Raghunathan PL, Guthrie C. RNA unwinding in U4/U6 snRNPs requires ATP hydrolysis and the DEIH-box splicing factor Brr2. Curr Biol 1998; 8(15):847-855.
193. Konarska MM, Sharp PA. Association of U2, U4, U5 and U6 small nuclear ribonucleoproteins in a spliceosome-type complex in absence of precursor RNA. Proc Natl Acad Sci USA 1988; 85(15):5459-5462.

194. Mattaj IW, Habets WJ, van Venrooij WJ. Monospecific antibodies reveal details of U2 snRNP structure and interaction between U1 and U2 snRNPs. EMBO J 1986; 5(5):997-1002.
195. Maroney PA, Romfo CM, Nilsen TW. Functional recognition of 5' splice site by U4/U6.U5 tri-snRNP defines a novel ATP-dependent step in early spliceosome assembly. Mol Cell 2000; 6(2):317-328.
196. Tardiff DF, Rosbash M. Arrested yeast splicing complexes indicate stepwise snRNP recruitment during in vivo spliceosome assembly. RNA 2006; 12(6):968-979.
197. Behzadnia N, Hartmuth K, Will CL et al. Functional spliceosomal A complexes can be assembled in vitro in the absence of a penta-snRNP. RNA 2006; 12(9):1738-1746.
198. Kornblihtt AR, de la Mata M, Fededa JP et al. Multiple links between transcription and splicing. RNA 2004; 10(10):1489-1498.
199. Neugebauer KM. On the importance of being co-transcriptional. J Cell Sci 2002; 115(Pt 20):3865-3871.
200. Nilsen TW. Spliceosome assembly in yeast: one ChIP at a time? Nat Struct Mol Biol 2005; 12(7):571-573.
201. Lacadie SA, Rosbash M. Cotranscriptional spliceosome assembly dynamics and the role of U1 snRNA:5'ss base pairing in yeast. Mol Cell 2005; 19(1):65-75.
202. Gornemann J, Kotovic KM, Hujer K, Neugebauer KM. Cotranscriptional spliceosome assembly occurs in a stepwise fashion and requires the cap binding complex. Mol Cell 2005; 19(1):53-63.
203. Bruzik JP, Maniatis T. Enhancer-dependent interaction between 5' and 3' splice sites in trans. Proc Natl Acad Sci USA 1995; 92(15):7056-7059.
204. Chiara MD, Reed R. A two-step mechanism for 5' and 3' splice-site pairing. Nature 1995; 375(6531):510-513.
205. Lim SR, Hertel KJ. Commitment to splice site pairing coincides with A complex formation. Mol Cell 2004; 15(3):477-483.
206. Lallena MJ, Chalmers KJ, Llamazares S et al. Splicing regulation at the second catalytic step by Sex-lethal involves 3' splice site recognition by SPF45. Cell 2002; 109(3):285-296.
207. Chen S, Anderson K, Moore MJ. Evidence for a linear search in bimolecular 3' splice site AG selection. Proc Natl Acad Sci USA 2000; 97(2):593-598.
208. Smith CW, Porro EB, Patton JG et al. Scanning from an independently specified branch point defines the 3' splice site of mammalian introns. Nature 1989; 342(6247):243-247.
209. Chiara MD, Palandjian L, Feld Kramer R et al. Evidence that U5 snRNP recognizes the 3' splice site for catalytic step II in mammals. EMBO J 1997; 16(15):4746-4759.
210. Luukkonen BG, Seraphin B. The role of branchpoint-3' splice site spacing and interaction between intron terminal nucleotides in 3' splice site selection in Saccharomyces cerevisiae. EMBO J 1997; 16(4):779-792.
211. Burgess SM, Guthrie C. Beat the clock: paradigms for NTPases in the maintenance of biological fidelity. Trends Biochem Sci 1993; 18(10):381-384.
212. Kress TL, Guthrie C. Molecular biology. Accurate RNA siting and splicing gets help from a DEK-hand. Science 2006; 312(5782):1886-1887.
213. Burgess SM, Guthrie C. A mechanism to enhance mRNA splicing fidelity: the RNA-dependent ATPase Prp16 governs usage of a discard pathway for aberrant lariat intermediates. Cell 1993; 73(7):1377-1391.
214. Kandels-Lewis S, Seraphin B. Involvement of U6 snRNA in 5' splice site selection. Science 1993; 262(5142):2035-2039.
215. Lesser CF, Guthrie C. Mutations in U6 snRNA that alter splice site specificity: implications for the active site. Science 1993; 262(5142):1982-1988.
216. McPheeters DS. Interactions of the yeast U6 RNA with the pre-mRNA branch site. RNA 1996; 2(11):1110-1123.
217. Collins CA, Guthrie C. Allele-specific genetic interactions between Prp8 and RNA active site residues suggest a function for Prp8 at the catalytic core of the spliceosome. Genes Dev 1999; 13(15):1970-1982.
218. Siatecka M, Reyes JL, Konarska MM. Functional interactions of Prp8 with both splice sites at the spliceosomal catalytic center. Genes Dev 1999; 13(15):1983-1993.
219. Dagher SF, Fu XD. Evidence for a role of Sky1p-mediated phosphorylation in 3' splice site recognition involving both Prp8 and Prp17/Slu4. RNA 2001; 7(9):1284-1297.
220. Villa T, Guthrie C. The Isy1p component of the NineTeen complex interacts with the ATPase Prp16p to regulate the fidelity of pre-mRNA splicing. Genes Dev 2005; 19(16):1894-1904.
221. Mayas RM, Maita H, Staley JP. Exon ligation is proofread by the DExD/H-box ATPase Prp22p. Nat Struct Mol Biol 2006; 13(6):482-490.
222. Soares LM, Zanier K, Mackereth C et al. Intron removal requires proofreading of U2AF/3' splice site recognition by DEK. Science 2006; 312(5782):1961-1965.

Chapter 3

Relating Alternative Splicing to Proteome Complexity and Genome Evolution

Yi Xing and Christopher Lee*

Abstract

Prior to genomics, studies of alternative splicing primarily focused on the function and mechanism of alternative splicing in individual genes and exons. This has changed dramatically since the late 1990s. High-throughput genomics technologies, such as EST sequencing and microarrays designed to detect changes in splicing, led to genome-wide discoveries and quantification of alternative splicing in a wide range of species from human to Arabidopsis.[1,2] Consensus estimates of AS frequency in the human genome grew from less than 5% in mid-1990s to as high as 60-74% now.[3] The rapid growth in sequence and microarray data for alternative splicing has made it possible to look into the global impact of alternative splicing on protein function and evolution of genomes. In this chapter, we review recent research on alternative splicing's impact on proteomic complexity and its role in genome evolution.

Alternative Splicing and Proteomic Complexity

Alternative splicing is a ubiquitous regulatory mechanism of protein function.[4] In the central dogma of molecular biology, one gene makes one protein, which carries out a specific function. In 1978, Walter Gilbert hypothesized that variations in splicing, i.e., alternative splicing, can generate functionally distinct protein products from a single gene.[5] Nowadays, it is widely accepted that this new paradigm of "one gene, many proteins" is how we should think about gene function and regulation in higher eukaryotes.

When an alternative splicing event occurs within the protein-coding region of a gene, it can modify the protein product in a variety of ways. If the length of the alternatively spliced region is an exact multiple of three nucleotides, alternative splicing will insert or remove a peptide segment without affecting the rest of the protein. Such alternative splicing events are often referred to as "frame-preserving" alternative splicing events (see Fig. 1). By contrast, if the length of the alternatively spliced region is not a multiple of three nucleotides, alternative splicing will shift the downstream reading frame. Such alternative splicing events are referred to as "frame-switching" alternative splicing events (Fig. 1). Genome-wide analyses of alternative splicing in human and other eukaryotes show that 40% of alternative splicing events are frame-preserving, similar to the percentage expected by random chance (~40%).[6] Interestingly, this percentage is much higher in evolutionarily conserved alternative exons.[7-10] For exons observed to be alternatively spliced in multiple species, 50-70% are frame-preserving. A recent analysis of mouse splicing microarray data indicates that 53% of tissue-specific exons in mouse are frame-preserving[11] (also see[12]). Taken together, these data suggest

*Corresponding Author: Christopher Lee—Department of Chemistry, University of California, Los Angeles, California 90095, USA. Email: leec@chem.ucla.edu

Alternative Splicing in the Postgenomic Era, edited by Benjamin J. Blencowe and Brenton R. Graveley.

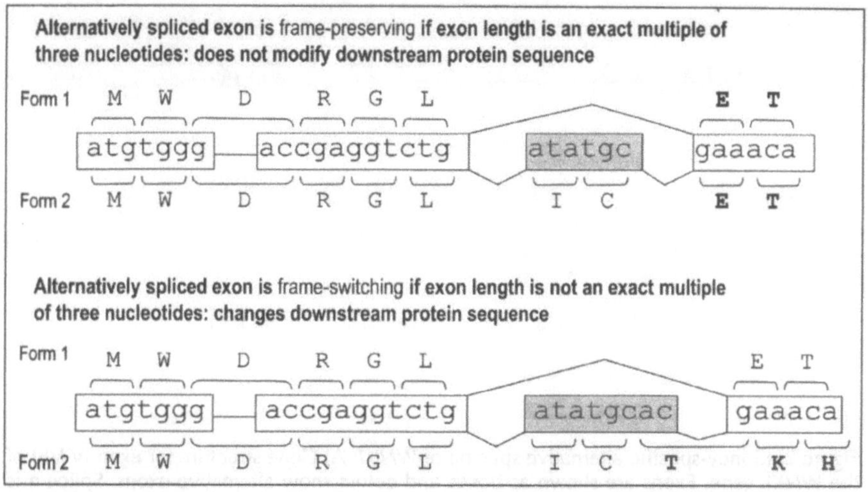

Figure 1. Exon length determines whether an alternatively spliced single-exon skip is frame-preserving or frame-switching. We define an alternatively spliced exon as frame-preserving if its length is an exact multiple of three nucleotides, as its alternative splicing will not alter the protein reading frame of subsequent exons (top). Reproduced from: Resch A et al. Nucleic Acids Res 2004; 32:1261-1269;[8] with permission of Oxford University Press.

that evolutionarily ancient "functional" alternative splicing events tend to add or delete a modular protein-coding unit while keeping the rest of the protein intact. It is worth noting that a large number of frame-switching alternative splicing events introduce premature termination codons (PTCs) into the transcript isoforms.[13] The mRNA nonsense-mediated decay (NMD) pathway is activated, leading to the degradation of the premature transcripts[13,14] (refer to chapter by Lareau et al).

Alternative splicing can cause large insertions or deletions within the protein product, or subtle changes as small as a few amino acids. Wang and colleagues show that the length of alternatively spliced regions follows an approximate power-law distribution.[15] Statistical analyses of alternative cassette exons show that they are significantly shorter than constitutive exons,[16] and the occurrence of alternative splicing is much more frequent in micro-exons (i.e., exons less than 30bp).[17] In an extreme situation, alternative splicing at NAGNAG 3' splice sites shifts the site of exon ligation by 3nt, causing the insertion or deletion of a single amino acid to the resulting protein isoforms.[18] However, it is not clear whether such a single amino acid indel event is functionally relevant, or simply reflects the inherent noise and stochasticity in splice site recognition.[19]

Alternative Splicing of Protein Domains and Functional Sites

In many cases, bioinformatic analyses of full-length protein isoform sequences can suggest the functional consequence of a specific alternative splice form. Figure 2 shows one such example. Xu and colleagues identified a novel alternative splice form in the serine/threonine kinase gene *WNK1*.[20] *WNK1* and its kidney-specific homolog *WNK4* express proteins that localize to the distal renal tubules of the kidney and play an important role in maintaining salt balance of the body.[21] Interestingly, *WNK1* expresses a novel splice form specifically in the kidney that contains a novel exon upstream of exon 5. This novel kidney-specific splice form disrupts the N-terminal kinase domain encoded by the non-kindey *WNK1* isoform. Based on such bioinformatics evidence, Xu and colleagues hypothesized that alternative splicing of *WNK1* down-regulates *WNK1* kinase activity in the kidney, allowing the *WNK1* activity to be replaced by *WNK4*. Misregulation of this splicing event can lead to elevated kinase activity and might be responsible

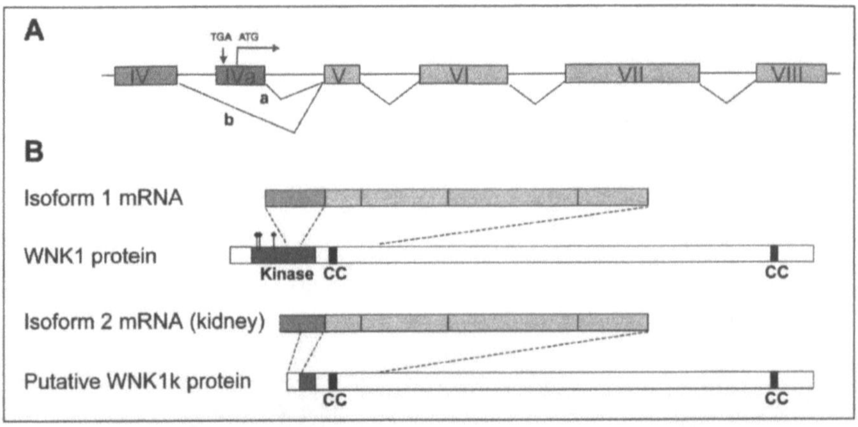

Figure 2. Kidney-specific alternative splicing of *WNK1*. A) Gene structure for exon IV-VIII of the *WNK1* gene. Exons are shown as boxes and colors show alternative exons. Splice a is specific to kidney. The putative in-frame stop codon TGA and start codon ATG are indicated. B) The two alternative forms of *WNK1* mRNA inferred from the expressed sequence data and the schematic representation of *WNK1* protein sequences. The conserved kinase domain, two coiled-coil (CC) domains and the corresponding protein regions of mRNA forms are indicated. Three amino acids (K233, C250, D368) that are required for the kinase activity of *WNK1* are marked by flags on the *WNK1* protein. Xu Q et al. Nucleic Acids Res 2002; 30:3754-3766;[20] with permission of Oxford University Press.

for pseudohypoaldosteronism type II (PHAII) hypertension.[20] These bioinformatics predictions were followed by a series of experimental investigations, which established the existence and functional importance of *WNK1* alternative splicing.[22-26]

A number of studies have performed sequence-analyses of full-length protein isoforms to investigate the global impact of alternative splicing on protein domains. Kriventseva and colleagues extracted 4084 protein isoforms of 1780 genes from the SWISS-PROT database and mapped the alternatively spliced regions onto protein domain annotations in Pfam, SMART and PROSITE. Their analysis indicates that alternative splicing tends to remove or insert a protein domain in its entirety (33% of alternatively spliced regions, compared to 16% expected by a random model).[27] Similarly, short alternative splicing events within protein domains tend to target functional residues more frequently than expected by random chance. These data suggest that natural selection favors the use of alternative splicing in creating functional diversity of the proteome. Liu and colleagues collected 932 human genes with multiple RefSeq transcripts in the NCBI Locuslink database.[28] Twenty-four CDD (Conserved Domain Database) domains had a strong bias towards alternatively spliced regions. These domains are involved in diverse biological processes such as apoptosis, calcium signaling, transcriptional regulation, etc. Resch and colleagues constructed a full-length protein isoform database from EST sequences, consisting of 13384 full-length isoforms of 4422 human genes.[6,29] They identified fifty CDD domains as being preferentially targeted by alternative splicing, including many well-known protein interaction domains such as KRAB domain and ankyrin repeats. In the Kruppel family of transcription factors, alternative splicing has a strikingly strong tendency toward disrupting its protein interaction domain (KRAB repressor domain), while leaving the DNA binding domain intact. Alternative splicing of these transcription factors acts like a "switch", converting a transcriptional repressor into an activator by turning off key protein-protein interactions. Preferential domain targeting by alternative splicing was also observed in mouse transcription factors.[30] Although exon skipping commonly leads to the disruption of protein domains, Hiller and colleagues show that it can also create new domains by joining two nonconsecutive exons together.[31]

Another subject of extensive studies is the alternative splicing of transmembrane segments. The release of functional fragments from membrane anchorage via proteolysis is a well-known regulatory mechanism of transmembrane proteins.[32] Interestingly, alternative splicing can produce a similar effect, by removing the exon coding for the transmembrane segment. For example, alternative splicing of *IL-6R* can create a soluble isoform that activates *IL-6R* signaling in cells with no endogenous *IL-6R* expression.[33] Genome-wide analyses of alternative splicing in human and mouse genes indicate that alternative splicing-mediated removal of transmembrane segments is prevalent in genes encoding single-pass membrane proteins.[34,35] Many membrane-bound receptors generate soluble isoforms by alternative splicing. The soluble isoform can act as a diffuse signal for activating signaling in other cells (such as IL-6R), or as an antagonist to the signaling pathway in the original cells by competing for ligand. Davis and colleagues used a membrane organization annotation pipeline to analyze 8032 mouse genes with multiple protein isoforms in the FANTOM3 database.[36] They classified protein isoforms into soluble intracellular proteins, soluble secreted proteins, type I membrane proteins, type II membrane proteins and multi-spanning membrane proteins. The conversion between soluble proteins and membrane proteins was found to be the most frequent type of conversion between different protein isoforms of the same gene.[36]

These large-scale analyses of protein isoforms (also see refs. 37,38) shed light on the role of alternative splicing in regulating protein function. Other mechanisms of gene regulation, such as transcriptional regulation and micoRNA-mediated translational repression, change the quantity of the final protein product. In contrast, alternative splicing can create a new protein with very different functions by altering its key functional regions such as globular domains and subcellular localization signals. From an evolutionary point of view, alternative splicing is like nature's protein engineering experiment to test the functionality of novel protein forms (see details below).

Alternative Splicing and Protein Structure

In contrast to the large amount of sequence data available for alternatively spliced proteins, there is very limited structural information for alternatively spliced proteins. So far 3D structures for only a handful of protein isoform pairs have been solved (see a summary in ref. 39). In two cases (*EDA-A* and *AdGST-1*), alternative splicing inserts less than ten amino acids into folded regions, causing a slight structural rearrangement. In three other cases, alternative splicing alters protein segments that are intrinsically disordered and the structured regions of these three proteins remain unaltered. Romero and colleagues collected 46 alternatively spliced proteins having experimentally characterized disordered regions.[39] 34% of total residues on these proteins are located in disordered regions. The percentage is significantly higher for alternatively spliced regions, with 57% of the residues being disordered. The same trend was observed in a larger dataset containing proteins with computationally predicted disordered regions. Disordered regions tend to be enriched for protein functional sites such as binding sites for SH3 domain or substrate sites for posttranslational modification. Therefore the high preference of alternative splicing in disordered regions provides a means to expand the functional diversity of proteins without disrupting the protein structure. In fact, since the frequencies of alternative splicing and protein intrinsic disorders both increase dramatically in higher eukaryotes, Romero and colleagues suggest that expansions of these two phenomenons might be evolutionarily linked.[39]

A few studies have used structure prediction tools to investigate the impact of alternative splicing on protein structure. Wang and colleagues collected structures of 1209 alternatively spliced proteins, including 351 from PDB and 858 from computational structure predictions. They show that alternative splicing events tend to occur in coiled regions and are usually located at the surface of proteins.[15] Wen and colleagues focused on 57 cases of very-short alternative splicing (VSAS) events with less than 17aa insertions or deletions and computationally predicted the secondary structures of these protein isoforms. In 2/3 of the cases, alternative splicing inserts a fragment with a predicted secondary structure different from those of its flanking sequences. This result suggests that short alternative splicing events might have an unexpectedly significant impact on the protein products.[40] This is indeed the case in the alternative splicing of the Piccolo gene.[41]

Alternative splicing of Piccolo inserts nine amino acids into its C2A domain. Structural modeling predicted this protein segment to elongate a surface loop without affecting the Calcium binding site of Piccolo. However, NMR structure determination of the longer isoform indicated a very surprising 3D structure. The nine-amino-acid insertion displaced a beta-strand in the core of the protein, triggering a large conformational change and altering the Calcium binding activity of Piccolo. This example underscores the importance of using experimental approaches to study the impact of alternative splicing on protein structure.

Alternative Splicing and Genome Evolution

Over the last few years, another major theme has emerged: the effect of alternative splicing on various processes of genome evolution, ranging from the small scale (individual nucleotide mutations) to the large scale (e.g., exon creation and loss). In this connection, it is significant that Gilbert's early proposal of the alternative splicing hypothesis explicitly suggested that it could make important contributions to genome and proteome evolution.[5] Shortly after alternative splicing entered its "genomic era" (via large-scale EST analyses), bioinformatics researchers began to examine this intriguing hypothesis using genome-wide alternative splicing databases.

Several parallel lines of evidence suggest that alternative splicing is associated with reduced selection pressure (and thus more rapid evolutionary change) than is observed in constitutively spliced regions. For example, Sorek and coworkers showed that Alu sequences, which normally are considered disruptive of protein coding regions, are in fact found in protein-coding exons, but only in alternatively spliced exons.[42] They argued that Alu elements would not be tolerated in constitutive exons due to negative selection pressure against disrupting the protein product, but evidently are tolerated in alternative exons (which are only included in a fraction of the transcripts synthesized from the gene). Similarly, a comparative genomics study of exon conservation found that alternatively spliced exons show much more rapid rates of turnover during mammalian evolution (exon creation and loss) than do constitutive exons.[43] Specifically, the exon inclusion level (the fraction of transcripts of a gene which contain a given exon, estimated from EST data) made a crucial difference: constitutive and major-form exons (those included in the majority of transcripts) were strongly conserved (e.g., 98% are conserved between mouse and human), but minor-form exons (those included in only a minority of transcripts) were highly divergent (about three-quarters are expressed as exons in mouse but not in human, or vice-versa and indeed have poor homology to the orthologous gene in human). These results suggested that minor-form alternatively spliced exons undergo much more rapid creation and loss during mammalian genome evolution.

A number of studies have analyzed such "genome-specific" exons. Pan et al designed a DNA microarray to detect alternative splicing and to measure exon inclusion levels quantitatively.[44] Comparing the microarray-measured inclusion levels vs. exon conservation patterns between mouse and human, they found that only a small fraction (about a quarter) of major-form exons were "genome-specific", but that a large fraction (over 70%) of minor-form exons were "genome-specific". Studies of conservation of constitutive exons vs. alternative exons in insects[45] obtained similar results, indicating that this pattern may be a general feature of alternative splicing in genome evolution. Pan and coworkers also showed that alternative splicing introduces another new form of plasticity in genome evolution, which they referred to as "species-specific alternative splicing". Whereas constitutive exons display little evolutionary change between human and mouse (e.g., 98% are evolutionarily conserved between the two species), a substantial fraction (11%) of exons that are alternatively spliced in one species are found to be constitutive in the other species.[46] It is notable that such a major change (skipping of an entire exon) has evolved at such a high frequency (11% of cases) during mammalian evolution; for purposes of comparison, this is about the same rate as is observed at the opposite extreme, for very minor changes—single nucleotide mutations, most of them silent—in human vs. mouse exons. Again, the implication is that alternative splicing is associated with rapid evolution of large-scale changes in gene structure, which could have large effects on gene products and functions.

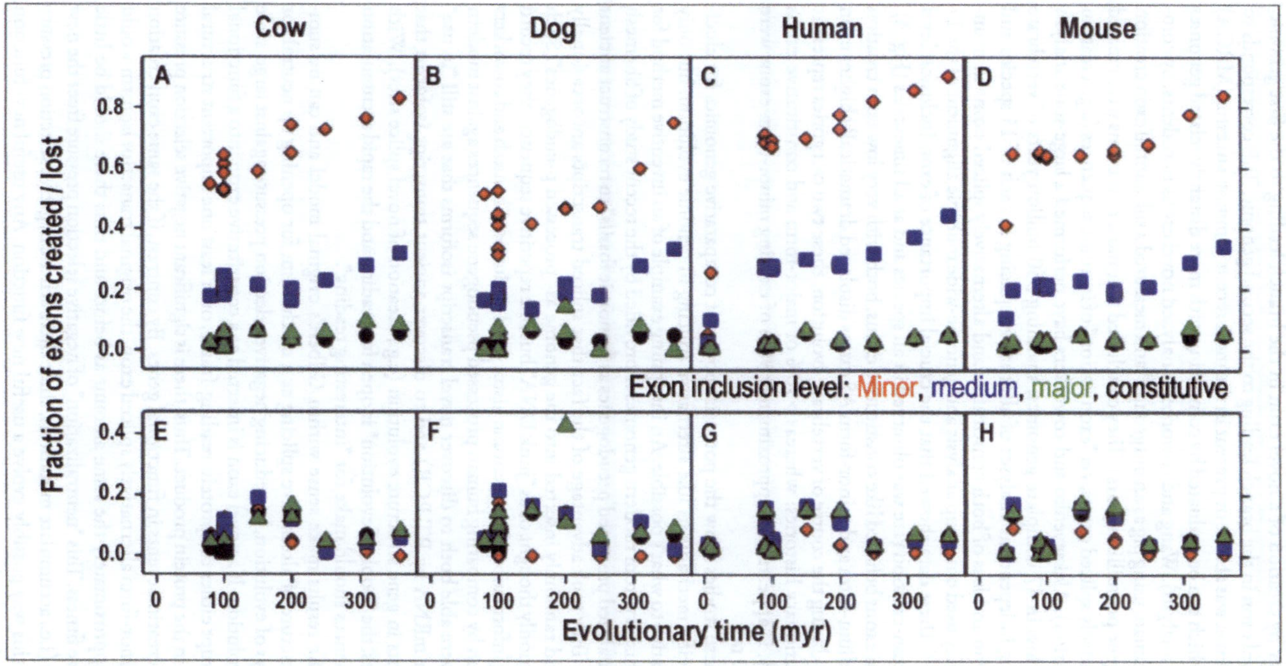

Figure 3. Fraction of modern exons created or lost versus evolutionary time. The fraction of minor-form exons (red diamonds), medium-form exons (blue squares), major-form exons (green triangles) and constitutive exons (black circles) (A-D) created or (E-H) lost as a function of evolutionary divergence time. Alternatively spliced exons and inclusion levels were identified independently from EST data for four different mammals: cow; dog; human; mouse; exon creation and loss rates were then estimated against each of 16 animal genomes by using the UCSC 17-genome alignment and outgroup analysis (see text). The amount of loss and creation in constitutive and major form exons is the same indicated by strong overlap of points on the graph. In the alternative set the fraction of created exons is anticorrelated with inclusion level. The amount of loss is similar for all times and inclusion categories. Similar trends are observed in data from different source organisms, plotted in separate panels. From: Alekseyenko A, et al. RNA 2007; 13:661-670;[49] with permission from the RNA Society.

Recent studies have sought to distinguish exon creation vs. exon loss events during genome evolution.[47,48] When an exon is present in one modern genome but absent from another, this can be explained either by creation of a novel exon in the branch leading to the first genome, or by loss of an ancestral exon in the branch leading to the second genome. This corresponds to inferring whether the exon was absent or present in the most recent common ancestor (MRCA) of the two genomes, which can be evaluated by comparing with more distantly related genomes (so-called "outgroup" analysis). Wang and coworkers analyzed novel exons in rodents, by comparison of mouse vs. human, using pig as an outgroup,[48] and measured a substantial exon creation rate (2.71×10^{-3} per gene per million years). They concluded that most new exons were created as minor-form alternatively spliced exons, via "exonization" of intronic sequences (e.g., creation of splice sites by mutation). Alekseyenko and coworkers have performed a large-scale analysis of exon creation and loss in 17 complete genomes, spanning 350 million years of vertebrate evolution,[49] combining independent analyses of alternative splicing in each of 15 species and analysis of exon creation and loss of both constitutive and alternatively spliced exons in mammals, fish and birds. They used outgroup analysis and genome-wide multiple alignment of the 17 genomes by UCSC.[50,51] These data showed that the crucial importance of exon-inclusion level for determining the exon-creation rate was observed in all species and at all timescales (Fig. 3). In general, major-form exons behaved like constitutive exons, both with very low exon creation rates. By contrast, medium-form and minor-form AS exons displayed dramatically higher exon creation rates. Indeed during the course of vertebrate evolution, these two categories represent almost opposite evolutionary histories: whereas >90% of major-form and constitutive exons are older than 350 my, it appears that approximately 90% of existing minor-form exons were created in the last 350 my.

Taken together, these studies show the potential power of comparative genomics for elucidating the role of specific mechanisms like alternative splicing in genome evolution, but only scratch the surface relative to what's possible. An intriguing example of an inventive method for mining evolutionary history from modern genomes is provided by the recent study of Shemesh and coworkers, who analyzed processed pseudogenes as "genomic fossils" to reconstruct ancient transcript isoforms.[52] They took advantage of the fact that spliced transcripts are occasionally reverse-transcribed and randomly inserted into the genome as "processed pseudogenes". Such pseudogenes are commonly thought of as "junk DNA", but like repetitive sequences they record potentially interesting information about genome history and evolution. Shemesh and coworkers mined this information by comparing human processed pseudogene sequences against modern gene structures and were able both to discover novel transcript isoforms that are still "in use" (detectable in human mRNA by RT-PCR) and to discover ancient transcript isoforms that reveal important events in gene structure evolution (e.g., creation of novel splice sites). With such inventive analyses, the "book of evolution" is open for reading and the rapid accumulation of many genome sequences should make for "interesting reading".

All of these specific results make sense within Gilbert's original model and can be summarized by a simple picture of alternative splicing as a mechanism for opening up neutral (or near-neutral) pathways of evolution, by reducing negative selection pressure against large-scale changes in genome evolution. If a novel exon is inserted as a constitutive exon into a functional gene, it is likely to disrupt either the protein reading-frame, or at least one important structural or functional element in the protein product. Thus, there is significant negative selection pressure against random exon creation events in functional genes. By contrast, if the same exon creation was introduced as a minor-form alternatively spliced exon, the original transcript isoform would still be produced in approximately the same amount as before and thus there should be little impact on reproductive fitness. This "neutralization" of negative selection pressure frees the new exon to evolve rapidly (i.e., accumulate mutations, since there is little negative selection pressure against them) and in this way possibly evolve a useful new function. Any useful new function could be fixed by positive selection.

Alternative Splicing and Amino Acid Mutation Selection Pressure

This model predicts that alternatively spliced exons should be associated with a relaxation of negative selection pressure against not only exon creation, but a wide variety of mutational processes. One area that has received much study is the accumulation of amino acid mutations. Many groups have compared amino acid mutation selection pressure metrics (dn/ds, or Ka/Ks) for alternatively spliced exons vs. constitutive exons and all of them have found marked increases, implying reduced negative selection pressure in alternatively spliced exons.[48,53-61] Again, as for exon creation, this effect appears to be strongly linked to exon inclusion level: major-form exons show Ka/Ks levels similar to constitutive exons, whereas minor-form exons show the largest change.[57,58] This pattern may be general. For example, negative selection pressure against another type of evolutionary change (introduction of premature termination codons) also appears to be relaxed in minor-form alternative transcripts as compared with major-form alternative transcripts.[62]

These results suggest the intriguing idea of evolutionary "hotspots" within a protein sequence: one portion of the coding region, provided by an alternatively spliced exon, apparently can have a much higher rate of amino acid mutations than the rest of the protein. A classic example is *BRCA1*.[54,63] Hurst and Pal first reported the finding that *BRCA1* has a Ka/Ks "hotspot" (in the region between codons 200-300, Ka/Ks rises to peak values of 2.0-8.0, indicating positive selection). Ordinarily, Ks is assumed to represent the neutral background mutation rate and thus should be approximately constant within a gene, while the Ka value is expected to reflect varying levels of amino acid selection pressure in different regions of the protein. Surprisingly, in this region of *BRCA1*, the opposite pattern was observed: Ka remained relatively constant, while Ks plunged several-fold, resulting in the observed increase in the Ka/Ks ratio. Subsequently, Orban and Olah pointed out that this region is alternatively spliced (exon skipping of exons 9 and 10 and alternative 5' splicing of exon 11) and that the region of lowest Ks overlaps two putative exonic splice enhancers (ESE) elements.[63] They proposed that this result might reflect purifying selection on regulation of alternative splicing in this region.

Alternative Splicing and RNA Selection Pressure

This surprising result, along with clues from genome-wide comparative genomics studies, suggested that relaxation of selection pressure was not the complete story. Alternative splicing implies regulation: whereas constitutive exons can be spliced in an unregulated manner (i.e., always spliced), alternative exons may require regulatory elements either to activate or suppress their splicing in specific tissues or circumstances. The functional importance of such regulatory elements in turn would create a new level of selection pressure, this time on the RNA sequence (as opposed to the amino acid sequence). We will refer to this as "RNA selection pressure". The elucidation of evidence for RNA selection pressure and analysis of its relationship with alternative splicing have been an important focus of recent research. Two rather different types of studies have converged: comparative genomics analysis of highly conserved genomic regions; and studies of alternatively spliced exons and their flanking introns.

Initial comparisons of the mouse and human genomes uncovered one major surprise: the existence of "ultraconserved" regions of complete sequence identity, extending over hundreds of nucleotides in length,[64] and shown by further analysis to be almost completely conserved in chicken and dog as well. Intriguingly, most of these regions were not in protein-coding exons (which are commonly thought of as the most conserved regions of sequence in the genome), but rather in introns. The finding of a statistical association of these regions with genes involved in RNA binding and splicing regulation and their proximity to alternatively spliced exons, led Haussler and coworkers to suggest that such regions could be involved in regulation of alternative splicing. The authors also found much larger numbers of smaller ultraconserved regions: over 5000 of length 100 nt or greater and tens of thousands at lower cutoffs. Thus ultraconserved regions could play an important part in the regulation of a significant fraction of genes.

Sironi and coworkers analyzed the distribution of multispecies conserved sequences (MCS) across annotated introns.[65,66] They found that MCS density was inversely correlated with intron size and correlated with proximity to conserved alternative splicing events. Furthermore, mapping of experimentally validated intronic splice regulatory elements found that about half correspond to MCS regions identified by comparative genomics. Thus these MCS not only appear to be strongly linked with alternative splicing regulation, but may actually identify the majority of functional regulatory sites.

Approaching the problem from the opposite direction, studies of alternatively spliced exons reached similar conclusions.[10,56,58,67-70] For example, Sugnet et al[10] reported that exon skipping was associated with substantially higher conservation in the flanking introns (approximately 70-85% identity for alternative exons, vs. 60-65% for constitutive exons), for at least 100 nt on either side of the alternative exon. They also observed a slight increase in percent identity within the exon itself (approximately 90% identity for alternative exons, vs. 80-85% identity for constitutive exons). They also examined alternative-5' and alternative-3' splicing patterns and again found increased intronic conservation, but only on one side, in the intron immediately adjacent to the alternative splice site. Similar results have been reported by others.[56,58,69,70] It has proved useful to disentangle the effects of amino acid selection pressure (e.g., as measured by Ka/Ks) from pure "RNA selection pressure" (e.g., as measured at synonymous sites by Ks)[56,58-61,63] (for a review considering the possible coupling of Ks effects and Ka/Ks measurements see ref. 71). Analyzed this way, the strength of RNA selection pressure associated with alternative splicing is surprisingly high, ranging from two- to six-fold reductions in Ks for minor-form exons,[58] compared with constitutive exons (even within the same gene). These differences could not be explained by codon usage bias or GC composition. The fact that the strength of this selection pressure correlates directly with the strength of the splicing reaction (as measured by the exon inclusion level), suggests that it is associated with regulation of alternative splicing.

It is striking that a very different selection pressure metric, protein frame-preservation, is also strongly anti-correlated with exon inclusion levels: it is highest for minor-form exons and decreases to near-random for major-form and constitutive exons.[8] Furthermore, frame-preservation also takes a similar amount of time to evolve: in human vs. chimp comparisons (5 my), frame-preservation is only slightly higher for minor-form exons compared with major-form or constitutive exons; for mouse vs. rat (40 my), the frame-preservation ratio is four-fold higher for minor-form exons and somewhat higher in human vs. mouse comparisons (90 my).[58] Thus the time course of evolution of frame-preservation selection pressure roughly parallels that of Ks selection pressure. A common-sense interpretation is that whereas relaxation of negative selection pressure occurs immediately upon the creation of an alternatively spliced exon, fixation of a new, positive function for the exon (as indicated by selection for protein modularity and RNA selection pressure on splice regulatory motifs) takes significant time.

Connecting RNA Selection Pressure to Real Splicing Regulation

While evidence of RNA selection pressure associated specifically with alternatively spliced exons is suggestive, the questions of identifying the actual splice regulatory elements and proving that they are functional are nontrivial. A number of groups have attacked this research problem. For example, Fairbrother et al performed statistical analyses to find motifs that were specifically enriched in exons with weak splice sites, compared with exons with strong splice sites.[72] They identified ten distinct motifs from this analysis as putative ESEs and tested them in a splicing reporter construct based on rescue of splicing of *SXN* exon 2. All ten showed some enhancement of exon inclusion that in all but one case was significantly diminished by a point mutation predicted to disrupt the ESE motif. Recently, Fairbrother and coworkers have further validated these ESE motifs using human SNP data.[73] They showed that their ESE predictions have a slight tendency to cluster near splice sites (a maximum of 0.13 ESEs/nt in the region 10-20 nt from the splice site, vs. a background density of 0.12 more than 70 nt from a splice site) and that within this region there is also a significant decrease in SNP density ($5\text{-}6 \times 10^{-4}$ SNP/nt within the region 10-20 nt from the splice site, vs. 7×10^{-4} SNP/nt more than 30 nt from a splice site). By using the chimp genome to identify which SNP allele was ancestral at each site, they were also able to show evidence of negative selection against SNPs that overlap the putative ESEs: about 20% fewer SNPs were observed to overlap the predicted ESEs than expected under a neutral model; this was true regardless of whether the SNP was predicted to disrupt the putative ESE, or not. This effect was strongest in the 20 nt around a splice site, where there was an approximately 40% odds-ratio reduction for SNPs that disrupted predicted ESEs. Overall a subset of 57 ESE hexamers were identified as conserved based on the SNP analysis. Another study of the same predicted ESEs showed that SNPs overlapping these motifs were least common at exon synonymous sites (6.69% of SNPs overlapped ESEs), relative to nonexonic regions (where the ESE hexamers are presumably not functional; 7.20-7.68%).[74] The authors calculated that approximately 13% fewer SNPs predicted to disrupt ESEs were observed at synonymous sites than expected under a neutral model. Intriguingly, the authors observed no evidence of negative selection against SNPs overlapping ESEs at nonsynonymous sites; this matches the observation from comparative genomics studies that Ka in alternative exons shows no decrease in alternative exons (relative to constitutive exons), whereas Ks was markedly reduced.[56,58]

Parmley et al have evaluated these ESE motifs using comparisons of the human, chimp and mouse genomes.[75] They found again that within the predicted ESEs, Ks was 5- 35% lower than in non-ESE exonic sequence (with the strongest effect near splice sites) and that this effect does not appear to be due to skewed CpG or skewed nucleotide distribution. However, the authors found that evidence of a link between putative ESE conservation and alternative splicing was less clear. First, the putative ESEs do not appear to explain the reduced Ks in alternative exons; non-ESE regions in alternative exons also showed a more than two-fold reduction in Ks. Second, alternative exons show no increase in predicted ESE density compared with constitutive exons. Of course, it is possible that there genuinely is an increase in functional ESEs in alternative exons, but that this is obscured by a large fraction of false positives (random matches to the ESE hexamer sequences that are not functional). Parmley et al's alternative exon data suggest that Ks is slightly lower in predicted ESEs than in non-ESE regions of the same alternative exons, but this does not fully resolve the question.

Kabat and coworkers have also used comparative genomics approaches to identify candidate splice regulatory motifs in introns flanking alternatively spliced exons, in nematode worms.[76] By finding regions of high nucleotide conservation between *C. elegans* and *C. briggsae*, they located 147 alternative exons with apparently conserved regulatory sites and identified short sequence motifs in these sites by statistical scoring. They were able to validate several of these via an in vivo splicing reporter assay. As additional animal and plant genome sequences become available, large-scale comparative genomics analyses such as this should be able to reveal many new splicing regulatory elements, but experiments will often be required to ascertain their functional significance.

Future Challenges

Currently, alternative splicing research is only at the early stages of connecting genome-wide evidence of RNA selection to elucidating the detailed regulatory mechanisms that control "programs" of alternative splicing. Some basic questions need to be answered. How many different regulatory elements control each alternative exon? If anything, comparative genomics data (such as Ks measurements) indicate "too much" conservation, implying that a large fraction of alternative exon sequence is under RNA selection pressure. For example, a six-fold reduction in Ks for minor-form exons (compared with constitutive exons) implies that 5/6 of the synonymous sites in the exon are constrained. Thus virtually the whole exon would be considered a "required splice regulatory element" for its own splicing, a somewhat puzzling scenario. Of course, these regions of strong conservation are almost always considerably larger than the alternative exon itself, extending 100 nt or more into the flanking introns on both sides.[10,58,67,68] Indeed, many of these alternative exons are part of ultraconserved regions of 200 nt or more of absolute conservation between human, mouse and rat.[64] One possible interpretation is that such large regions of strong conservation are due to the overlap of many, many ESE/ESS/ISE/ISS sites required for each alternative exon. This raises the challenge of how to decipher the combinatorial "program" of multiple binding sites controlling each splicing event.

Another possibility is that entirely different mechanisms, such as RNA secondary structure formation, play an important role in regulating splicing. RNA secondary structures are typically much larger (in sequence length) than splicing factor binding sites. Recently, some very striking examples of RNA secondary structure involvement in splicing regulation have been reported for the *DSCAM* gene (refer to chapter by Park and Graveley).[77,78] Secondary structure can be important for splicing regulation in mammalian genes as well. For example, in *FGFR2* two intronic elements (IAS2 and ISAR) are required for activation of exon IIIb splicing and not only form a stem-loop structure, but can be replaced by other stem-loop sequences without loss of activity.[79,80] Alternative splicing of type II procollagen is developmentally regulated, with exon 2 (encoding a cysteine-rich von Willebrand factor C-like domain) included only in chondrocyte precursors. McAlinden et al identified a 13 bp stem + 12 nt loop structure immediately adjacent to exon 2's 5' splice site, that is essential for this switching event.[81] They confirmed the existence of the stem-loop structure using RNase digestion experiments. The stem sequence is absolutely conserved from mammals to fish and appears to work in conjunction with a weak 5' splice site, which is also strongly conserved through evolution. Another stem structure involved in splicing regulation has been reported in *CFTR*.[82] Applying secondary structure analysis to large-scale comparative genomics data is an important goal towards assessing how common such mechanisms are in regulating alternative splicing throughout the genome.

Acknowledgement

This work was supported by grants from the NIH (U54 RR021813) and DOE (DE-FC02-02ER63421), and a Dreyfus Foundation Teacher-Scholar Award to C.L.

References

1. Modrek B, Lee C. A genomic view of alternative splicing. Nature Genet 2002; 30:13-19.
2. Blencowe BJ. Alternative splicing: new insights from global analyses. Cell 2006; 126:37-47.
3. Johnson JM, Castle J, Garrett-Engele P et al. Genome-wide survey of human alternative pre-mRNA splicing with exon junction microarrays. Science 2003; 302:2141-2144.
4. Black DL. Protein diversity from alternative splicing: a challenge for bioinformatics and postgenome biology. Cell 2000; 103:367-370.
5. Gilbert W. Why genes in pieces? Nature 1978; 271:501.
6. Resch A, Xing Y, Modrek B et al. Assessing the impact of alternative splicing on domain interactions in the human proteome. J. Proteome Res 2004; 3:76-83.
7. Sorek R, Shamir R, Ast G et al. How prevalent is functional alternative splicing in the human genome? Trends Genet 2004; 20:68-71.
8. Resch A, Xing Y, Alekseyenko A et al. Evidence for a subpopulation of conserved alternative splicing events under selection pressure for protein reading frame preservation. Nucleic Acids Res 2004; 32:1261-1269.
9. Philipps DL, Park JW, Graveley BR et al. A computational and experimental approach toward a priori identification of alternatively spliced exons. RNA 2004; 10:1838-1844.
10. Sugnet CW, Kent WJ, Ares Jr. M et al. Transcriptome and genome conservation of alternative splicing events in humans and mice. Pac Symp Biocomput 2004; 66-77.
11. Xing Y, Lee C. Protein modularity of alternatively spliced exons is associated with tissue-specific regulatoiin of alternative splicing. PLoS Genet 2005; 1:e34.
12. Sugnet CW, Srinivasan K, Clark TA et al. Unusual intron conservation near tissue-regulated exons found by splicing microarrays. PLoS Comput Biol 2006; 2:e4.
13. Lewis BP, Green RE, Brenner SE et al. Evidence for the widespread coupling of alternative splicing and nonsense-mediated mRNA decay in humans. Proc Natl Acad Sci USA 2003; 100:189-192.
14. Hillman RT, Green RE, Brenner SE et al. An unappreciated role for RNA surveillance. Genome Biol 2004; 5:R8.
15. Wang P, Yan B, Guo JT et al. Structural genomics analysis of alternative splicing and application to isoform structure modeling. Proc Natl Acad Sci USA 2005; 102:18920-18925.
16. Sorek R, Shemesh R, Cohen Y et al. A non-EST-based method for exon-skipping prediction. Genome Res 2004; 14:1617-1623.
17. Volfovsky N, Haas BJ, Salzberg SL et al. Computational discovery of internal micro-exons. Genome Res 2003; 13:1216-1221.
18. Hiller M, Huse K, Szafranski K et al. Widespread occurrence of alternative splicing at NAGNAG acceptors contributes to proteome plasticity. Nat Genet 2004; 36:1255-1257.

19. Chern TM, van Nimwegen E, Kai C et al. A simple physical model predicts small exon length variations. PLoS Genet, 2006; 2:e45.
20. Xu Q, Modrek B, Lee C et al. Genome-wide detection of tissue-specific alternative splicing in the human transcriptome. Nucleic Acids Res 2002; 30:3754-3766.
21. Wilson FH, Disse-Nicodeme S, Choate KA et al. Human hypertension caused by mutations in WNK kinases. Science 2001; 293:1107-1112.
22. Wade JB, Fang L, Liu J et al. WNK1 kinase isoform switch regulates renal potassium excretion. Proc Natl Acad Sci USA 2006; 103:8558-8563.
23. Subramanya AR, Yang CL, Zhu X et al. Dominant-negative regulation of WNK1 by its kidney-specific kinase-defective isoform. Am J Physiol Renal Physiol 2006; 290:F619-624.
24. Lazrak A, Liu Z, Huang CL et al. Antagonistic regulation of ROMK by long and kidney-specific WNK1 isoforms. Proc Natl Acad Sci USA 2006; 103:1615-1620.
25. Delaloy C, Lu J, Houot AM et al. Multiple promoters in the WNK1 gene: one controls expression of a kidney-specific kinase-defective isoform. Mol Cell Biol 2003; 23:9208-9221.
26. O'Reilly M, Marshall E, Speirs HJ et al. WNK1, a gene within a novel blood pressure control pathway, tissue-specifically generates radically different isoforms with and without a kinase domain. J Am Soc Nephrol 2003; 14:2447-2456.
27. Kriventseva EV, Koch I, Apweiler R et al. Increase of functional diversity by alternative splicing. Trends Genet 2003; 19:124-128.
28. Liu S, Altman RB. Large scale study of protein domain distribution in the context of alternative splicing. Nucleic Acids Res 2003; 31:4828-4835.
29. Xing Y, Resch A, Lee C et al. The Multiassembly Problem: reconstructing multiple transcript isoforms from EST fragment mixtures. Genome Res 2004; 14:426-441.
30. Taneri B, Snyder B, Novoradovsky A et al. Alternative splicing of mouse transcription factors affects their DNA-binding domain architecture and is tissue specific. Genome Biol 2004; 5:R75.
31. Hiller M, Huse K, Platzer M et al. Creation and disruption of protein features by alternative splicing—a novel mechanism to modulate function. Genome Biol 2005; 6:R58.
32. Arribas J, Borroto A. Protein ectodomain shedding. Chem Rev 2002; 102:4627-4638.
33. Peters M, Muller AM, Rose-John S et al. Interleukin-6 and soluble interleukin-6 receptor: direct stimulation of gp130 and hematopoiesis. Blood 1998; 92:3495-3504.
34. Xing Y, Xu Q, Lee C et al. Widespread production of novel soluble protein isoforms by alternative splicing removal of transmembrane anchoring domains. FEBS Lett 2003; 555:572-578.
35. Cline MS, Shigeta R, Wheeler RL et al. The effects of alternative splicing on transmembrane proteins in the mouse genome. Pac Symp Biocomput 2004; 17-28.
36. Davis MJ, Hanson KA, Clark F et al. Differential use of signal peptides and membrane domains is a common occurrence in the protein output of transcriptional units. PLoS Genet 2006; 2:e46.
37. Takeda J, Suzuki Y, Nakao M et al. Large-scale identification and characterization of alternative splicing variants of human gene transcripts using 56,419 completely sequenced and manually annotated full-length cDNAs. Nucleic Acids Res 2006; 34:3917-3928.
38. Nakao M, Barrero RA, Mukai Y et al. Large-scale analysis of human alternative protein isoforms: pattern classification and correlation with subcellular localization signals. Nucleic Acids Res, 2005; 33:2355-2363.
39. Romero PR, Zaidi S, Fang YY et al. Alternative splicing in concert with protein intrinsic disorder enables increased functional diversity in multicellular organisms. Proc Natl Acad Sci USA 2006; 103:8390-8395.
40. Wen F, Li F, Xia H et al. The impact of very short alternative splicing on protein structures and functions in the human genome. Trends Genet 2004; 20:232-236.
41. Garcia J, Gerber SH, Sugita S et al. A conformational switch in the Piccolo C2A domain regulated by alternative splicing. Nat Struct Mol Biol 2004; 11:45-53.
42. Sorek R, Ast G, Graur D et al. Alu-containing exons are alternatively spliced. Genome Res 2002; 12:1060-1067.
43. Modrek B, Lee C. Alternative splicing in the human, mouse and rat genomes is associated with an increased rate of exon creation/loss. Nature Genet 2003; 34:177-180.
44. Pan Q, Shai O, Misquitta C et al. Revealing global regulatory features of Mammalian alternative splicing using a quantitative microarray platform. Mol Cell 2004; 16:929-941.
45. Malko DB, Makeev VJ, Mironov AA et al. Evolution of exon-intron structure and alternative splicing in fruit flies and malarial mosquito genomes. Genome Res 2006; 16:505-509.
46. Pan Q, Bakowski MA, Morris Q et al. Alternative splicing of conserved exons is frequently species-specific in human and mouse. Trends Genet 2005; 21:73-77.
47. Cusack BP, Wolfe KH. Changes in alternative splicing of human and mouse genes are accompanied by faster evolution of constitutive exons. Mol Biol Evol 2005; 22:2198-2208.

48. Wang W, Zheng H, Yang S et al. Origin and evolution of new exons in rodents. Genome Res 2005; 15:1258-1264.
49. Alekseyenko A, Kim N., Lee C et al. Global analysis of exon creation vs. loss and the role of alternative splicing, in 17 vertebrate genomes. RNA 2007; 13:661-670.
50. Blanchette M, Kent WJ, Riemer C et al. et al. Aligning multiple genomic sequences with the threaded blockset aligner. Genome Res 2004; 14:708-715.
51. Hinrichs AS, Karolchik D, Baertsch R et al. The UCSC Genome Browser Database: update 2006. Nucleic Acids Res 2006; 34:D590-598.
52. Shemesh R, Novik A, Edelheit S et al. Genomic fossils as a snapshot of the human transcriptome. Proc Natl Acad Sci USA 2006; 103:1364-1369.
53. Iida K, Akashi H. A test of translational selection at 'silent' sites in the human genome: base composition comparisons in alternatively spliced genes. Gene 2000; 261:93-105.
54. Hurst LD, Pal C. Evidence for purifying selection acting on silent sites in BRCA1. Trends Genet 2001; 17:62-65.
55. Filip LC, Mundy NI. Rapid evolution by positive Darwinian selection in the extracellular domain of the abundant lymphocyte protein CD45 in primates. Mol Biol Evol 2004; 21:1504-1511.
56. Baek D, Green P. Sequence conservation, relative isoform frequencies and nonsense-mediated decay in evolutionarily conserved alternative splicing. Proc Natl Acad Sci USA 2005; 102:12813-12818.
57. Xing Y, Lee C. Assessing the application of Ka/Ks ratio test to alternatively spliced exons. Bioinformatics 2005; 21:3701-3703.
58. Xing Y, Lee C. Evidence of functional selection pressure for alternative splicing events that accelerate evolution of protein subsequences. Proc Natl Acad Sci USA 2005; 102:13526-13531.
59. Chen FC, Wang SS, Chen CJ et al. Alternatively and constitutively spliced exons are subject to different evolutionary forces. Mol Biol Evol 2006; 23:675-682.
60. Ermakova EO, Nurtdinov RN and Gelfand MS. Fast rate of evolution in alternatively spliced coding regions of mammalian genes. BMC Genomics 2006; 7:84.
61. Plass M, Eyras E. Differentiated evolutionary rates in alternative exons and the implications for splicing regulation. BMC Evol Biol 2006; 6:50.
62. Xing Y, Lee C. Negative selection pressure against premature protein truncation is reduced by alternative splicing and diploidy. Trends Genet 2004; 20:472-475.
63. Orban TI, Olah E. Purifying selection on silent sites—a constraint from splicing regulation? Trends Genet 2001; 17:252-253.
64. Bejerano G, Pheasant M, Makunin I et al. Ultraconserved elements in the human genome. Science 2004; 304:1321-1325.
65. Sironi M, Menozzi G, Comi GP et al. Analysis of intronic conserved elements indicates that functional complexity might represent a major source of negative selection on noncoding sequences. Hum Mol Genet 2005; 14:2533-2546.
66. Sironi M, Menozzi G, Comi GP et al. Fixation of conserved sequences shapes human intron size and influences transposon-insertion dynamics. Trends Genet 2005; 21:484-488.
67. Sorek R, Ast G. Intronic sequences flanking alternatively spliced exons are conserved between human and mouse. Genome Res 2003; 13:1631-1637.
68. Kaufmann D, Kenner O, Nurnberg P et al. In NF1, CFTR, PER3, CARS and SYT7, alternatively included exons show higher conservation of surrounding intron sequences than constitutive exons. Eur J Hum Genet 2004; 12:139-149.
69. Itoh H, Washio T, Tomita M et al. Computational comparative analyses of alternative splicing regulation using full-length cDNA of various eukaryotes. RNA 2004; 10:1005-1018.
70. Zheng CL, Fu XD, Gribskov M et al. Characteristics and regulatory elements defining constitutive splicing and different modes of alternative splicing in human and mouse. RNA 2005; 11:1777-1787.
71. Xing Y, Lee C. Can RNA selection pressure distort the measurement of Ka/Ks? Gene 2006; 370:1-5.
72. Fairbrother WG, Yeh RF, Sharp PA et al. Predictive identification of exonic splicing enhancers in human genes. Science 2002; 297:1007-1013.
73. Fairbrother WG, Holste D, Burge CB et al. Single Nucleotide Polymorphism-Based Validation of Exonic Splicing Enhancers. PLoS Biol 2004; 2:E268.
74. Carlini DB, Genut JE. Synonymous SNPs Provide Evidence for Selective Constraint on Human Exonic Splicing Enhancers. J Mol Evol 2005.
75. Parmley JL, Chamary JV, Hurst LD et al. Evidence for Purifying Selection Against Synonymous Mutations in Mammalian Exonic Splicing Enhancers. Mol Biol Evol 2006.
76. Kabat JL, Barberan-Soler S, McKenna P et al. Intronic alternative splicing regulators identified by comparative genomics in nematodes. PLoS Comput Biol 2006; 2:e86.
77. Graveley BR. Mutually exclusive splicing of the insect Dscam pre-mRNA directed by competing intronic RNA secondary structures. Cell 2005; 123:65-73.

78. Kreahling JM, Graveley BR. The iStem, a long-range RNA secondary structure element required for efficient exon inclusion in the Drosophila Dscam pre-mRNA. Mol Cell Biol 2005; 25:10251-10260.
79. Baraniak AP, Lasda EL, Wagner EJ et al. A stem structure in fibroblast growth factor receptor 2 transcripts mediates cell-type-specific splicing by approximating intronic control elements. Mol Cell Biol 2003; 23:9327-9337.
80. Muh SJ, Hovhannisyan RH, Carstens RP. A Nonsequence-specific double-stranded RNA structural element regulates splicing of two mutually exclusive exons of fibroblast growth factor receptor 2 (FGFR2). J Biol Chem 2002; 277:50143-50154.
81. McAlinden A, Havlioglu N, Liang L et al. Alternative splicing of type II procollagen exon 2 is regulated by the combination of a weak 5' splice site and an adjacent intronic stem-loop cis element. J Biol Chem 2005; 280:32700-32711.
82. Hefferon TW, Groman JD, Yurk CE et al. A variable dinucleotide repeat in the CFTR gene contributes to phenotype diversity by forming RNA secondary structures that alter splicing. Proc Natl Acad Sci USA 2004; 101:3504-3509.

Chapter 4

Complex Alternative Splicing

Jung Woo Park and Brenton R. Graveley*

Abstract

Alternative splicing is a powerful means of controlling gene expression and increasing protein diversity. Most genes express a limited number of mRNA isoforms, but there are several examples of genes that use alternative splicing to generate hundreds, thousands and even tens of thousands of isoforms. Collectively such genes are considered to undergo complex alternative splicing. The best example is the ***Drosophila Down syndrome cell adhesion molecule (Dscam)*** gene, which can generate 38,016 isoforms by the alternative splicing of 95 variable exons. In this review, we will describe several genes that use complex alternative splicing to generate large repertoires of mRNAs and what is known about the mechanisms by which they do so.

Introduction

Alternative splicing affords eukaryotes with the opportunity to produce multiple proteins from a single gene.[1-3] This allows organisms to maximize the coding capacity of their genomes. To illustrate this, let us consider two different ways in which evolution can produce two highly related proteins starting from a single gene. Take, for instance, a representative human gene that contains ~10 exons, produces a single mRNA isoform and encompasses ~28,000 bp.[4,5] At least two different scenarios can give rise to a new variant of the encoded protein that contains an additional 10 amino acids. First, the gene could be duplicated and diverge such that either a pre-existing exon is extended by 30 nucleotides or a new 30 nt exon is created. Alternatively, a single 30 nucleotide cassette exon could be inserted into, or arise within an intron in the original gene allowing for two mRNA isoforms that either contain or lack the exon to now be produced. While these two scenarios have a similar outcome—the production of a new protein 10 amino acids longer than the original protein—they can have drastically different consequences on the size of the genome. Gene duplication requires expanding the genome by at least 28,000 bp. In contrast, creating the same protein by simply adding an alternative exon to the original gene would only increase the size of the genome by ~30 nt. The difference in efficiency between gene duplication or the evolution of new alternative exons becomes more pronounced as the number of new isoforms increases. For example, in the same amount of genome space required to generate a single new isoform of our hypothetical gene by gene duplication, hundreds of new isoforms could be created by evolving new alternative exons. Thus, alternative splicing is an extremely economical means of increasing protein diversity.

Alternative splicing is prevalent in metazoan genomes. For example, current estimates suggest that at least 42% of *Drosophila* genes[6] and over two thirds of mouse and human genes[7] encode alternatively spliced pre-mRNAs. These numbers have been increasing at a brisk pace over the past several years and are likely to still be underestimates as many low abundance, tissue-specific or

*Corresponding Author: Brenton R. Graveley—Department of Genetics and Developmental Biology, University of Connecticut Health Center, 263 Farmington Avenue, Farmington, CT 06030-3301. Email:graveley@neuron.uchc.edu

Alternative Splicing in the Postgenomic Era, edited by Benjamin J. Blencowe and Brenton R. Graveley.

developmentally regulated isoforms almost certainly remain to be characterized. Thus, it is now fair to say that the majority of metazoan genes encode alternatively spliced pre-mRNAs. Alternative splicing is clearly the rule, not the exception.

An issue that is quite distinct from the number of genes that encode alternatively spliced pre-mRNAs is the number of isoforms generated per gene. Figure 1 depicts the number of mRNA isoforms per gene in *D. melanogaster*, which is perhaps the best annotated metazoan genome. There are several conclusions that can be drawn from this graph. First, many genes encode only a single mRNA isoform. Second, very few genes encode a large number of mRNA isoforms—the greatest number of mRNA isoforms encoded by a gene in this dataset is 26 for the *longitudinals lacking (lola)* and *modifier of mdg4 (mod (mdg4))* genes. Thus, the number of genes encoding multiple mRNA isoforms decreases as the number of isoforms increases, yet it would appear that alternative splicing is rarely used to create a tremendously diverse set of mRNA isoforms from a single gene.

There are, however, many caveats that should be kept in mind when interpreting this dataset as it significantly underestimates both the number of mRNA isoforms expressed per gene and the number of genes encoding multiple mRNAs. The current annotation of the *D. melanogaster* genome (version 4.1) lists only 21.7% of the genes as encoding more than one mRNA isoform. This is because this dataset includes only annotated mRNA isoforms most of which have been manually curated. Despite the overall high quality of the annotation, many mRNA isoforms that are well-documented in the literature are not present in this annotation of the genome. Moreover, recent microarray analyses suggests that at least 42% of genes encode multiple mRNA isoforms and few, if any of the isoforms detected by this method are present in this dataset.[6] As the annotation and our understanding of the entire repertoire of mRNAs expressed by the *Drosophila* genome improves, the slope of this graph will change such that both the number of genes that express multiple isoforms and the number of isoforms per gene will increase.

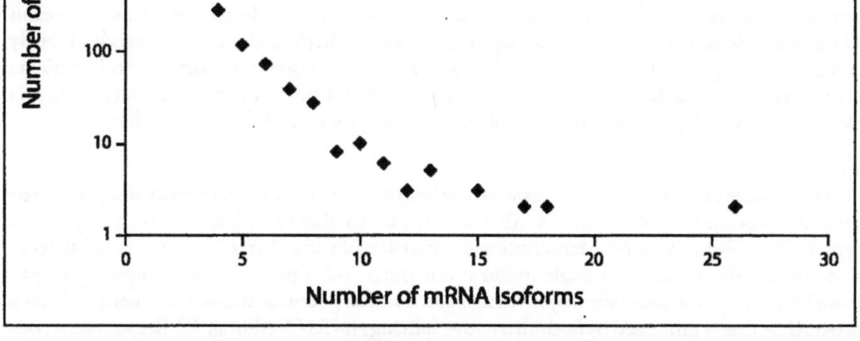

Figure 1. Number of distinct mRNA isoforms per gene in *Drosophila melanogaster*. The number of distinct mRNA isoforms annotated for each gene in version 4.1 of the *D. melanogaster* genome was determined. The number of genes is plotted as a function of the number of annotated mRNA isoforms derived from each gene. Note that the Y axis is represented using the logarithmic scale.

In this review we will consider in detail a few genes that express a large repertoire of mRNA isoforms and represent the outliers of this graph. These genes represent what we refer to as complex alternative splicing events. Genes in this class typically have both a complex genomic organization and appear to use unique mechanisms in the expression of their mRNA repertoire. Throughout this review, we will focus on *Drosophila* genes as they represent some of the most unusual and best characterized complex alternative splicing events.

An Overview of Alternative Splicing—From Simple to Complex

As revealed in the plot shown in Figure 1, there are thousands of *Drosophila* genes that encode pre-mRNAs that are alternatively spliced to generate only two mRNA isoforms. A classic example of this is *doublesex* (*dsx*) which functions as a key regulatory gene in the sex-determination pathway (Fig. 2).[8] In males, the *dsx* pre-mRNA is spliced to include exons 1-3, 5 and 6 and to skip exon 4. In contrast, in females, the same pre-mRNA is spliced to include exons 1-4 and a poly(A) site within exon 4 is used. The male and female-specific mRNAs encode male- and female-specific DSX proteins that function as transcription factors to regulate the expression of genes that control the sexual differentiation pathway.[9] This is an excellent example where alternative splicing is used to create two mRNA isoforms that encode proteins that function as a binary switch to control an extremely important aspect of biology. While many genes that encode only two mRNA isoforms are of obvious interest and biological importance, for the remainder of this review, we will describe genes and regulatory mechanisms that generate truly phenomenal numbers of mRNA isoforms by virtue of alternative splicing.

Mhc

An excellent example of a gene that undergoes complex alternative splicing in *Drosophila* is the *Myosin heavy chain* (*Mhc*) gene which encodes a protein that plays a critical role in the function of muscle cells.[10] The *Mhc* gene contains 30 exons, 17 of which are alternatively spliced (Fig. 2). With the exception of exon 18, which is represented at the genomic level by a single alternative exon, the other alternatively spliced exons are organized into 5 separate clusters which contain from 2 to 5 exons each. The alternative exons within each cluster are included in the mRNA in a mutually exclusive manner—only one exon from each cluster is included in the final mRNA. The consequence of this gene organization is that 480 different mRNAs can by generated from this gene by alternative splicing.

A few RNA sequence elements have been identified that play roles in controlling the splicing of the exon 11 cluster of *Mhc*, which contains five exons (exons 11a to 11e).[11] The regulatory elements were all initially identified based on their evolutionary conservation and are called conserved intronic elements (CIEs). One of these elements, CIE3, is located in the the last intron of the exon 11 cluster. Deletion of CIE3 results in skipping of exon 11e in the indirect flight muscle of the fly and exon 10 is spliced directly to exon 12. However, the splicing of the other 4 exon 11 variants is unaffected. Though the mechanism by which this element functions is entirely unknown, it is clear that the CIE3 plays an important role in regulating splicing of the exon 11 cluster.

Para

Even greater numbers of distinct proteins can be synthesized from other *Drosophila* genes when other RNA processing events, such as RNA editing, are combined with alternative splicing. One type of RNA editing is the posttranscriptional conversion of adenosine to inosine—a modification that can alter the identity of a single amino acid in the encoded protein. The *Drosophila paralytic* (*para*) gene, which encodes the major voltage-gated action potential sodium channel, produces a pre-mRNA that is processed by both alternative splicing and RNA editing.[12,13] The *para* gene contains 13 alternative exons and can potentially synthesize 1,536 different mRNAs utilizing alternative splicing alone (Fig. 2). However, at least 11 adenosines are edited to inosine in *para* transcripts. Considering both RNA editing and alternative splicing, 1,032,192 different *para* transcripts can theoretically be synthesized from this single gene. These examples serve to illustrate how alternative splicing and in some cases RNA editing, can significantly expand protein diversity.

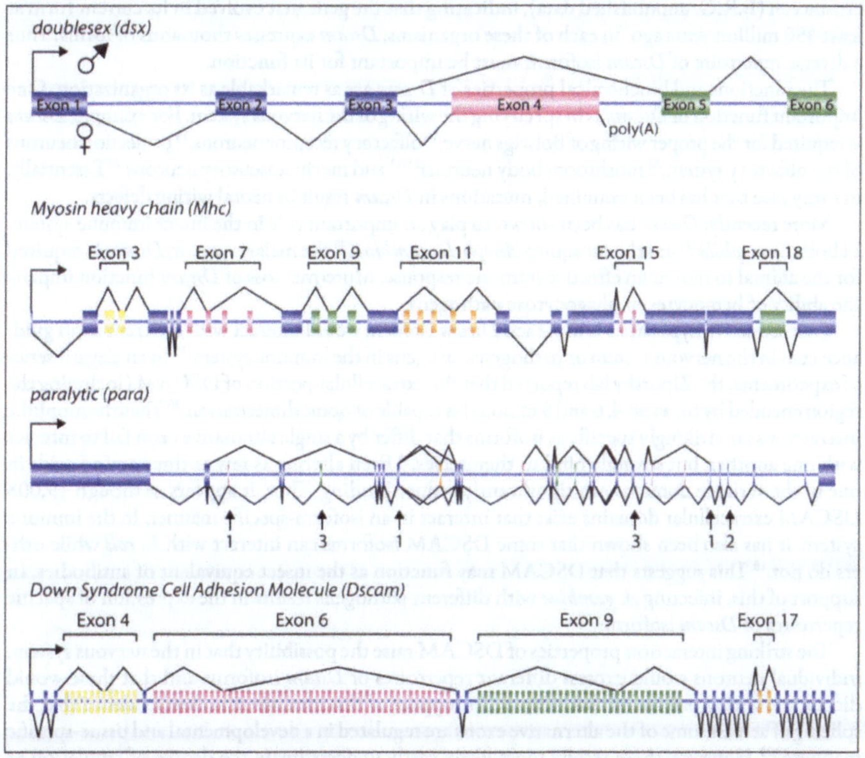

Figure 2. Variation in the complexity of alternative splicing in *Drosophila*. A) The *doublesex* pre-mRNA generates two distinct isoforms in a sex-specific manner. B) The *Mhc* gene can produce 480 different mRNAs. C) The *para* gene undergoes both alternative splicing and RNA editing (the sites and number of RNA editing events are indicated by the arrows). As a result, 1,032,192 different *para* mRNAs can potentially be synthesized. D) The *Dscam* gene can generate 38,016 different mRNAs by virtue of alternative splicing alone.

Dscam

The *Dscam* gene is by far the most extreme case in which alternative splicing alone can generate an extraordinarily diverse repertoire of mRNAs and proteins. This gene, which is essential in *Drosophila*, contains 115 exons, 95 of which are alternatively spliced. The alternative exons are organized into four distinct clusters—the exon 4, 6, 9 and 17 clusters—that contain 12, 48, 33 and 2 variable exons each (Fig. 2).[14,15] Importantly, the exons within each cluster are alternatively spliced in a mutually exclusive manner. As a result, *Dscam* potentially encodes 38,016 different isoforms. The *Drosophila* DSCAM protein is most similar to the human Down syndrome Cell Adhesion Molecule Protein. However, the human gene does not appear to undergo any of the alternative splicing that occurs in *Drosophila*.

The *Drosophila* DSCAM protein contains an extracellular domain, composed of 10 immunoglobulin domains and four fibronectin type III domains, connected to a transmembrane domain and an intracellular domain.[14,15] Alternative splicing of the exon 4, 6 and 9 clusters alters the sequence of three of the Ig domains, while alternative splicing of the exon 17 cluster alters the transmembrane domain. These splicing events have important consequences on the function of the protein. *Dscam* is conserved in all insects[16-18] and has recently been identified in *Daphnia pulex*, a crustacean (B.R.G. unpublished data), indicating that the gene first evolved in its current form at least 450 million years ago. In each of these organisms, *Dscam* expresses thousands of forms. Thus a diverse repertoire of *Dscam* isoforms must be important for its function.

The functions and biochemical properties of *Dscam* are as remarkable as its organization. One important function of *Dscam*, is in specifying the wiring of the nervous system. For example, *Dscam* is required for the proper wiring of Bolwigs nerve,[14] olfactory receptor neurons,[19] projection neurons of the olfactory system,[20] mushroom body neurons[21-23] and mechanosensory neurons.[24] Essentially, in every case that has been examined, mutations in *Dscam* result in neural wiring defects.

More recently, *Dscam* has been shown to play an important role in the insect immune system. In both *Drosophila*[18] and the mosquito *Anopheles gambiae*,[25] the malaria vector, *Dscam* is required for the animal to mount an effective immune response. Moreover, loss of *Dscam* function impairs the ability of hemocytes to phagocytose pathogens.

One attractive hypothesis is that each *Dscam* isoform would interact with different axon guidance cues in the nervous system or pathogens/antigens in the immune system.[15] In an elegant series of experiments, the Zipursky lab reported that the extracellular portion of DSCAM (including the region encoded by the exon 4, 6 and 9 clusters) is capable of homodimerization.[26] These homophilic interactions are strikingly specific as isoforms that differ by a single alternative exon fail to interact with one another, but strongly bind to themselves.[26] Even altering as few as three amino acids in one of the variable domains can significantly reduce binding. Thus, it appears as though 19,008 DSCAM extracellular domains exist that interact in an isoform-specific manner. In the immune system, it has also been shown that some DSCAM isoforms can interact with *E. coli* while others do not.[18] This suggests that DSCAM may function as the insect equivalent of antibodies. In support of this, infecting *A. gambiae* with different pathogens results in the expression of specific repertoires of *Dscam* isoforms.[25]

The striking interaction properties of DSCAM raise the possibility that in the nervous system, individual neurons would express different repertoires of *Dscam* isoforms and that these would dictate the wiring pattern of that neuron. In support of this idea, there is clear evidence that the splicing of at least some of the alternative exons are regulated in a developmental and tissue-specific manner.[27,28] However, these results most likely vastly underestimate the degree of regulation as they were performed on either entire animals or large, complex tissues. It will be necessary to analyze *Dscam* expression at single cell resolution to adequately address this issue. Some attempts have been made to determine the number of isoforms expressed in individual cells by combining single-cell RT-PCR with microarray analysis.[23,28] These studies suggest that individual neurons express a limited collection of isoforms (less than 50). However, this issue again could benefit from a detailed analysis at the single cell level in the fly. Nonetheless, it is clear that individual neurons express different collections of isoforms.

Recent evidence also suggests that the precise collection of isoforms expressed in an individual cell is critical to specify the wiring pattern of the neuron.[24] To address this, two *Dscam* alleles were generated that each lacked 5 of the exon 4 variants—one lacked exons 4.2 through 4.6 while the other lacked exon 4.4 through 4.8. This reduces the *Dscam* repertoire from 38,016 to 22,176 potential isoforms. They found that axonal targeting of mechanosensory neurons was disrupted in flies homozygous for both of these alleles. More importantly, however, the pattern of axon branching was different for these two alleles, yet highly consistent in multiple individuals carrying each allele. Because the repertoire of potential isoforms that can be expressed from the two alleles is different, yet the overall number of isoforms is identical, these results provide strong evidence that the sequences encoded by the exon 4 variants have nonredundant functions. In other words, the identity of the isoforms expressed in an individual neuron is critical to specify the correct wiring pattern. The implication of this is that the splicing of *Dscam* must be precisely regulated.

Mechanisms Used for Mutually Exclusive Splicing in *Dscam*

One of the most intriguing aspects of *Dscam* is the fact that the exons within each variable cluster are spliced in a strictly mutually exclusive manner. Work on other genes has revealed several mechanisms that are used to ensure that the splicing of pairs of alternative exons is strictly mutually exclusive. First, the splice sites in the intron separating the two alternative exons can be spatially arranged such that when splicing factors recognize one splice site, they prevent the binding of splicing factors to the other splice site through steric hinderance. This mechanism has been shown to occur in the mammalian α-*tropomyosin*[29] and α-*actinin*[30] genes. Alternatively, splicing of the two exons would be mutually exclusive if the intron separating the exons is too small to be efficiently spliced. For example, in *Drosophila*, introns smaller than 59 nucleotides cannot be removed by the spliceosome.[31] Second, mutually exclusive splicing can also be ensured if a gene has a unique arrangement of splice sites that are recognized by the major (which consists of U1, U2, U4, U6 and U5 snRNPs) and minor (which consists of the U11, U12, U4atac, U6atac and U5 snRNPs) spliceosomes.[32] Because, neither spliceosome can remove an intron containing a mixture of major and minor splice sites,[33] the splicing of such genes is by definition, mutually exclusive. The human stress-activated protein kinase (*JNK* 1) gene contains an alternatively spliced region with this type of organization.[34] Finally, the splicing of two cassette exons may not be truly mutually exclusive, but rather may simply appear to be so. This could occur if the two alternative exons are not a multiple of three nucleotides. If neither or both exons are included, premature termination codons will be introduced into the mRNA and those isoforms will be subject to degradation by the nonsense mediated decay pathway.[35] As a result, mRNAs containing one and only one exon will be stable and the splicing of such pre-mRNAs will appear to be mutually exclusive, though in reality it is not.

Importantly, none of these mechanisms can explain how the exon 4, 6 and 9 clusters of *Dscam* are spliced in a mutually exclusive manner. First, even if steric hinderance could prevent adjacent exons from being spliced together, it would be unable to prevent non-adjacent exons from being spliced. Second, none of the introns separating the variable exons are below the size limit for *Drosophila* introns and again, this size limit would not effect non-adjacent exons. Third, none of the splice sites in *Dscam* conform to the minor spliceosome consensus sequence. Finally, the NMD mechanism cannot operate. The exons within the exon 4 cluster are all multiples of 3 such that the reading frame would not change regardless of how many exons are included. For the exon 6 and 9 clusters, while the inclusion of more than one exon would result in a frameshift, the inclusion of multiples of three exons (i.e., 4 exons, 7 exons, etc.) would not. However, such products have not been observed. Thus, it would appear that a novel mechanism(s) must exist to ensure that the splicing of the exon 4, 6 and 9 clusters of *Dscam* occurs in a mutually exclusive manner. Surprisingly, in each cluster that has been studied, it has been shown that RNA secondary structures play a critical role in the mutually exclusive splicing.

Mutually Exclusive Splicing of the Exon 4 Cluster

Our laboratory recently identified sequences located in the intron between exon 3 and the first exon 4 variant that plays an important role in the splicing of the exon 4 cluster. These sequences were shown to form an RNA secondary structure we call the inclusion stem, or iStem, that is required for the efficient inclusion of all 12 variable exons in the exon 4 cluster.[36] The iStem is formed by basepairing interactions between a 20 nt sequence located immediately downstream of the 5' splice site of exon 3 with a second 20 nt sequence located 300 nt downstream (Fig. 3A). The structural aspects of the iStem are supported by experiments showing that mutations in either half of the iStem that disrupt basepairing result in skipping of all 12 exon 4 variants and that compensatory mutations that restore the structure, but not the sequence, also restore the function of the iStem. The iStem is conserved in all 12 *Drosophila* species that have been sequenced, though the precise sequence varies considerably and there are several examples of compensatory changes. Finally, several observations suggest that only the base-paired portion of the iStem is critical for it's function. First, deleting large portions of the intervening sequence have little effect on the function of the iStem. More importantly, however, is the fact that the sequence and the distance between the regions that engage in basepairing interactions is highly variable among all *Drosophila* species. Thus, the iStem is a structural rather than a sequence-specific element that governs inclusion of all of the exons within a single cluster.

The fact that the structure, but not the sequence of the iStem is required for its function raises some interesting issues regarding the mechanisms by which it may act. One possibility is that the RNA structure of the iStem somehow promotes exon inclusion in a manner that does not require additional protein factors. Alternatively, a protein, or protein complex interacts with the iStem in a sequence nonspecific manner—perhaps a double-stranded RNA binding protein, or an RNA helicase—and this complex somehow promotes exon inclusion. An intriguing property of the iStem is that despite the fact that it controls the inclusion of all of the exons within the exon 4 cluster, it does not play a significant role in determining which variable exon is selected. Thus, although the mechanism by which the iStem functions is not known, it is clear that the iStem is a novel type of regulatory element that simultaneously controls the splicing of multiple alternative exons.

In addition to the iStem, a second RNA secondary structure exists in the exon 4 cluster and is located within the intron between exon 4.12 and exon 5. Like the iStem, this RNA structure promotes the inclusion of all 12 exon 4 variants and is therefore called the iStem2 (Kreahling and Graveley, unpublished). The iStem2 is also evolutionarily conserved, consists of two distantly located sequences that basepair with one another and mutations in either half of the iStem2 results in skipping of the exon 4 variants. Interestingly, the iStem and iStem2 appear to function together, as simultaneously disrupting both structures more strongly affects exon inclusion than does either structure alone.

These results clearly show that RNA secondary structures play an important role in the mutually exclusive splicing of the exon 4 cluster of the *Dscam* pre-mRNA. As we shall see, this will be a recurring theme as we consider the other alternatively spliced clusters in *Dscam*.

Mutually Exclusive Splicing of the Exon 6 Cluster

RNA secondary structures also play an important function in the mutually exclusive splicing of the *Dscam* exon 6 cluster. While searching the entire *Dscam* gene for sequence elements that were conserved in all *Drosophila* species and could therefore potentially function as binding sites for splicing regulatory proteins, two classes of sequence elements in the exon 6 cluster were identified—the docking site, which is located in the intron downstream of constitutive exon 5 and the selector sequences, which are located upstream of each of the 48 exon 6 variants.[16,37] The most striking aspect of these elements is that each selector sequence is complementary to a portion of the docking site (Fig. 3B). As a result, the interaction between a selector sequence and the docking site juxtaposes one and only one, alternative exon to the upstream constitutive exon. Moreover, because each selector sequence interacts with the docking site, only one selector sequence at a time can bind to the docking site. The mutually exclusive nature of the docking site-selector sequence interactions immediately suggests that the formation of these competing RNA structures is a central component of the mechanisms guaranteeing that only one exon 6 variant is included in each *Dscam* mRNA.[16,37]

Though the formation of these structures remain to be experimentally proven, the evidence that they do is extremely compelling. First, these sequence elements are highly conserved—both the docking site and the selector sequences are conserved in all 20 of the sequenced arthropod genomes and therefore first evolved at least 450 million years. Second, each exon 6 variant in all of the sequenced insect genomes contains an upstream selector sequence. Third, each selector sequence has the potential to basepair with the docking site. Fourth, several double compensatory mutations have been identified in the honeybee (*Apis mellifera*) *Dscam* gene.[16] The docking site in *A. mellifera* contains two nucleotides that are different from the docking sites of all other insects. However, several of the *A. mellifera* selector sequence have the potential to form base pairing interactions with the variable docking site nucleotides. Finally, the high degree of conservation of the docking site is expected of an element that interacts with a large number of elements as mutations in the docking site would affect the splicing of all of the exon 6 variants. Thus, all evidence to date strongly suggests that these RNA secondary structures form and that they play a key role in the mutually exclusive splicing mechanism of the exon 6 cluster.

The exact mechanism by which the docking site-selector sequence interaction promotes the splicing of the adjacent exon is unknown. Moreover, it is not known whether these interactions serve to simply ensure that only one exon 6 variant is included or whether they play a direct role in selecting which of the 48 exons are included. For example, it is possible that selector sequences that have a high affinity for the docking site will be included more efficiently than exons having selector sequences with weaker binding affinities. Current data argues against this idea as there is no correlation between the ΔG of the interaction of each selector sequence with the docking site and the frequency of inclusion observed in microarray[28] and sequencing[14,18,19,23] experiments. However, this could be complicated by factors such as distance, varying elongation rates of RNA polymerase along the gene and the potential existence of other RNA structures that would alter the actual length or possibly the accessibility of regions within the exon 6 cluster pre-mRNA. Nonetheless, the exon 6 cluster clearly employs a novel and elegant mechanism to ensure that only one exon is included.

Mutually Exclusive Splicing of the Exon 17 Cluster

Although nothing is known about how splicing of the exon 9 cluster of *Dscam* occurs in a mutually exclusive manner, there are indications that RNA secondary structures again play a key role in this process in the exon 17 cluster. Anastassiou and colleagues[37] identified four conserved sequence elements (designated A, A', B and B' in Fig. 3C) in the intron between exons 16 and 17.1 that have the potential to form intriguing RNA secondary structures. These four sequences form two sets of complementary sequences—A can basepair with A' and B can basepair with B'. Though not as obviously important as the docking site-selector sequence interaction, these sequences potentially function to ensure that only one of the two exon 17 variants is included. On one hand, when the A-A' stem forms, exon 17.1 would be preferentially included. In contrast, when the B-B' stem forms, the 3' splice site of exon 17.1 would be occluded and as a result, exon 17.2 would be included. These interactions are not mutually exclusive as both could potentially form at the same time. As a result, it will be important to functionally analyze the role of these sequence elements in the splicing of the exon 17 cluster. Nonetheless, the conservation, location and structures of these elements do suggest that they play some role in this process.

Lessons from Dscam

There are two striking aspects of the mechanisms of mutually exclusive splicing that have been uncovered by studying *Dscam*. First, in each cluster that has been studied, RNA secondary structures have been identified that play key roles in mutually exclusive splicing. Second, the mechanisms by which each of these structures functions appears to be distinct. This is particularly surprising as it suggests that there are multiple solutions to the common problem of how only one exon is selected for inclusion when three or more exons can be chosen from. Although *Dscam* is the most extreme case of such a genomic organization, it is not the sole member of this class of genes—there are at least 9 other *Drosophila* genes (*Mhc, ATPalpha, GluClalpha, slo, heph, TepII, wupA, 14-3-3zeta* and *Pfk*) that contain at least one cluster of three or more mutually exclusive exons. The fact that there are multiple mechanisms by which pre-mRNAs with three or more alternative exons can be spliced in a mutually exclusive manner suggests that it is relatively easy to evolve a way to negotiate this problem. Moreover, the fact that several *Drosophila* genes have such an organization, combined with the genomic economy of exon duplication events suggests that this is a robust evolutionary strategy for generating multiple related proteins. It is therefore striking that there are no known genes in humans, or any other vertebrates for that matter, that contain more than two exons that are spliced in a mutually exclusive manner. Why is this? Have alternative mechanisms for generating protein diversity such as VDJ recombination become so successful as to plunge multiple exon mutually exclusive splicing into extinction in the vertebrate lineage? Is the vertebrate spliceosome incapable of negotiating pre-mRNAs with such a configuration? Answers to these questions will likely provide insight into the mechanism of genome evolution.

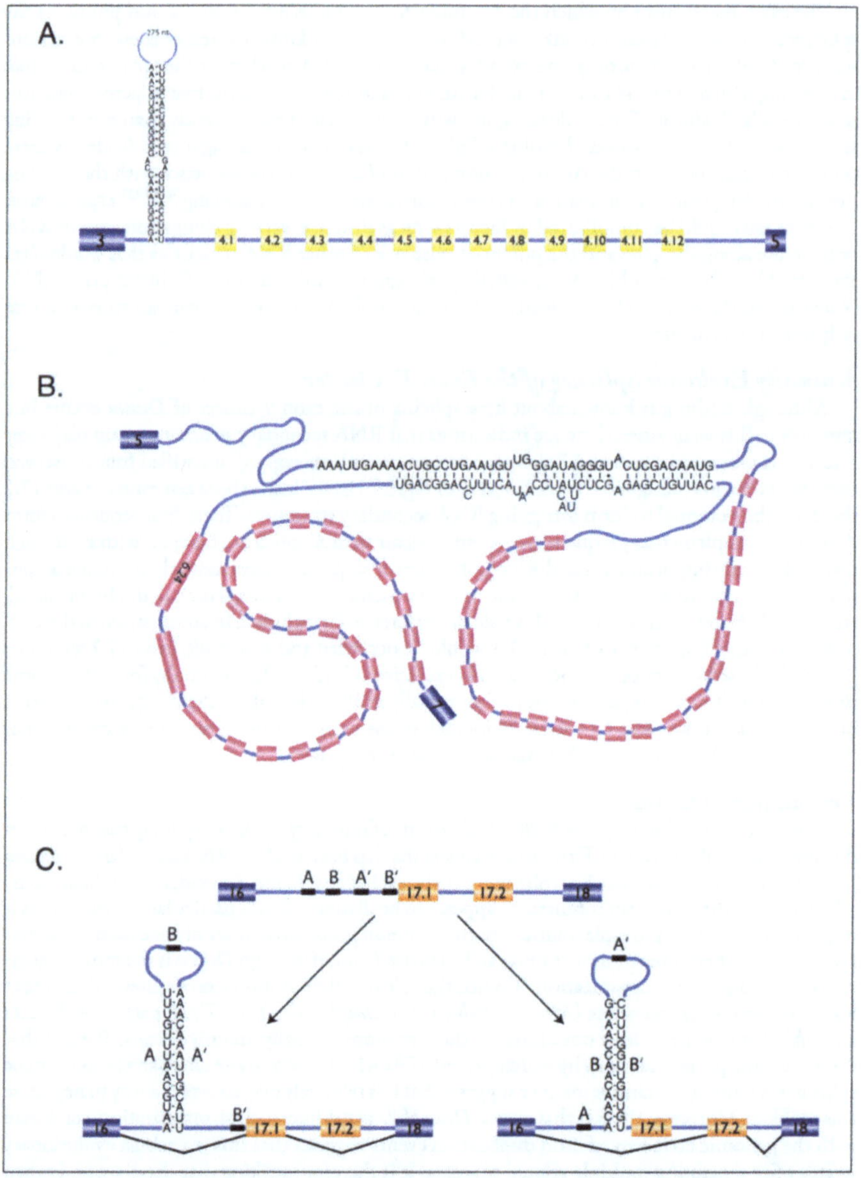

Figure 3. RNA secondary structures important for mutually exclusive splicing in the *Drosophila Dscam* pre-mRNA. A) The iStem functions to promote the inclusion of any one of the 12 exon 4 variants. In the absence of the iStem, the exon 4 variants are excluded. B) The docking site-selector sequence interactions function to prevent the inclusion of multiple exon 6 variants. Due to the fact that only one selector sequence can interact with the docking site at a time, only one exon 6 variant can be included. C) Four conserved sequence elements (represented by the boxes labeled A, A', B and B') have the potential to form RNA structures that may prevent the inclusion of both exon 17 variants in the same mRNA.

The second take home message from these findings is that each of these structures were identified by comparative genomics which was made possible by the availability of multiple genomes from closely related species. The only exception is the iStem. In this case, the 3' half of the iStem was initially identified by mutagenesis studies. However, the discovery that it was part of an RNA secondary structure rather than a protein binding site eluded detection until the sequences of several *Drosophila* genomes were available. As the ability to perform comparative genomics studies across multiple genomes has only recently become possible, it is likely that many RNA secondary structures that play an important role in alternative splicing will be discovered over the next several year and we anticipate that such elements will turn out to be quite ubiquitous.

Trans-Splicing

Two genes in the *Drosophila* genome mentioned earlier, ***lola*** and ***mod(mdg4)***, take the concept of complexity to a whole new level. These two genes utilize trans-splicing to generate mRNAs. Trans-splicing is a mechanism by which two different RNA molecules are spliced together to generate a single mRNA. Trans-splicing occurs frequently in organisms such as nematodes, trypanosomes and planarians.[38-40] In these cases noncoding spliced leader RNAs (SL RNAs) are spliced to the 5' end of the majority of the mRNAs in these organisms. This serves to both add a 5' cap to the mRNAs and to process polycistronic pre-mRNAs into monocistronic mRNAs. However, as the trans-splicing that has been characterized in *Drosophila* involves the joining of RNAs containing coding sequence, these two processes are quite distinct.

The first clear example of a biologically relevant trans-spliced coding gene was ***mod(mdg4) (modifier of mdg4)***, which encodes a BTB-domain containing transcription factor.[41] This was realized immediately upon sequencing the ***mod (mdg4)*** genomic locus and analyzing its organization. ***Mod (mdg4)*** encodes 28 distinct mRNA isoforms that are generated by the splicing of four common exons (1-4) located at the 5' end of the gene to a collection of alternative exons at the 3' end of the gene (Fig. 4). The striking observation was that seven of the isoforms contained alternative exons transcribed from the opposite strand than the strand used to transcribe the common exons![42,43] Thus, the splicing of these transcripts must occur via trans-splicing. This was rigorously proven by placing one of the trans-spliced exons on a separate chromosome from the common exons and showing that functional mRNAs were generated containing both sequences.[42] Importantly, ***mod(mdg4)*** is also trans-spliced in other ***Diptera*** and ***Lepidoptera*** species though the number and location of the variable exons differs between species.[41,44]

The second gene shown to be trans-spliced is ***lola (longitudinals lacking)***, which also encodes a BTB domain-containing transcription factor essential for axon guidance.[45,46] The ***lola*** gene is 60 kb in length and contains four alternative promoters and 32 exons that are alternatively spliced to generate 80 mRNAs[47] (Fig. 4). Each transcript contains one of the first four exons generated by alternative promoter use, the constant exons 5-8 and one set of the variable exons at the 3' end. Unlike ***mod(mdg4)***, the fact that ***lola*** undergoes trans-splicing was not obvious from the organization of the gene. Rather, this insight came from interallelic complementation studies.[48] Horiuchi and colleagues found ***lola*** alleles with mutations in the common exons are homozygous lethal, as are alleles with mutations in the variable exons. However, what was most striking was the observation that an allele with a common exon mutation can be complemented by alleles with variable exon mutations. Moreover, the idea that this occurred via trans-splicing was verified by showing that chimeric mRNAs were generated in the complemented animals that contained SNPs specific to each allele.[48] As with ***mod(mdg4)***, at least some of the variable exons are known to be transcribed from their own promoters.[48]

The identification of genes that use trans-splicing as their mechanism of mRNA synthesis raises numerous interesting issues. The first is why is trans-splicing used instead of cis-splicing? Perhaps this represents yet another means of ensuring mutually exclusive splicing—as the region of the pre-mRNA that is trans-spliced is consumed during the reaction, it is not possible to include multiple variable exons between two constant exons. Another important question is how does this type of trans-splicing occur? How are the two pre-mRNAs joined together and what specifies

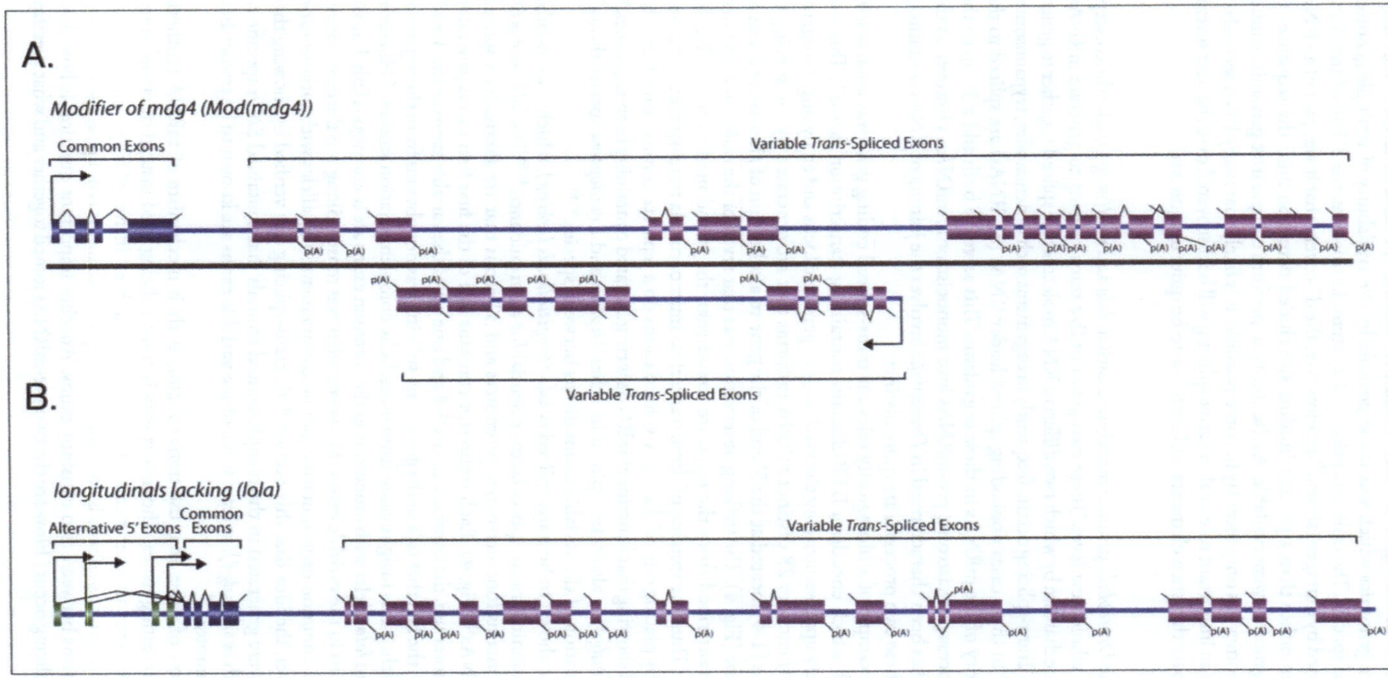

Figure 4. Trans-spliced genes in *Drosophila*. A) The *modifier of mdg4 mod(mdg4)* gene contains numerous variable 3' exons that are joined to a set of four common exons. Several of the variable exons are encoded on the opposite strand than the common exons and therefore are joined together by trans-splicing. B) The *longitudinals lacking (lola)* gene also contains numerous variable 3' exons that are joined to a set of common exons. In this case, the variable exons are encoded on the same strand of as the common exons. However, interallelic complementation experiments indicate that the *lola* mRNAs are synthesized by trans-splicing.

which transcripts are to be joined together? Why are the *lola* common exons not spliced to the *mod (mdg4)* variable exons and vice-versa? The observation that a *mod (mdg4)* variable exon placed on a different chromosome than the common exons suggests that this is not due to confining the transcripts to a specific nuclear location. What prevents the trans-spliced precursors from being spliced to other pre-mRNAs in the cell? Finally, as the genomic organization of *lola* does not obviously suggest trans-splicing as the dominant synthetic mechanism, one must wonder how many other genes are trans-spliced? Does this phenomenon occur outside of insects? There are several examples of mammalian genes that undergo trans-splicing.[49,50] However, in most cases, this occurs in only a small subset of mRNAs that are synthesized and may therefore represent noise rather than being biologically significant.[51] Nonetheless, *mod (mdg4)* and *lola* add another intriguing example of complex alternative splicing mechanisms that raise more questions than answers.

Conclusions

The examples that have been described in this chapter serve to illustrate the fact that alternative splicing can be extraordinarily complex and is used to generate a vast repertoire of isoforms. For example, the *Drosophila* proteome potentially contains at least 1,084,500 members taking into account the diversity generated by the *Mhc, Dscam, para, lola* and *mod (mdg4)* genes alone. Clearly a wide variety of regulatory mechanisms are used to precisely control the expression of these mRNAs in space and time and the fidelity of their synthesis. As our exploration of the transcriptomes of various organisms expands, additional examples of complex alternative splicing events will most certainly be unveiled. Understanding how these remarkable events are controlled and their functional roles in biology will be an exciting and consuming endeavor for years to come.

Acknowledgements

Work in our laboratory is funded by grants from the National Institutes of Health, The Raymond and Beverly Sackler Fund for the Arts and Sciences and the State of Connecticut Stem Cell Initiative.

References

1. Black DL. Protein diversity from alternative splicing: a challenge for bioinformatics and postgenome biology. Cell 2000; 103:367-370.
2. Blencowe BJ. Alternative splicing: new insights from global analyses. Cell 2006; 126:37-47.
3. Graveley BR. Alternative splicing: increasing diversity in the proteomic world. Trends Genet 2001; 17:100-107.
4. Lander ES, Linton LM, Birren B et al. Initial sequencing and analysis of the human genome. Nature 2001; 409:860-921.
5. Venter JC, Adams MD, Myers EW et al. The sequence of the human genome. Science 2001; 291:1304-1351.
6. Stolc V, Gauhar Z, Mason C et al. A gene expression map for the euchromatic genome of Drosophila melanogaster. Science 2004; 306:655-660.
7. Johnson JM, Castle J, Garrett-Engele P et al. Genome-wide survey of human alternative pre-mRNA splicing with exon junction microarrays. Science 2003; 302:2141-2144.
8. Lopez AJ. Alternative splicing of pre-mRNA: developmental consequences and mechanisms of regulation. Annu Rev Genet 1998; 32:279-305.
9. Burtis KC, Baker BS. Drosophila doublesex gene controls somatic sexual differentiation by producing alternatively spliced mRNAs encoding related sex-specific polypeptides. Cell 1989; 56:997-1010.
10. George EL, Ober MB, Emerson CPJ. Functional domains of the Drosophila melanogaster muscle myosin heavy-chain gene are encoded by alternatively spliced exons. Mol Cell Biol 1989; 9:2957-2974.
11. Standiford DM, Sun WT, Davis MB et al. Positive and negative intronic regulatory elements control muscle-specific alternative exon splicing of Drosophila myosin heavy chain transcripts. Genetics 2001; 157:259-271.
12. Palladino MJ, Keegan LP, O'Connell MA et al. A-to-I pre-mRNA editing in Drosophila is primarily involved in adult nervous system function and integrity. Cell 2000; 102:437-449.
13. Thackeray JR, Ganetzky B. Conserved alternative splicing patterns and splicing signals in the Drosophila sodium channel gene para. Genetics 1995; 141:203-214.
14. Schmucker D et al. Drosophila Dscam is an axon guidance receptor exhibiting extraordinary molecular diversity. Cell 2000; 101:671-684.

15. Zipursky SL, Wojtowicz WM, Hattori D. Got diversity? Wiring the fly brain with Dscam. Trends Biochem Sci 2006; 31:581-588.
16. Graveley BR. Mutually exclusive splicing of the insect Dscam pre-mRNA directed by competing intronic RNA secondary structures. Cell 2005; 123:65-73.
17. Graveley BR, Kaur A, Gunning D et al. The organization and evolution of the dipteran and hymenopteran Down syndrome cell adhesion molecule (Dscam) genes. RNA 2004; 10:1499-1506.
18. Watson FL, Püttmann-Holgado R, Thomas F et al. Extensive diversity of Ig-superfamily proteins in the immune system of insects. Science 2005; 309:1874-1878.
19. Hummel T, Vasconcelos ML, Clemens JC et al. Axonal targeting of olfactory receptor neurons in Drosophila is controlled by Dscam. Neuron 2003; 37:221-231.
20. Zhu H, Hummel T, Clemens JC et al. Dendritic patterning by Dscam and synaptic partner matching in the Drosophila antennal lobe. Nat Neurosci 2006; 9:349-355.
21. Wang J, Ma X, Yang JS et al. Transmembrane/juxtamembrane domain-dependent Dscam distribution and function during mushroom body neuronal morphogenesis. Neuron 2004; 43:663-672.
22. Wang J, Zugates CT, Liang IH et al. Drosophila Dscam is required for divergent segregation of sister branches and suppresses ectopic bifurcation of axons. Neuron 2002; 33:559-571.
23. Zhan XL, Clemens JC, Neves G et al. Analysis of Dscam diversity in regulating axon guidance in Drosophila mushroom bodies. Neuron 2004; 43:673-686.
24. Chen BE, Kondo M, Garnier A et al. The molecular diversity of Dscam is functionally required for neuronal wiring specificity in Drosophila. Cell 2006; 125:607-620.
25. Dong Y, Taylor HE, Dimopoulos G. AgDscam, a hypervariable immunoglobulin domain-containing receptor of the Anopheles gambiae innate immune system. PLoS Biol 2006; 4:e229.
26. Wojtowicz WM, Flanagan JJ, Millard SS et al. Alternative splicing of Drosophila Dscam generates axon guidance receptors that exhibit isoform-specific homophilic binding. Cell 2004; 118:619-633.
27. Celotto AM, Graveley BR. Alternative splicing of the Drosophila Dscam pre-mRNA is both temporally and spatially regulated. Genetics 2001; 159:599-608.
28. Neves G, Zucker J, Daly M et al. Stochastic yet biased expression of multiple Dscam splice variants by individual cells. Nat Genet 2004; 36:240-246.
29. Smith CW, Nadal-Ginard B. Mutually exclusive splicing of alpha-tropomyosin exons enforced by an unusual lariat branch point location: implications for constitutive splicing. Cell 1989; 56:749-758.
30. Southby J, Gooding C, Smith CW. Polypyrimidine tract binding protein functions as a repressor to regulate alternative splicing of alpha-actinin mutally exclusive exons. Mol Cell Biol 1999; 19:2699-2711.
31. Kennedy CF, Kramer A, Berget SM. A role for SRp54 during intron bridging of small introns with pyrimidine tracts upstream of the branch point. Mol Cell Biol 1998; 18:5425-5434.
32. Patel AA, Steitz, JA. Splicing double: insights from the second spliceosome. Nat Rev Mol Cell Biol 2003; 4:960-970.
33. Sharp PA, Burge CB. Classification of introns: U2-type or U12-type. Cell 1997; 91:875-879.
34. Letunic I, Copley RR, Bork P. Common exon duplication in animals and its role in alternative splicing. Hum Mol Genet 2002; 11:1561-1567.
35. Jones RB, Wang F, Luo Y et al. The nonsense-mediated decay pathway and mutually exclusive expression of alternatively spliced FGFR2IIIb and -IIIc mRNAs. J Biol Chem 2001; 276:4158-4167.
36. Kreahling JM, Graveley BR. The iStem, a long-range RNA secondary structure element required for efficient exon inclusion in the Drosophila Dscam pre-mRNA. Mol Cell Biol 2005; 25:10251-10260.
37. Anastassiou D, Liu H, Varadan V. Variable window binding for mutually exclusive alternative splicing. Genome Biol 2006; 7:R2.
38. Blumenthal T. Trans-splicing and polycistronic transcription in Caenorhabditis elegans. Trends Genet 1995; 11:132-136.
39. Nilsen TW. trans-splicing: an update. Mol Biochem Parasitol 1995; 73:1-6.
40. Zayas RM, Bold TD, Newmark PA. Spliced-leader trans-splicing in freshwater planarians. Mol Biol Evol 2005; 22:2048-2054.
41. Krauss V, Dorn R. Evolution of the trans-splicing Drosophila locus mod(mdg4) in several species of Diptera and Lepidoptera. Gene 2004; 331:165-176.
42. Dorn R, Reuter G, Loewendorf A. Transgene analysis proves mRNA trans-splicing at the complex mod(mdg4) locus in Drosophila. Proc Natl Acad Sci USA 2001; 98:9724-9729.
43. Labrador M, Mongelard F, Plata-Rengifo P et al. Molecular biology: Protein encoding by both DNA strands. Nature 2001; 409:1000.
44. Gabler M, Volkmar M, Weinlich S et al. Trans-splicing of the mod(mdg4) complex locus is conserved between the distantly related species Drosophila melanogaster and D virilis. Genetics 2005; 169:723-736.
45. Crowner D, Madden K, Goeke S et al. Lola regulates midline crossing of CNS axons in Drosophila. Development 2002; 129:1317-1325.

46. Giniger E, Tietje K, Jan LY et al. Lola encodes a putative transcription factor required for axon growth and guidance in Drosophila. Development 1994; 120:1385-1398.
47. Ohsako T, Horiuchi T, Matsuo T et al. Drosophila lola encodes a family of BTB-transcription regulators with highly variable C-terminal domains containing zinc finger motifs. Gene 2003; 311:59-69.
48. Horiuchi T, Giniger E, Aigaki T. Alternative trans-splicing of constant and variable exons of a Drosophila axon guidance gene, lola. Genes Dev 2003; 17:2496-2501.
49. Finta C, Zaphiropoulos PG. Intergenic mRNA molecules resulting from trans-splicing. J Biol Chem 2002; 277:5882-5890.
50. Wu Q, Maniatis T. A striking organization of a large family of human neural cadherin-like cell adhesion genes. Cell 1999; 97:779-790.
51. Tasic B, Nabholz CE, Baldwin KK et al. Promoter choice determines splice site selection in protocadherin alpha and gamma pre-mRNA splicing. Mol Cell 2002; 10:21-33.

Chapter 5

Technologies for the Global Discovery and Analysis of Alternative Splicing

John A. Calarco,[†] Arneet L. Saltzman,[†] Joanna Y. Ip and Benjamin J. Blencowe*

Abstract

During the past ~20 years, studies on alternative splicing (AS) have largely been directed at the identification and characterization of factors and mechanisms responsible for the control of splice site selection, using model substrates and on a case by case basis. These studies have provided a wealth of information on the factors and interactions that control formation of the spliceosome. However, relatively little is known about the global regulatory properties of AS. Important questions that need to be addressed are: which exons are alternatively spliced and under which cellular contexts, what are the functional roles of AS events in different cellular contexts, and how are AS events controlled and coordinated with each other and with other levels of gene regulation to achieve cell- and development- specific functions. During the past several years, new technologies and experimental strategies have provided insight into these questions. For example, custom microarrays and data analysis tools are playing a prominent role in the discovery and analysis of splicing regulation. Moreover, several non-microarray-based technologies are emerging that will likely further fuel progress in this area. This review focuses on recent advances made in the development and application of high-throughput methods to study AS.

Introduction

Recent studies employing custom microarrays and associated data analysis tools have shed light upon the possible extent of AS regulation associated with specific cell and tissue types, both under physiologically normal and disease-associated contexts. Related studies have also yielded novel insights into the global contributions of individual factors to splicing regulation. As a result, particular genes and cellular functions that are likely regulated in a coordinated fashion by individual or combinations of specific splicing factors, as well as features of the sequence "code" underlying regulated splicing events, have been identified. The enormous amount of data generated by such high-throughput approaches has led to a wealth of opportunities for follow-up studies directed at exploring not only additional global aspects of AS, but also the functions of a myriad of individual exons in gene regulation and the mechanisms involved.

In this review, we will survey the development and application of different high-throughput approaches that have been employed to study AS on a large-scale. We will assess what has been learned from these technological advances as well as how we envisage their further development

[†] Co-First Authors
*Corresponding Author: Benjamin J. Blencowe—Banting and Best Department of Medical Research and Department of Molecular and Medical Genetics. Terrence Donnelly Center for Cellular and Biomolecular Research, 160 College Street, University of Toronto, Toronto, Canada, M5S 3E1. Email: b.blencowe@utoronto.ca

Alternative Splicing in the Postgenomic Era, edited by Benjamin J. Blencowe and Brenton R. Graveley.

contributing to our knowledge of AS. We will also survey some of the emerging technologies for their potential to address questions that cannot currently be addressed by microarray-based systems. We are clearly at an exciting juncture in the field, where the merging of the new technologies with more conventional approaches in splicing research promises to facilitate new insight into the nature of the complexity and regulation of AS in multicellular organisms.

Microarray-Based Methods for the Large-Scale Discovery and Characterization of Alternative Splicing Events

Tiling and Splice Junction Microarrays

One of the first indications that DNA microarrays could be used to detect AS in mRNA transcripts came from studies applying 'tiling' probe arrays, i.e., microarrays with closely spaced or overlapping probes that span entire chromosomes or genomes (Fig. 1A).[1-3] Initially employed were tiling microarrays fabricated using the ink-jet technology developed at Rosetta Inpharmatics.[4] These microarrays contained 60mer oligonucleotide probes with sequences overlapping by 10 nucleotides corresponding to the nonrepetitive portions of chromosome 22. The microarrays were hybridized with cDNA corresponding to 69 different cell or tissue types.[1] In the same study, microarrays containing single probes for all annotated exons in the human genome were also employed (Fig. 1B). The initial aim of these strategies was to identify transcribed regions of the genome and to confirm and extend knowledge of the exon/intron architecture of annotated genes. In addition to providing such insights, evidence for the differential usage of certain exons in transcripts was apparent from comparisons of the hybridization efficiencies of probes to specific exons relative to neighboring exon probes.

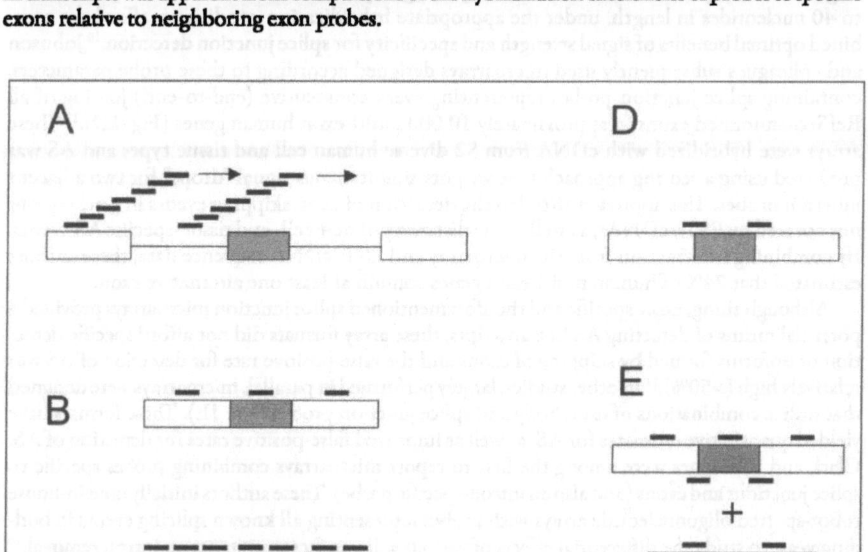

Figure 1. Microarray formats for monitoring AS. The diagram shows various microarray formats employing oligonucleotide probes (thick lines), which are typically anchored to glass slides. The probe arrangements are shown in relation to a single cassette-type alternative exon (gray rectangle) flanked by intron sequences and constitutive exons (white rectangles). The following probe configurations are illustrated: (A) genomic tiling; (B) single probe per exon; (C) multiple probes per exon; (D) junction-specific probes for included exons; (E) combinations of exon and junction specific probes for included and skipped exons.

Another tiling-based approach employed 25mer oligonucleotide microarrays fabricated using a photolithography-based process developed by Affymetrix.[3] In this study, the oligonucleotides were spaced at 35 nucleotide intervals spanning human chromosomes 21 and 22 and the data generated from hybridizing cDNA from 11 different tumor and fetal cell lines also provided evidence of possible differential exon inclusion levels in transcripts.[3] Subsequent re-analysis of the data from this study using tailored statistical approaches provided evidence for a higher degree of exon skipping events than previously predicted using the same array platform.[5] A more direct analysis of microarray data for the purpose of AS detection subsequently emerged from profiling 1600 rat genes in ten different tissues using a custom-designed Affymetrix array containing probes largely concentrated in the 3'-UTR regions of genes (Fig. 1C).[2] In this study, the authors concluded that approximately 17% of the UTR regions displayed evidence of AS.

In more recent studies, exon-specific and tiling microarrays have been used for the analysis of AS in various organisms and in different developmental contexts.[6-8] For example, Stolc and colleagues used a microarray that incorporated probes to all exons and approximately 40% of the predicted splice junctions in *Drosophila melanogaster*.[6] The results from profiling different *Drosophila* developmental stages predicted the existence of 5440 previously unidentified splice isoforms and it was estimated that approximately 40% of genes in the fruit fly contain one or more AS events.

Many of the studies cited above utilized cDNA labeling procedures that depended on oligo-dT priming and therefore 3'-end detection bias was an issue. To address this issue, Johnson and colleagues developed a full-length RNA amplification and labeling protocol and used this in conjunction with ink-jet fabricated microarrays containing 38 mer splice junction-specific oligonucleotide probes.[9] Careful controls by this group and others indicated that probes of 36 to 40 nucleotides in length, under the appropriate hybridization conditions, afford the combined optimal benefits of signal strength and specificity for splice junction detection.[10] Johnson and colleagues subsequently used microarrays designed according to these probe parameters, containing splice junction probes representing every consecutive (end-to-end) joining of all RefSeq-annotated exons in approximately 10,000 multi-exon human genes (Fig. 1D).[11] These arrays were hybridized with cDNA from 52 diverse human cell and tissue types and AS was predicted using a scoring approach that weights simultaneous signal "drops" for two adjacent junction probes. This approach afforded the detection of exon skipping events in gene regions not covered by ESTs/cDNAs, as well as the discovery of new cell- and tissue-specific AS events. By combining information from the microarray and EST/cDNA sequence data, these authors estimated that 74% of human multi-exon genes contain at least one alternative exon.

Although tiling, exon-specific and the aforementioned splice junction microarrays provided a potential means of detecting AS in transcripts, these array formats did not afford specific detection of isoforms formed by skipping of exons and the false-positive rate for detection of AS was relatively high (>50%).[11] In other studies, largely performed in parallel, microarrays were designed that utilize combinations of exon body and splice junction probes (Fig. 1E). These formats have yielded quantitative estimates for AS, as well as improved false-positive rates for detection of AS. Clark and colleagues were among the first to report microarrays combining probes specific to splice junctions and exons (and also an intron-specific probe). These authors initially used in-house robot-spotted oligonucleotide arrays with probes representing all known splicing events in budding yeast to study the differential effects of various splicing factor mutants on intron removal.[10] Since budding yeast contains few AS events, the primary aim of this study was to assess whether the ~200 single intron-containing genes are spliced with different efficiencies in different splicing factor mutant strains.

Fiber Optic-Based Detection Microarrays and DASL

A different microarray system was developed by Fu and colleagues working in collaboration with Illumina.[12] The method described by these authors involves a series of steps including RNA/cDNA-mediated annealing, selection, extension and ligation (RASL or DASL) followed by hybridization to arrays and monitoring using a fiber optic-based detection system (Fig. 2).[12,13] Specifically,

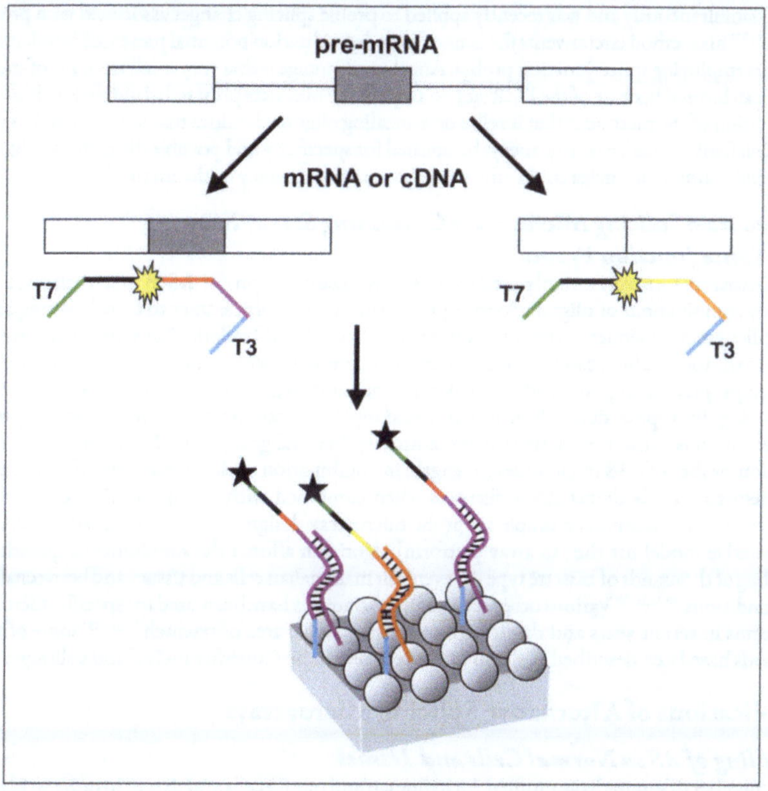

Figure 2. Monitoring of AS using RASL or DASL (RNA/cDNA-mediated annealing, selection, extension and ligation), followed by hybridization to fiber optic-based detection microarrays. Ligation products are generated when a common oligonucleotide (black line) anneals to an exon sequence directly flanking a splice junction shared by two mRNA isoforms and splice isoform-specific oligonucleotides (red and yellow lines) anneal on the other side of the junction. Following annealing and ligation (yellow explosion) of the oligonucleotides (common and isoform-specific) the ligation products are amplified by PCR using universal T7 and T3 primer sites, included on the ends of the annealing oligos. Biotin (black star) is incorporated during the amplification process through one of the PCR primers and the amplified products are then hybridized through unique bar-code sequences (pink and orange lines) that are complementary to sequences located on the fiber-optic detection microarray.

this technique makes use of a common oligonucleotide designed to anneal to the exon sequence flanking a splice junction shared by two mRNA or cDNA isoforms and two specific oligonucleotides (each flanked by a specific bar-code sequence that will hybridize to the fiber-optic-coupled detection array) designed to anneal to alternative exon sequences unique to each splice isoform that flank the common exon. Hence, only one of the two specific oligonucleotides should anneal to each isoform. The annealed oligonucleotides (common and isoform-specific) are ligated together and amplified by PCR using universal primer sites at the ends of the oligos. Biotin is incorporated during the amplification process through one of the PCR primers and the amplified products are then hybridized to the fiber optic array through the unique bar-code sequence and detected through the use of labeled streptavidin. The RASL/DASL technique has been multiplexed for monitoring thousands of AS events simultaneously and was recently applied to profile splicing changes associated with prostate cancer.[13] This method circumvents the issue of cross-hybridization potential presented by microarray formats employing splice-junction probes. Another advantage is that very small amounts of starting RNA can be used because of the PCR step to amplify the products prior to hybridization. However, a limitation of the method is that it relies on annealing oligonucleotides that are restricted to splice junction flanks, which may not always be optimal for specificity and possible differences in ligation or amplification efficiencies could affect the quantitative accuracy of the method.

Alternative Splicing Microarrays Combining Sets of Exon and Splice Junction Probes

Microarray formats currently used by most investigators to profile AS in metazoan organisms employ combinations of oligonucleotide probes that are complementary to exon body sequences and splice junction sequences representing both "included" and "excluded" or skipped isoforms.[14-25] As with the initial tiling, exon or splice-junction microarrays mentioned above, these formats typically employ sets of oligonucleotides anchored at one end on glass slides. Although some strategies employing this type of design have incorporated multiple overlapping 25 mer probes targeted to splice junctions to improve measurement sensitivity,[24] several groups have found that single splice junction probes of ~38 nucleotides in length, in combination with the exon-specific probes, are sufficient to reliably detect AS differences when combined with the appropriate data analysis tools.[15,18,21,23] For example, a simple six-probe microarray design in conjunction with GenASAP (Generative model for the AS array platform) algorithm affords the simultaneous quantitative profiling of thousands of cassette type AS events in mammalian cells and tissues and between different conditions.[20,21,26-29] Various other data analysis strategies have been used for specific microarray platforms in recent years and this is currently a very active area of research.[2,5,17,25] Some of these methods have been described in detail in a recent review by Cuperlovic-Culf and colleagues.[30]

Applications of Alternative Splicing Microarrays

Profiling of AS in Normal Cells and Tissues

A number of groups have profiled AS in human and mouse cells and tissues in order to identify and characterize tissue- and developmental stage-specific splicing events.[11,17,21,23,24,28,31] Results from these studies indicate that many AS events occur in a tissue-regulated manner, analogous to the regulation of gene expression at the transcriptional level. Johnson and colleagues identified numerous splicing differences between human tissues and cell lines[11] and work from our group and the groups of Ares and Lee have uncovered numerous new tissue-specific AS events in mouse tissues.[17,21,23] Huang and colleagues profiled AS in various developmental stages of the testes and mature spermatozoa using cDNA microarrays and validated by sequencing 74 of these developmentally-regulated AS events.[31] Approximately 83-85% of these AS events are predicted to alter the reading frames of the resulting transcripts. Recently, our group together with the group of Lynch has profiled AS changes that occur upon activation of the Jurkat T-cell line.[27] Approximately 10% of the alternative exons represented on the microarray undergo significant changes in inclusion level and many of these differences occurred in transcripts of genes known to have important roles in T-cell function. These recent studies have demonstrated the utility of AS microarrays for the identification of novel tissue- and development stage-specific AS events as well as quantitative information on the inclusion levels of the corresponding exons. As such, they have provided substantial amounts of new information about AS that is prime for future characterization.

Splicing-sensitive microarrays have also been employed to address biological questions in other model systems. Ner-Gaon and Fluhr used whole genome arrays to assess the extent to which intron retention, the most common class of AS event in *Arabidopsis thaliana*, occurs in this species.[8] These investigators estimate that approximately 8% of all transcripts undergo intron retention and that the vast majority of these AS events alter the reading frame of the affected transcripts. This result implies that intron retention evolved as an important mechanism to regulate the function and activity of certain genes in *Arabidopsis*.

Another study by McIntyre and colleagues profiled sex-specific splicing differences in various laboratory lines of *Drosophila melanogaster*.[7] Regulation of specific AS events in transcripts such as those from the *doublesex* and *sex-lethal* genes have been shown to be important in the sex determination pathway in *Drosophila*.[32,33] However, it was not known prior to this study what proportion of transcripts undergo AS in a sex-specific manner. Surprisingly, the work of McIntyre and colleagues indicated that between 11% and 24% of transcripts known to have multiple variants are predicted to show sex-biased splicing patterns. This study also found that a large proportion of previously identified splicing regulators[34] are differentially expressed between male and female flies, suggesting that these differences could be responsible, at least in part, for the large number of AS differences observed. Future experiments should help establish whether a causal connection exists between the differential regulation of these splicing factors and sex-related splicing changes.

In addition to surveying large numbers of AS events, some groups have designed focused microarrays for profiling single pre-mRNA transcripts with complex patterns of AS. In one such study, Nagao and colleagues designed microarrays with exon, intron and exon-junction probes to study previously identified and potentially cryptic splice variants of the *PTCH* gene, the human homolog of the *Drosophila* gene patched.[19] Protein products from the *PTCH* locus are important components of the hedgehog signaling pathway and mutations in this gene have been linked to patients with autosomal dominant nevoid basal cell carcinoma syndrome (NBCCS). In the study by Nagao and colleagues it was found that transcript variants containing exon 12B are more abundant in the brain and heart than in the other tissues surveyed. Inclusion of this exon introduces a premature termination codon that is predicted to result in a either a nonfunctional or dominant-negative acting protein, potentially impacting brain and heart development. In addition, microarray experiments using RNA from cell lines derived from NBCCS patients identified novel isoforms generated by the usage of cryptic splice sites. Another recent study employing a focused array profiled documented AS events in transcripts of genes encoding all of the known human phosphodiesterases.[14] Tissue-specific AS was detected in the transcripts of specific phosphodiesterase genes, thus providing a more detailed picture of the expression landscape for this important class of enzymes. Together, the results of the studies summarized above demonstrate the utility of focused AS arrays for monitoring known splice variants and for the detection of aberrant variants associated with diseases.

Alternative Splicing Patterns Associated with Cancer

Transcript level profiles derived from microarray data have been used to establish molecular 'signatures' that facilitate distinguishing normal from cancerous tissues, the subtyping of different breast cancers and leukemic cell types and the prediction of whether tumours respond to specific drugs, as well as prognostic outcomes.[33] Since disease-associated mutations very often target the splicing process and given increasing evidence for links between alterations in splicing patterns and disease,[35,36] it has been suggested that information from splicing-sensitive microarrays may further improve the accuracy of identifying disease-related signatures.[37] Evidence has been provided in one AS profiling study that this may indeed be the case (see below). It has also been proposed that the identification of disease-associated splice isoform changes will lead to the development of aberrant isoform-specific drugs that do not interfere with isoforms from the same gene that encode normal gene functions.[38]

A number of groups have microarray-profiled various tumour tissues and tumour-derived cell lines in recent years.[13,18,22] Relogio and colleagues profiled approximately 100 AS events implicated in cancers in normal lymphocytes and Hodgkin's lymphoma cell lines from patients with differing severities of the disease.[22] In parallel, the expression patterns of 86 splicing regulators were also monitored to assess the extent to which trans-acting factors might be deregulated in cancer. Alternative splicing differences were detected between the normal lymphocytes and Hodgkin's lines and a correlation was observed between AS differences and tumor severity. This study also revealed transcript abundance differences and ectopic expression of various splicing factors, including the neuronal Nova proteins, which were associated with increased inclusion of exon 6a in transcripts of *JNK2*, an important kinase implicated in Hodgkin's lymphomas.

In another recent study, prostate tumour samples and cell lines were profiled for AS and transcript level differences.[13] Li, Fu and colleagues used the DASL-PCR/fiber-optic array platform described above to profile changes in transcripts from approximately 1500 genes implicated in prostate cancer. These investigators also developed a data analysis method to assign similarity scores based both on splicing and transcript levels and performed hierarchical clustering using these similarity criteria. This 'two-dimensional' clustering technique was used to distinguish between cancerous and noncancerous samples and also between the severity and grade of individual prostate tumors. The results indicated that profiling splicing differences together with transcript level differences can afford a refined method to characterize tumors and potentially also to discover examples of transcription-coupled splicing changes linked to prostate cancer progression. Subsequent analysis of the data from this study indicated that the splice variant profiling data was more predictive of prostate cancer origin than was the transcription profiling data,[39] thus supporting the idea that splice isoform "signatures" have considerable potential for detection and disease characterization.

Li, Lin and colleagues have also recently used exon and splice junction microarrays to profile AS in transcripts from 64 genes in human breast cancer cell lines and normal mammary epithelial cells.[18] To investigate the influence of cell culturing environments on tumor formation and progression in vivo, the authors compared profiles from cultured cells grown in two dimensional and three dimensional environments and also from xenografted tumours grown in nude mice. Changes in AS were observed in transcripts from specific genes such as *MYL6*, *DDR1* and *CD44* when comparing breast cancer lines with normal lines. It was also found that the different culturing conditions affected AS patterns, with three-dimensional culturing of breast cancer cells in matrigel substrate showing stronger similarities to the AS patterns in xenografted tumors, than to the AS patterns obtained from the cell lines grown in two-dimensional culture.

Cancer-associated splice isoforms identified in the studies summarized above may help shed light on important molecular changes underlying the development and progression of cancers, in addition to serving as potentially useful cancer biomarkers. All three of the studies mentioned above indicate that alterations in trans-acting splicing factors likely underlie many of the splice isoform changes associated with cancers and that the elucidation of the nature and relevance of these changes may also lead to the identification of valuable cancer biomarkers.

Global Regulation by Trans-Acting Splicing Factors and AS Networks

Several groups have used microarrays to investigate the global activities of trans-acting splicing factors.[10,15,24] In the study by Clark et al various strains of *Saccharomyces cerevisiae* with temperature-sensitive mutations in pre-mRNA processing factors were surveyed for effects on differential splicing efficiency, using the microarrays described above.[10] The use of different indices for monitoring intron accumulation and splicing efficiency allowed these investigators to separate mutants that differentially affect pre-mRNA levels and splicing efficiency. Interestingly, it was found that many mutants affecting intron accumulation did not affect splicing efficiency and vice versa, indicating that the two processes are controlled in many cases by different sets of factors.

Blanchette et al investigated the targets of *Drosophila melanogaster* splicing regulators using custom-designed AS microarrays combining exon body and junction probes.[15] These authors used double-stranded RNA interference in cultured *Drosophila* SL2 cells to knockdown expression of dASF/SF2, B52/dSRp55, hrp48 and PSI. Labeled cDNA prepared from RNA isolated from the cells was hybridized to an AS array with probes representing 2,931 unique transcripts. Each splicing factor was found to affect the AS of different numbers and subsets of exons, with dASF/SF2 affecting the most exons (~300) and PSI affecting the least (~48 exons). Subsequent analysis of the intronic regions flanking the dASF/SF2- and B52/dSRp55-regulated exons identified overrepresented sequences that resemble the SELEX-defined motifs of these factors. A relatively small degree of overlap was observed between the exon targets of these splicing factors and few exons displayed opposite splicing patterns when dASF/SF2 and hrp48 (the *Drosophila* homolog of hnRNP A1) were knocked down. Mammalian ASF/SF2 and hnRNP-A1 proteins were previously shown to have opposing activities in the regulation of AS in the context of several analyzed transcripts (refer to chapters by Lin and Fu, and Marinez-Contreras et al) and these antagonistic functions were thought to occur on a more general level. However, the results of Blanchette et al suggest that the control of AS events could require the combinatorial action of numerous different splicing regulators, including as yet unknown combinations of both identified and unidentified factors.

Recent work from Ule et al made use of an Affymetrix microarray combining exon and splice junction probes to survey splicing differences between wild-type mice and mice carrying deletions in the genes encoding the neural-specific splicing factors Nova1 and Nova2.[24] These authors identified in their microarray data a set of AS events differentially regulated between neocortex and thymus tissues that change in level in the absence of Nova2. All of the Nova2-dependent AS events in neocortex were validated by RT-PCR assays, allowing the authors to estimate that Nova2 regulates approximately 7% of AS events that are differentially spliced in brain versus thymus tissues. Importantly, it was found that a significant fraction of the Nova2-regulated alternative exons belong to genes that have functions associated with the synapse. Furthermore, it was found that ~74% of the proteins corresponding to the Nova-regulated list of genes were previously reported to interact with one another. These data strongly suggested that Nova proteins function to regulate the splicing of a network of genes that function at the synapse (Fig. 3A). Recent work from our lab involving AS microarray profiling of diverse mouse tissues has provided additional evidence supporting the concept of regulated splicing networks that operate to control specific functions in the central nervous system (CNS).[28] A set of ~150 alternative exons displaying differential inclusion levels in CNS tissues relative to the other profiled tissues was identified and it was found that the genes containing these exons are significantly enriched in annotations associated with functions including GTP-regulated signalling, vesicle-mediated transport, cytoskeletal organization and biogenesis, as well as in nervous system-specific functions. In another recent study we used a human-specific AS microarray to identify a set of AS events that display differential regulation upon activation of the Jurkat T-cell line. Of the AS events that were analyzed further, >50% were also detected in activated, normal CD4+ and CD8+ T-cell lymphocytes. Interestingly, genes containing the differentially-regulated alternative exons were significantly enriched in annotations associated with cell cycle control.[27]

These results resemble those from earlier studies from microarray profiling of transcript levels using conventional microarrays, in which subsets of genes with cell and tissue coregulated transcript levels are often functionally related and operate in the same processes and pathways.[40-42] However, importantly, the sets of genes that display cell/tissue coregulation at the level of AS do not overlap extensively with the sets of genes coregulated at the level of transcription and in general overlapping yet distinct functional categories are enriched in each set. These studies extended previous observations from microarray profiling indicating that different subsets of genes are regulated in a tissue-specific manner at the levels of AS and transcription (Fig. 3B).[17,21] Taken together with the results of Ule and colleagues, as well as recent microarray-based evidence for gene function coordination at the levels of mRNA stability (e.g., by miRNAs) and export,[43,44] it is increasingly apparent that AS represents one of several layers of regulation acting on different subsets of genes to confer cell- and tissue-specific functions in mammals (Fig. 3C). Future research in gene expression will undoubtedly explore the mechanisms by which the splicing machinery controls functionally-related sets of genes and how these networks operate in the context of networks controlled at other levels of gene expression to define cell- and tissue-specific activities.

General Global Regulatory Features of AS

Alternative splicing microarray data has afforded new opportunities to discover and analyze global regulatory properties of splicing that operate in a cell- and tissue-type *independent* manner. One area of investigation has been to determine the impact of the evolutionary origin of exons on their inclusion levels. The results of our initial profiling experiments using ten diverse adult mouse tissues showed that weakly included exons are overrepresented by genome-specific AS events (i.e., exons present in the mouse genome but not in the human genome). This observation confirmed earlier predictions based entirely on the analysis of aligned cDNA and EST sequences by Modrek

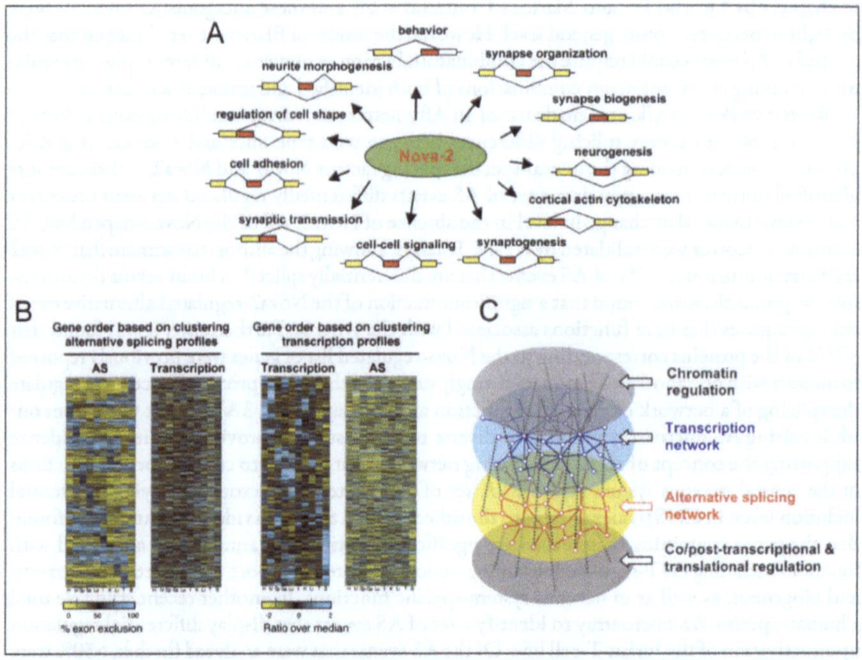

Figure 3. Global regulatory features of AS revealed by microarray profiling. A) Darnell and colleagues compared splicing patterns between neocortex from wild type and mutant mice lacking Nova-2. Genes with Nova-2 dependent AS differences are predominantly associated with functions linked to synapse activity.[24] The diagram illustrates different types of Nova-2 regulated AS events monitored on the microarray (which include alternative 5'/3' splice site selection events as well as cassette alternative exons) and a subset of the Biological Process annotation categories (assigned using the Gene Ontology system) associated with genes containing alternative exons that display a significant change in inclusion level between the neocortex of wild type and Nova-2$^{-/-}$ mice. B) Different subsets of mouse genes are regulated at the AS and transcriptional levels. Genes are shown clustered based on AS levels (left-most panel), as determined by AS profiling using a mouse-specific microarray,[21] and the same set of genes is shown in the same order in the adjacent panel. Genes are also shown clustered based on transcript levels (third panel from left), as measured using probes specific for constitutive exons flanking each profiled alternative exon and the same set of genes is shown in the same order in the adjacent panel. (Adapted from: Blencowe BJ. Cell 126(1):37-47;[37] ©2006 with permission of Elsevier.) C) Data from microarray AS profiling experiments, including those described above, suggest that different subsets of genes may comprise "layers" of networks that coordinate specific cellular functions. The dotted lines indicate interconnections that serve to integrate these and the other layers of gene regulation shown. (Adapted from: Pan Q et al. Mol Cell 16(6):929-941;[21] ©2004 with permission from Elsevier.)

and coworkers.[45] The AS microarray data also revealed that the majority of alternative exons at steady state are highly included in mouse tissues and that highly included alternative exons are often represented by conserved exons that undergo species-specific AS (i.e., in this case conserved exons that are constitutively spliced in human but alternatively spliced in mouse). It was also found that more highly included exons, such as those involving conserved exons that are spliced in a species-specific manner, more often have the potential to alter conserved and functionally-defined protein domains.[21,46,47] Interestingly, alternative exons that are conserved between human and mouse were found to have more variable inclusion levels in different tissues and also to less often have the potential to disrupt conserved protein domains. In agreement with these findings, further analysis of our dataset by Xing et al[48] revealed that cassette alternative exons conserved between mouse and human, which represent less than 20% of the total number of cassette alternative exons, more often undergo differential regulation or "switching" between tissues. Taken together, a conclusion from these studies is that overall selection pressure has acted to avoid major alterations to conserved domain-encoding regions of transcripts, although there are clear exceptions to this trend in which AS can trigger major structural and functional alterations by deleting or including conserved protein domains. An important goal of future studies will be to elucidate the functions of specific regulated AS events that affect regions of transcripts that do not overlap defined or well conserved protein domain coding sequences.

Assessing the Functional Relevance of AS Events

An important general question is the extent to which splice isoforms detected in transcript sequence and microarray data have important biological functions. In particular, how often are detected splice isoforms the products of infrequent splicing events that are present at a low steady state levels and do not confer a function that is under significant selection pressure? Relevant to this question is the interesting observation that ~40% of AS events detected in alignments of cDNA and EST data have the potential to introduce one or more premature termination codons (PTCs). The resulting splice variants are therefore potential substrates for mRNA turnover by the nonsense mediated mRNA decay (NMD) pathway (refer to chapter by Lareau et al).[49-51] Based on this finding it was proposed that AS coupled to NMD may act on a frequent basis to control gene expression.[50] Indeed, experimental evidence supports an important role for this regulatory mechanism, termed "RUST" (regulatory unproductive splicing and translation). For example, it has been shown in several cases to operate in the auto-regulation of expression of RNA binding proteins. In these cases the RNA binding proteins, which include members of the SR and hnRNP families of splicing factors, can bind to their own pre-mRNAs resulting in the activation of an AS event that introduces a PTC. The PTC can be inserted by inclusion or skipping of the exon (i.e., the exon itself may contain a PTC or the PTC is introduced via a frame-shift), resulting in subsequent degradation of the PTC-containing isoform by NMD (refer to chapter by Lareau et al for specific examples).

Using AS microarray profiling, we have assessed the extent to which AS coupled NMD may function as a regulatory mechanism.[20] The microarray data revealed that PTC-containing isoforms are generally present at low steady state abundance in diverse mammalian cells and tissues. Knockdown of the essential NMD factor Upf1 by RNAi in HeLa cells showed that only a small fraction (~10%) of PTC-containing isoforms significantly increase in level when NMD is blocked. This result therefore indicated that most PTC-containing isoforms are present at low levels independently of NMD, whereas a small fraction of these isoforms may be involved in Upf1-depedent regulation. It was also observed that PTC-containing splice isoforms are more often species-specific than conserved between human and mouse. This observation was consistent with the results of previous analyses of transcript sequences showing that conserved AS events are more likely to preserve reading frames than are species-specific AS events.[46,47] Taken together, a conclusion that can be drawn from these studies is that most PTC-containing isoforms may not be functionally significant or under significant selection pressure. However, as mentioned above, a small subset of AS events function to regulate transcript levels via NMD. Interestingly, many of the PTC-introducing AS events that display pronounced Upf1-dependent regulation are conserved between humans and mouse and a subset of these events are found in genes associated with splicing and other steps in pre-mRNA processing (A.L.S. and B.J.B., unpublished results). Thus, in agreement with previous studies performed on a gene-by-gene basis, AS-coupled NMD events detected by microarray profiling experiments appear to often function in the regulation of the levels of factors involved in the post-transcriptional control of gene expression.

Elucidating Features of the Splicing Regulatory "Code" via Microarray Profiling

Recent microarray profiling experiments have been performed on larger numbers of AS events and in a wider spectrum of tissues.[23,28] In these studies clustering of tissue-specific microarray profiled AS events coupled with sequence analyses of the exons and flanking introns has resulted in the identification of motifs that are significantly over-represented in groups of tissue-regulated AS events.

Recent work from Sugnet and colleagues[23] used this approach on data generated from profiling 22 mouse tissues using an Affymetrix microarray and identified two new motifs, one associated with muscle-specific AS that resembles a branchpoint-like sequence and the other associated with brain-specific AS. Recent work from our group has resulted in the identification of a set of motifs that are highly correlated with regulation of AS in central nervous system (CNS) tissues.[28] Interestingly, many of the motifs that correlate strongly with inclusion of alternative exons in CNS tissues relative to the other profiled tissues are pyrimidine-rich and are primarily located in the intron regions flanking the alternative exons. These motifs represent potential binding sites for neural/brain-specific PTB (nPTB/brPTB), which has been documented to promote increased inclusion of alternative exons in transfected reporters in cells of neuronal origin. Additional motifs that do not resemble the binding sites of characterized splicing factors were also identified; different subsets of these motifs correlate strongly with increased inclusion and/or exclusion of exons in CNS tissues. These studies thus have the potential to identify important elements of the splicing code underlying the control of tissue-regulated AS. An important direction for future studies will be to expand this approach to define groups of AS events that are coregulated in other tissues and cell-types, such that additional motifs and sequence features that may underlie regulation (e.g., RNA folding) can be identified on a more systematic basis. Obviously, parallel studies will be required to experimentally validate these sequence elements and also to identify the factors that bind to them.

Comparing Global Patterns of AS Between Species

As mentioned above, less than 20% of cassette AS events appear to be conserved between human and mouse. The fact that most splice variants are different between these two mammalian species suggests an important role for AS in the evolution of species-specific characteristics. However, experiments directed at assessing the extent to which AS levels of orthologous exons diverge over time and how splicing levels compare between species that are separated by a shorter window of evolutionary time, have not been reported. We have recently investigated these questions by comparing AS patterns between corresponding tissues in human, chimpanzee and mouse. Using our human AS microarray we have profiled AS in human and chimpanzee frontal cortex and heart.[29] The majority of the AS events showed similar exon inclusion levels. However, 6-8% of AS events displayed pronounced differences in inclusion levels. These AS events were detected in transcripts from genes associated with diverse functions, including regulation of gene expression, signalling and predisposition to human lineage-specific diseases. These events could therefore represent splicing changes that could underlie a range of important functional and phenotypic differences between humans and chimpanzees.

AS Microarray Technology: The Way Forward?

Since their initial use a few years ago, it is evident that AS microarrays will become widespread tools for researchers studying global aspects of gene regulation. Clearly, if microarray designs and associated data analysis tools afford the effective surveying of transcriptional and splicing regulation in a single experiment, it makes little sense to focus on transcriptional regulation alone, given the likely importance of AS in virtually every aspect of metazoan biology. Widespread use of AS microarray technology in the next few years seems highly probable given that rapid progress is being made in generating new microarray formats with ever-increasing probe density. Moreover, there is continual progress in the development of software tools that maximize the power of data analysis such that reliable predictions for splicing changes or levels can be extracted from the raw data. The technology clearly has room for improvement, as is the case for any early stage technology. Nevertheless, the value of AS microarrays as both a discovery and analysis tool has been demonstrated and the current systems will continue to allow new and interesting insights to be generated concerning the role of AS regulation in different biological contexts.

A limitation of current AS microarray systems is that they do not provide information about the connectivity of exons in transcripts that contain multiple alternatively spliced regions. For example, if a gene containing two separate alternative exons is transcribed and each of the exons is detected by microarray profiling to be included 50% of the time, it is not possible to know whether the two exons are included together in 50% of the transcripts or whether, for example, 50% of the transcripts contain one exon and the other 50% of the transcripts contains the other exon. This problem becomes more acute in cases where a series of alternative exons belonging to a single gene may be spliced in different combinations in different cell types. Another limitation of most microarray systems is their sensitivity and accuracy in detection of splice changes in low abundance transcripts. Emerging technologies that may resolve some of these issues are discussed below.

Non Microarray-Based Methods for the Discovery and Characterization of Alternative Splicing Events

Applying Polonies to Quantify Splice Variants

The polymerase colony (polony) technology is a recently described approach affording the deconvolution of combinations of splicing events in mixtures of isoforms.[52,53] Polonies allow the amplification of an array of individual DNA molecules immobilized in a thin polyacrylamide gel coating a glass slide (see Fig. 4).[54-56] For AS profiling, the DNA is immobilized in the gel by covalent linkage of one PCR primer that is specific for a sequence flanking the alternatively spliced region of interest.[52,54] Different isoforms are then amplified and detected by the use of either fluorescently-labeled exon-specific probes, or single base extension (SBE) incorporating fluorescently-labeled nucleotides (Fig. 4).[52,55,56] Several exon probes can be used sequentially by denaturing, washing and rehybridizing the gel. The design is also amenable to 'multiplexing', since different fluorophore-labeled probes can be targeted to different exons and hybridized simultaneously.[52] Using this approach, single molecules are amplified into "polonies" and the combination of fluorescent signals in a polony can be used to infer the exon composition of the individual, amplified cDNAs.

The utility of the polony assay was initially demonstrated in applications in which different combinations of exon inclusion were detected in transcripts from the *MAPT* (Microtubule-Associated Protein Tau), *SMN* (Survival of Motor Neuron) and *CD44* genes. Similar to some of the AS microarray profiling systems described above, the polony technique depends on prior knowledge of transcript sequences for primer design. However, in addition to the main advantage that specific combinations of alternative exons/regions that occur within a single transcript can be discerned, the polony method is also highly sensitive and capable of detecting rare splice isoforms. Moreover, the method is quantitative, since the relative amounts of different isoforms can be measured by counting the number of arrayed polonies. However, a limitation to its use at present is its throughput, in part due to the relatively high cost of fabricating large numbers of exon-specific probes for multi-exon genes, as well as the limit to the number of denaturation cycles that can be performed on each polony gel.[52]

Other Primer Extension-Based Strategies

In addition to the polony methods, differential primer extension has been employed in other strategies to profile AS. A 'tag-microarray' minisequencing system originally developed for SNP genotyping was used for the multiplexed detection of 61 alternatively spliced transcripts from 19 genes in leukemia cell lines.[57] The approach used was somewhat similar to certain applications of the polony strategy described above, except that the primers were immobilized directly on

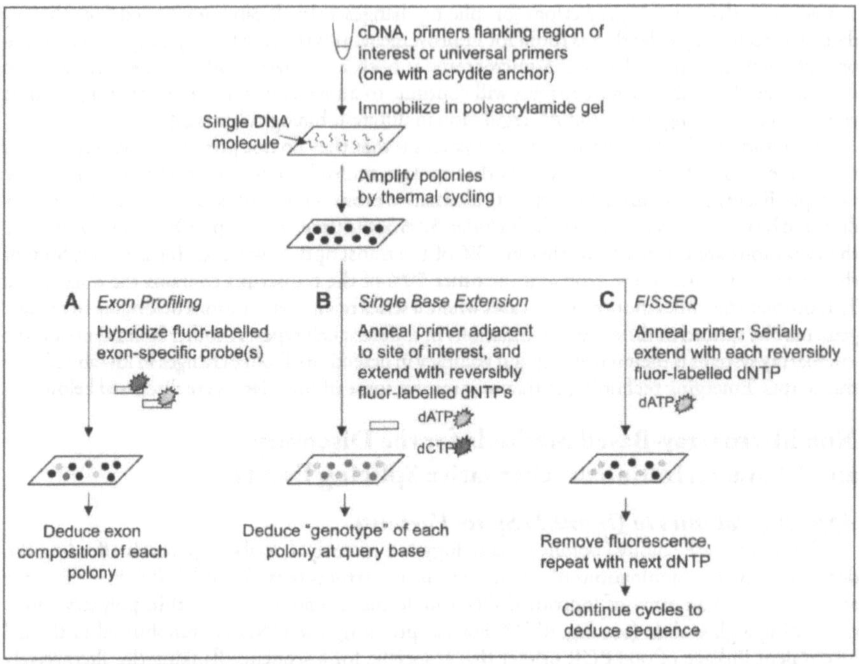

Figure 4. AS profiling on polonies. Individual cDNA molecules, reverse-transcribed from poly(A) + RNA, are immobilized in a thin polyacrylamide gel and then amplified into polonies by PCR. Three methods for interrogation of the transcript sequence are shown. A) In exon profiling, fluorophore-labeled exon probes are hybridized to the polonies to determine the combinations of exons in each polony. Different fluorophores can be used to examine several exons simultaneously. Following scanning to detect fluorescent signal, probes are denatured from the anchored DNA strands and washed away so that other probes may be hybridized.[52] B) To identify a particular nucleotide within a transcript, a primer complementary to the sequence immediately upstream is annealed to the polonies. Single base extension is then performed with fluorescently-labeled nucleotides to identify the base.[52,55] C) In FISSEQ, Fluorescence In Situ Sequencing, cycles of single base extension are performed with each nucleotide, allowing sequencing of up to 8 bases.[56] Figure adapted from references herein.

a glass slide rather than in a polyacrylamide matrix. The arrayed minisequencing primers were designed to anneal immediately adjacent to the splice junction and extended with fluorescent nucleotides.[57] This method, which has the potential to profile 200 transcripts in 80 samples on one slide[57] requires PCR amplification, which makes it relatively sensitive to the detection of low abundance transcripts, although it is only semi-quantitative in detecting the relative abundance of splice variants. A related arrayed primer extension (APEX) approach was used for detecting transcripts from variable exons of CD44 in a human breast tumour sample.[58] While this design also used primer extension on immobilized primers for detection of splice variants, it employed different primers for each splice junction sequence.

Another recently described method that permits high-throughput and automated profiling of AS involves primer extension using gene-specific RT-PCR and primer extension coupled to detection by MALDI-TOF MS (matrix-assisted laser desorption/ionization time of flight mass spectrometry).[59] Different products generated by primer extension that differ by one to a few bases at a site of AS are resolved and quantified by MALDI-TOF MS.[59] This method was applied to several genes in human tissue samples and was useful in detecting complex splicing patterns in *VEGF* (Vascular Endothelial Growth Factor) transcripts. Similar to other primer extension-based AS profiling methods, prior knowledge of the sequences involved is required for primer design. However, with the use of a 384 well-plate format, a throughput of a few thousand reactions per day could be achieved.[59]

Sequence-Based Detection of AS Events

While certain microarray designs incorporating tiling and splice junction oligonucleotides (discussed above) can be used for the de novo identification of AS events, many previous analyses have relied on the alignment of EST and cDNA sequences to genome sequence. Several databases of AS events have been constructed in this way. Lists of such databases for mammalian sequences have been published in several recent reviews,[60-63] and new resources are continuously being developed.[64] However, a recent analysis of sequences in the Mammalian Genome Collection (MGC) indicated that random cDNA sequencing is approaching saturation for the addition of new full-length cDNA clones to the MGC.[65] Therefore, the identification of novel transcripts and AS events that have been missed by systematic cDNA and EST sequencing will require the development of more efficient and directed strategies.[66] For example, several methods for generating cDNA libraries enriched for AS and for directed screening of existing libraries for novel transcripts have been described. In addition, combining new computational prediction approaches with large-scale experimental validation and the use of new high-throughput sequencing technologies will be discussed.

Generating "AS Libraries"

Several methods have been proposed to selectively clone alternatively spliced transcripts to produce "AS libraries". In the ASEtrap strategy,[67] adaptor sequences are ligated to the ends of fragmented double-stranded cDNA that has been denatured and renatured. The cDNA is then incubated with His-tagged *E. coli* SSBP (ssDNA-binding protein), which will bind to single-stranded 'loop regions' generated by renaturing of differentially alternatively spliced cDNAs. Affinity-purified SSBP-cDNA complexes are then either subjected to further cycles of enrichment or the cDNA is PCR amplified using the adaptor sequences and cloned into an AS library.[67] An ASEtrap library generated from placental cDNA had 10-fold enrichment for internal cassette-exon events when compared to a control library.[67] In a similar strategy called ASSET (AS sequence enriched tags),[68] instead of using SSBP, annealing of randomized biotinylated oligonucleotides was used to 'trap' the single-stranded loop regions.[68] ASSET led to the identification of previously unidentified splicing events in 436 genes.[68]

Another strategy called DATAS (Differential Analysis of Transcripts with Alternative Splicing)[38,69] has been proposed. In this approach, biotinylated cDNA generated from one condition or population of transcripts is hybridized with mRNA from a second condition or population to form RNA:DNA heteroduplexes. Single-stranded loop regions of mRNA will occur in regions where there are sequences inserted by AS. These RNA loops are isolated by RNaseH digestion of RNA in the RNA:DNA hybridized region and the biotinylated cDNA strand is then removed by magnetic capture on streptavidin beads. The isolated single-stranded RNA regions are then reverse transcribed and cloned into a library.[38,69] A disadvantage of this approach is that only the alternative region is cloned, whereas in previously mentioned techniques, the sequences flanking the alternative region are also included.

Directed Screening of Existing Libraries

Efficient and directed methods for screening cDNA libraries represents another approach for detection of alternatively spliced transcripts that are not present in EST collections.[70] For example, when 384 well-characterized genes not present in the MGC were targeted for full-length cDNA cloning, unexpected PCR products representing novel splice variants of five of these genes were uncovered, as well as potential splice variants for many more of the genes.[66]

A method for directed screening of plasmid cDNA libraries, SLIP (self-ligation of inverse PCR products), was used to produce full-length clones for 153 transcription factors not represented in the *Drosophila* Gene Collection (DGC).[70] In SLIP, a transcript is amplified from a pool of cDNA library plasmids, using a primer pair that target adjacent sequences in a known exon in the transcript and which direct extension in opposite directions.[70] The amplified sequence is then circularized and the library portion of the sequence is selectively digested with a methylation-sensitive restriction enzyme. The remaining cDNA product is cloned and sequenced. SLIP was successful for 68% of targets and able to recover full-length cDNA for rare and alternatively-spliced genes.[70] Unlike RT-PCR, which depends on accurate prediction of transcript ends for primer design, SLIP has the advantage of being able to recover sequences at the 5' and 3' ends of genes that are not present in annotated gene models.[70]

Another strategy for the screening of existing cDNA libraries is RecA-mediated exon profiling.[71] In this technique, target plasmids are probed using the homologous recombinase RecA, carrying a radioisotope-labeled oligonucleotide probe to the putative novel alternative exon, (i.e., identified by another method, such as a tiling array). The formation of a 'triplex' between the RecA carrying the oligonucleotide probe and a cDNA carrying the exon sequence is detected by electrophoretic mobility difference and autoradiography.[71] This strategy can be more cost-effective than library sequencing and was able to identify novel splice variants in four genes.[71]

Computational Prediction and Experimental Validation of Novel Alternative Exons

In addition to the identification of AS events by alignment of cDNA and EST sequences to genomes, sequence comparisons between genomes have been used to develop computational predictions of novel AS events. For example, using sequence features of human and mouse genes, two exon classification algorithms, ACESCAN[47] (ACE: alternative conserved exons) and a Non-EST-Based Method for Exon-Skipping Prediction[72] predicted many novel alternative exons conserved between human and mouse, a subset of which were verified by RT-PCR assays. Another probabilistic method, UNCOVER[73] (Unknown COnserved Variable Exon Recognition) also used sequence features to predict both skipped exons and retained introns in human and mouse. Other computational approaches combine comparative genomics and EST sequences, such as the cross-species EST-to-genome comparison algorithm ENACE,[74] which uses human, mouse and rat sequences to predict novel alternative exons and retained introns with no previous EST evidence in the same species.

In model organisms without sufficient EST coverage, alternative methods to aligning ESTs to the genome must be developed for AS prediction. For example, by examining conservation between *D. melanogaster* and *D. pseudoobscura*, putative alternative exons were identified, at least 40% of which were found to be alternatively spliced.[75] For the prediction of AS in *C. elegans*, a new gene-finding algorithm, LOCUS[76] (Length Optimized Characterization of Unknown Spliceforms), uses transcript length information to predict the form of alternatively spliced transcripts. The length information is derived from the *C. elegans* Orfeome project, in which transcripts were amplified and cloned using primers targeted to predicted first and last exons.[77,78] LOCUS could accurately predict the correct form of most alternatively spliced products on a test set of 151 alternatively spliced internal exons, including both cassette exons and alternative splice site usage.[76] However, the accuracy of LOCUS was decreased when used on the fly and human genomes, suggesting that the algorithm parameters need to be retrained and improved for applications in other species, perhaps by incorporating more complex scoring schemes, splicing enhancer/silencer sequences, or conservation information.[76]

High-Throughput Sequencing Technology

The demand in the academic and commercial sectors for cost-effective and high-throughput sequencing technology has led to several new sequencing strategies which have tremendous potential for the detection and monitoring of AS. These new technologies have been reviewed in depth in terms of their relative cost, throughput, accuracy and applications.[79-82] Most of these technologies currently yield considerably shorter sequence read-lengths per run compared to the read-lengths that can be achieved by conventional Sanger sequencing. However, they do offer advantages in terms of increased throughput and sensitivity.

The use of new, single molecule-based sequencing technologies for the sensitive detection of transcript diversity will allow the detection of novel transcript variants and the potential to examine the combinatorial patterns of splice site usage within transcripts. Several single molecule techniques have been described in which clonally amplified single DNA molecules are arrayed on a surface or in wells and sequenced in a highly parallel manner (Fig. 5A-C). For example, polonies (described above) can be used for amplifying single DNA molecules. A minisequencing technique, FISSEQ (Fluorescence In Situ SEQuencing),[56] can then be used to sequence up to 8 bases on each polony, using reversibly fluorophore-labeled nucleotide analogues (Fig. 4C).[56] In another polony-based approach, a combination of emulsion PCR amplification and sequencing on polonies using fluorophore-labeled probes allowed the resequencing of an *E. coli* genome (Fig. 5A).[83] Specific applications of polony technology to the profiling of AS have been discussed above.

In another technology referred to as "Massively Parallel Signature Sequencing (MPSS), sequences from the 3'-end of transcripts are generated and captured on paramagnetic beads (Fig. 5B).[84] Sequencing is performed by cycles of restriction digestion and ligation using fluorescently labeled probes. This method has been applied to the quantitative analysis of gene expression, since it allows the counting of mRNA molecules in a cell or tissue sample.[84] While the sequencing of short tags from the 3' end of transcripts would not be directly applicable to the profiling of AS, modifications to the sequence capture approach can be envisaged that would allow directed sequencing and quantification of alternatively spliced transcripts.

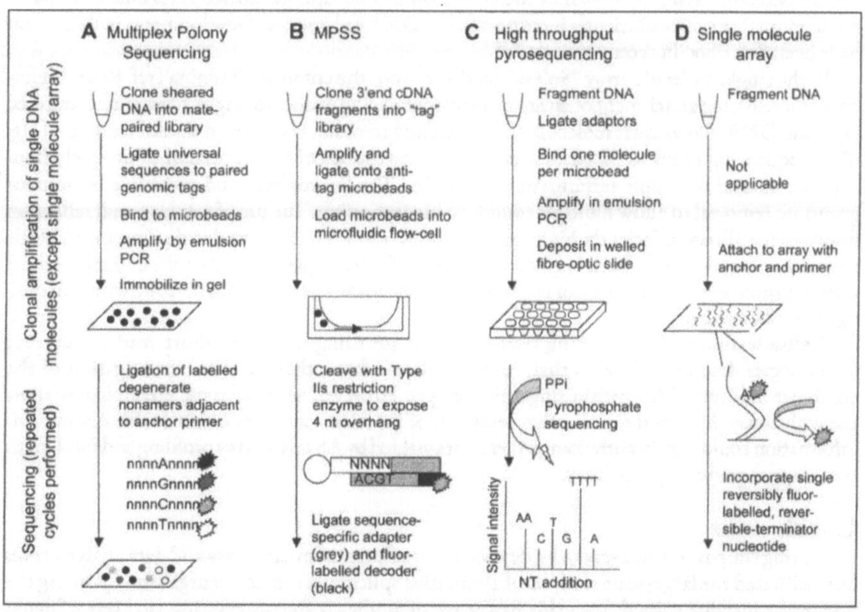

Figure 5. Overview of several high-throughput single-molecule-based sequencing techniques that can be applied to the profiling of AS. In multiplex polony sequencing[83] (A), MPSS[84] (Massively Parallel Signature Sequencing, Lynx/Illumina; B) and high throughput pyrosequencing[85] (454 Life Sciences™; C), various techniques are employed to clonally amplify single DNA molecules (top). In contrast, the single molecule array[82] (Solexa/Illumina; D) does not require clonal amplification. Separation of sequencing substrates using polony slides, microfluidics, welled fibre-optic slides, or high density arrays (A-D, respectively; middle) allows highly-parallel sequencing. Rounds of sequencing are then performed using methods based on the detection of specifically designed fluorescently-labelled (explosion) oligonucleotides (A,B) or single nucleotide triphosphates (D), or the detection of emitted light (lightning bolt, C).

A technology for high-throughput pyrosequencing of clonally amplified DNA (Fig. 5C) has recently been applied to the sequencing of the *Mycoplasma genitalium* genome.[85] In this approach, adapters are ligated to sheared genomic DNA, which is then singly coupled to beads and amplified by emulsion PCR.[86] Pyrosequencing of the amplified DNA on a single bead is then carried out in picolitre-size wells on a fiber-optic slide.[85] The release of pyrophosphate that accompanies successive nucleotide incorporations is then detected as light through enzymatic reactions.[87] This massively parallel pyrosequencing method can achieve a read length of 100 bases, the longest of the aforementioned technologies. When applied to transcriptome profiling of the LNCaP prostate cancer cell line treated with synthetic androgen,[88] this technique enabled quantitative measurement of transcript abundance, as well as identification of novel splice variants. 181,279 expressed sequence tags (ESTs) were generated from a small amount of fragmented cDNA and allowed the identification of 25 novel AS events.[88] In addition, 10,000 ESTs were mapped to genomic locations with no prior predicted transcription. However, there was some sequencing bias towards the ends of transcripts, which may be the result of incomplete cDNA fragmentation.[88]

The technologies described above involve the sequencing of clonally amplified single DNA molecules. However, new technologies that allow the sequencing of single molecules without amplification are emerging, offering advantages in terms of reduced amounts of starting material and a lack of possible amplification bias. Development of true single molecule sequencing technologies has been advanced by continuing progress in areas such as nanofluidic devices and sensitive optical detection systems, as well as the chemistry of fluorophore-labeled reversible terminator nucleotides that can be efficiently incorporated by DNA polymerase. Developments in these areas have been described in recent reviews.[89,90]

In the single molecule array "Solexa" platform from the company Illumina (ref. 82 and references therein), the attachment of an anchor and sequencing primer to fragmented, single stranded genomic DNA allows each molecule to be attached to a different site on a solid substrate (Fig. 5D). Sequencing of up to 30 bases is carried out by rounds of incorporation of uniquely-fluorophore-labeled reversible terminator nucleotides. The fluorophore and reversible terminator group are removed to allow multiple rounds of incorporation. The use of total internal reflection fluorescence allows sufficiently high signal-to-noise ratio for single molecule detection. While this technology has only been applied to genomic DNA sequencing, it could be modified for transcriptome sequencing in an analogous way to the massively parallel pyrosequencing technology described above.

A disadvantage of performing transcriptome profiling using the short-read sequencing technologies described above is that, in their current form, they preclude investigation of the combinatorial use of AS within single transcripts. However, some of these technologies allow the high-throughput and quantitative detection of junction regions associated with AS and this information could significantly complement data yielded by AS microarray profiling and full-length transcript sequencing procedures.

Conclusions

During the past several years, the development of custom microarrays and data analysis tools has facilitated the large-scale analysis of alternative splicing in metazoan organisms. Among the important insights gained from this first wave of studies is the observation that sets of genes regulated at the level of alternative splicing in a cell or tissue type-specific manner are primarily different from the sets of genes regulated in the same contexts at the level of transcription.[17,21,27] Moreover, the targets of a small number of alternative splicing factors have already been identified using alternative splicing profiling technology.[15,24] These studies have revealed that individual splicing factors such as the neural-specific protein Nova can regulate the alternative splicing of genes that are significantly associated with specific functions. The emerging picture from these and other recent studies is that alternative splicing, like transcription and other layers of gene regulation, is capable of coordinating the activities of genes that function in common processes and pathways. Such "networks" of alternative splicing-regulated genes await further delineation and characterization. For example, the expansion of datasets by profiling additional cell and tissue sources, as well as stages of differentiation and development, should facilitate the further identification of groups of genes that are regulated in a coordinated manner by alternative splicing. Such an expansion of data will provide a powerful basis from which to identify new features of the "regulatory code" underlying cell/tissue and differentiation stage-specific splicing decisions.[23] The knowledge of new motifs identified in this manner will in turn facilitate the identification of new trans-acting factors that regulate alternative splicing. As discussed in this chapter, these advances will undoubtedly benefit from ongoing improvements to the profiling capacity and accuracy of systems that can be used to monitor alternative splicing.

While this chapter has primarily focused on the developments and applications of systems for the large-scale analysis of alternative splicing, it does not discuss a major challenge that the field must face in the years ahead-namely the stark absence of information on the specific functions of alternatively spliced protein variants. As the data allowing the charting of new and existing splice variants rapidly expands, high throughput screens for assessing the functions of the splice variants will be required. It seems quite likely that RNAi and antisense technologies will be employed for the systematic targeting of splice variants associated with specific functions and pathways, for which there is an assayable readout. Similarly, given increasing evidence that the translated regions corresponding to many alternative exons occur on the surfaces of proteins,[91] an important goal will be to establish other high-throughput systems that permit protein-protein interactions involving collections of expressed splice variants to be assayed. Finally, as suggested previously,[37] we appear to be entering an era of "exonomics", in which exon-level resolution functional data will become increasingly important in order to obtain a more complete picture of the regulation and function of complex biological systems, such as the nervous system. The increased use of exon targeting strategies in a whole organism context will therefore also be critical in order to dissect the biological roles of individual exons, especially those that display conserved regulatory patterns.

Acknowledgements

We thank Brent Graveley for helpful comments on the manuscript. Our research funding is provided by the Canadian Institutes of Health Research, National Cancer Institute of Canada, and by Genome Canada through the Ontario Genomics Institute.

References

1. Shoemaker DD, Schadt EE, Armour CD et al. Experimental annotation of the human genome using microarray technology. Nature 2001; 409(6822):922-927.
2. Hu GK, Madore SJ, Moldover B et al. Predicting splice variant from DNA chip expression data. Genome Res 2001; 11(7):1237-1245.
3. Kapranov P, Cawley SE, Drenkow J et al. Large-scale transcriptional activity in chromosomes 21 and 22. Science 2002; 296(5569):916-919.
4. Hughes TR, Mao M, Jones AR et al. Expression profiling using microarrays fabricated by an ink-jet oligonucleotide synthesizer. Nat Biotechnol 2001; 19(4):342-347.
5. Kampa D, Cheng J, Kapranov P et al. Novel RNAs identified from an in-depth analysis of the transcriptome of human chromosomes 21 and 22. Genome Res 2004; 14(3):331-342.
6. Stolc V, Gauhar Z, Mason C et al. A gene expression map for the euchromatic genome of Drosophila melanogaster. Science 2004; 306(5696):655-660.
7. McIntyre LM, Bono LM, Genissel A et al. Sex-specific expression of alternative transcripts in Drosophila. Genome Biol 2006; 7(8):R79.
8. Ner-Gaon H, Fluhr R. Whole-genome microarray in Arabidopsis facilitates global analysis of retained introns. DNA Res 2006; 13(3):111-121.
9. Castle J, Garrett-Engele P, Armour CD et al. Optimization of oligonucleotide arrays and RNA amplification protocols for analysis of transcript structure and alternative splicing. Genome Biol 2003; 4(10):R66.
10. Clark TA, Sugnet CW, Ares Jr M. Genomewide analysis of mRNA processing in yeast using splicing-specific microarrays. Science 2002; 296(5569):907-910.
11. Johnson JM, Castle J, Garrett-Engele P et al. Genome-wide survey of human alternative pre-mRNA splicing with exon junction microarrays. Science 2003; 302(5653):2141-2144.

12. Yeakley JM, Fan JB, Doucet D et al. Profiling alternative splicing on fiber-optic arrays. Nat Biotechnol 2002; 20(4):353-358.
13. Li HR, Wang-Rodriguez J, Nair TM et al. Two-dimensional transcriptome profiling: identification of messenger RNA isoform signatures in prostate cancer from archived paraffin-embedded cancer specimens. Cancer Res 2006; 66(8):4079-4088.
14. Bingham J, Sudarsanam S, Srinivasan S. Profiling human phosphodiesterase genes and splice isoforms. Biochem Biophys Res Commun 2006; 350(1):25-32.
15. Blanchette M, Green RE, Brenner SE et al. Global analysis of positive and negative pre-mRNA splicing regulators in Drosophila. Genes Dev 2005; 19(11):1306-1314.
16. Fehlbaum P, Guihal C, Bracco L et al. A microarray configuration to quantify expression levels and relative abundance of splice variants. Nucleic Acids Res 2005; 33(5):e47.
17. Le K, Mitsouras K, Roy M et al. Detecting tissue-specific regulation of alternative splicing as a qualitative change in microarray data. Nucleic Acids Res 2004; 32(22):e180.
18. Li C, Kato M, Shiue L et al. Cell type and culture condition-dependent alternative splicing in human breast cancer cells revealed by splicing-sensitive microarrays. Cancer Res 2006; 66(4):1990-1999.
19. Nagao K, Togawa N, Fujii K et al. Detecting tissue-specific alternative splicing and disease-associated aberrant splicing of the PTCH gene with exon junction microarrays. Hum Mol Genet 2005; 14(22):3379-3388.
20. Pan Q, Saltzman AL, Kim YK et al. Quantitative microarray profiling provides evidence against widespread coupling of alternative splicing with nonsense-mediated mRNA decay to control gene expression. Genes Dev 2006; 20(2):153-158.
21. Pan Q, Shai O, Misquitta C et al. Revealing global regulatory features of mammalian alternative splicing using a quantitative microarray platform. Mol Cell 2004; 16(6):929-941.
22. Relogio A, Ben-Dov C, Baum M et al. Alternative splicing microarrays reveal functional expression of neuron-specific regulators in Hodgkin lymphoma cells. J Biol Chem 2005; 280(6):4779-4784.
23. Sugnet CW, Srinivasan K, Clark TA et al. Unusual Intron Conservation near Tissue-Regulated Exons Found by Splicing Microarrays. PLoS Comput Biol 2006; 2(1):e4.
24. Ule J, Ule A, Spencer J et al. Nova regulates brain-specific splicing to shape the synapse. Nat Genet 2005; 37(8):844-852.
25. Wang H, Hubbell E, Hu JS et al. Gene structure-based splice variant deconvolution using a microarray platform. Bioinformatics 2003; 19 (Suppl 1):i315-322.
26. Shai O, Morris QD, Blencowe BJ et al. Inferring global levels of alternative splicing isoforms using a generative model of microarray data. Bioinformatics 2006; 22(5):606-613.
27. Ip JY, Tong A, Pan Q et al. Global analysis of alternative splicing during T-cell activation. RNA 2007; 13:563-572.
28. Fagnani M, Barash Y, Ip JY et al. Functional coordination of alternative splicing in the mammalian central nervous system. Genome Biol 2007; 8:R108..
29. Calarco JA, Xing Y, Caceres M et al. Global analysis of alternative splicing differences between humans and chimpanzees. Genes Dev 2007; in press.
30. Cuperlovic-Culf M, Belacel N, Culf AS et al. Data analysis of alternative splicing microarrays. Drug Discov Today 2006; 11(21-22):983-990.
31. Huang X, Li J, Lu L et al. Novel development-related alternative splices in human testis identified by cDNA microarrays. J Androl 2005; 26(2):189-196.
32. MacDougall C, Harbison D, Bownes M. The developmental consequences of alternate splicing in sex determination and differentiation in Drosophila. Dev Biol 1995; 172(2):353-376.
33. Forch P, Valcarcel J. Splicing regulation in Drosophila sex determination. Prog Mol Subcell Biol 2003; 31:127-151.
34. Park JW, Parisky K, Celotto AM et al. Identification of alternative splicing regulators by RNA interference in Drosophila. Proc Natl Acad Sci USA 2004; 101(45):15974-15979.
35. Cartegni L, Chew SL, Krainer AR. Listening to silence and understanding nonsense: exonic mutations that affect splicing. Nat Rev Genet 2002; 3(4):285-298.
36. Caceres JF, Kornblihtt AR. Alternative splicing: multiple control mechanisms and involvement in human disease. Trends Genet 2002; 18(4):186-193.
37. Blencowe BJ. Alternative splicing: new insights from global analyses. Cell 2006; 126(1):37-47.
38. Bracco L, Kearsey J. The relevance of alternative RNA splicing to pharmacogenomics. Trends Biotechnol 2003; 21(8):346-353.
39. Zhang C, Li HR, Fan JB et al. Profiling alternatively spliced mRNA isoforms for prostate cancer classification. BMC Bioinformatics 2006; 7:202.
40. Eisen MB, Spellman PT, Brown PO et al. Cluster analysis and display of genome-wide expression patterns. Proc Natl Acad Sci USA 1998; 95(25):14863-14868.

41. Su AI, Wiltshire T, Batalov S et al. A gene atlas of the mouse and human protein-encoding transcriptomes. Proc Natl Acad Sci USA 2004; 101(16):6062-6067.
42. Zhang W, Morris QD, Chang R et al. The functional landscape of mouse gene expression. J Biol 2004; 3(5):21.
43. Keene JD, Lager PJ. Posttranscriptional operons and regulons co-ordinating gene expression. Chromosome Res 2005; 13(3):327-337.
44. Lim LP, Lau NC, Garrett-Engele P et al. Microarray analysis shows that some microRNAs downregulate large numbers of target mRNAs. Nature 2005; 433(7027):769-773.
45. Modrek B, Lee CJ. Alternative splicing in the human, mouse and rat genomes is associated with an increased frequency of exon creation and/or loss. Nat Genet 2003; 34(2):177-180.
46. Pan Q, Bakowski MA, Morris Q et al. Alternative splicing of conserved exons is frequently species-specific in human and mouse. Trends Genet 2005; 21(2):73-77.
47. Yeo GW, Van Nostrand E, Holste D et al. Identification and analysis of alternative splicing events conserved in human and mouse. Proc Natl Acad Sci USA 2005; 102(8):2850-2855.
48. Xing Y, Lee CJ. Protein modularity of alternatively spliced exons is associated with tissue-specific regulation of alternative splicing. PLoS Genet 2005; 1(3):e34.
49. Green RE, Lewis BP, Hillman RT et al. Widespread predicted nonsense-mediated mRNA decay of alternatively-spliced transcripts of human normal and disease genes. Bioinformatics 2003; 19 (Suppl 1): i118-121.
50. Lewis BP, Green RE, Brenner SE. Evidence for the widespread coupling of alternative splicing and nonsense-mediated mRNA decay in humans. Proc Natl Acad Sci USA 2003; 100(1):189-192.
51. Lejeune F, Maquat LE. Mechanistic links between nonsense-mediated mRNA decay and pre-mRNA splicing in mammalian cells. Curr Opin Cell Biol 2005; 17(3):309-315.
52. Zhu J, Shendure J, Mitra RD et al. Single molecule profiling of alternative pre-mRNA splicing. Science 2003; 301(5634):836-838.
53. Butz JA, Roberts KG, Edwards JS. Detecting changes in the relative expression of KRAS2 splice variants using polymerase colonies. Biotechnol Prog 2004; 20(6):1836-1839.
54. Mitra RD, Church GM. In situ localized amplification and contact replication of many individual DNA molecules. Nucleic Acids Res 1999; 27(24):e34.
55. Mitra RD, Butty VL, Shendure J et al. Digital genotyping and haplotyping with polymerase colonies. Proc Natl Acad Sci USA 2003; 100(10):5926-5931.
56. Mitra RD, Shendure J, Olejnik J et al. Fluorescent in situ sequencing on polymerase colonies. Anal Biochem 2003; 320(1):55-65.
57. Milani L, Fredriksson M, Syvanen AC. Detection of alternatively spliced transcripts in leukemia cell lines by minisequencing on microarrays. Clin Chem 2006; 52(2):202-211.
58. Kim H, Pirrung MC. Arrayed primer extension computing with variant mRNA splice forms. Multiple isoforms of CD44 in a human breast tumor. J Am Chem Soc 2002; 124(18):4934-4935.
59. McCullough RM, Cantor CR, Ding C. High-throughput alternative splicing quantification by primer extension and matrix-assisted laser desorption/ionization time-of-flight mass spectrometry. Nucleic Acids Res 2005; 33(11):e99.
60. Modrek B, Lee C. A genomic view of alternative splicing. Nat Genet 2002; 30(1):13-19.
61. Lareau LF, Green RE, Bhatnagar RS et al. The evolving roles of alternative splicing. Curr Opin Struct Biol 2004; 14(3):273-282.
62. Stamm S, Ben-Ari S, Rafalska I et al. Function of alternative splicing. Gene 2005; 344:1-20.
63. Zavolan M, van Nimwegen E. The types and prevalence of alternative splice forms. Curr Opin Struct Biol 2006; 16(3):362-367.
64. Holste D, Huo G, Tung V et al. HOLLYWOOD: a comparative relational database of alternative splicing. Nucleic Acids Res 2006; 34(Database issue):D56-62.
65. Gerhard DS, Wagner L, Feingold EA et al. The status, quality and expansion of the NIH full-length cDNA project: the Mammalian Gene Collection (MGC). Genome Res 2004; 14(10B):2121-2127.
66. Baross A, Butterfield YS, Coughlin SM et al. Systematic recovery and analysis of full-ORF human cDNA clones. Genome Res 2004; 14(10B):2083-2092.
67. Thill G, Castelli V, Pallud S et al. ASEtrap: a biological method for speeding up the exploration of spliceomes. Genome Res 2006; 16(6):776-786.
68. Watahiki A, Waki K, Hayatsu N et al. Libraries enriched for alternatively spliced exons reveal splicing patterns in melanocytes and melanomas. Nat Methods 2004; 1(3):233-239.
69. Schweighoffer F, Ait-Ikhlef A, Resink AL et al. Qualitative gene profiling: a novel tool in genomics and in pharmacogenomics that deciphers messenger RNA isoforms diversity. Pharmacogenomics 2000; 1(2):187-197.

70. Hoskins RA, Stapleton M, George RA et al. Rapid and efficient cDNA library screening by self-ligation of inverse PCR products (SLIP). Nucleic Acids Res 2005; 33(21):e185.
71. Hasegawa Y, Fukuda S, Shimokawa K et al. A RecA-mediated exon profiling method. Nucleic Acids Res 2006; 34(13):e97.
72. Sorek R, Shemesh R, Cohen Y et al. A non-EST-based method for exon-skipping prediction. Genome Res 2004; 14(8):1617-1623.
73. Ohler U, Shomron N, Burge CB. Recognition of unknown conserved alternatively spliced exons. PLoS Comput Biol 2005; 1(2):113-122.
74. Chen FC, Chen CJ, Ho JY et al. Identification and evolutionary analysis of novel exons and alternative splicing events using cross-species EST-to-genome comparisons in human, mouse and rat. BMC Bioinformatics 2006; 7:136.
75. Philipps DL, Park JW, Graveley BR. A computational and experimental approach toward a priori identification of alternatively spliced exons. RNA 2004; 10(12):1838-1844.
76. Agrawal R, Stormo GD. Using mRNAs lengths to accurately predict the alternatively spliced gene products in Caenorhabditis elegans. Bioinformatics 2006; 22(10):1239-1244.
77. Reboul J, Vaglio P, Rual JF et al. C. elegans ORFeome version 1.1: experimental verification of the genome annotation and resource for proteome-scale protein expression. Nat Genet 2003; 34(1):35-41.
78. Lamesch P, Milstein S, Hao T et al. C. elegans ORFeome version 3.1: increasing the coverage of ORFeome resources with improved gene predictions. Genome Res 2004; 14(10B):2064-2069.
79. Chan EY. Advances in sequencing technology. Mutat Res 2005; 573(1-2):13-40.
80. Metzker ML. Emerging technologies in DNA sequencing. Genome Res 2005; 15(12):1767-1776.
81. Shendure J, Mitra RD, Varma C et al. Advanced sequencing technologies: methods and goals. Nat Rev Genet 2004; 5(5):335-344.
82. Bentley DR. Whole-genome resequencing. Curr Opin Genet Dev 2006.
83. Shendure J, Porreca GJ, Reppas NB et al. Accurate multiplex polony sequencing of an evolved bacterial genome. Science 2005; 309(5741):1728-1732.
84. Brenner S, Johnson M, Bridgham J et al. Gene expression analysis by massively parallel signature sequencing (MPSS) on microbead arrays. Nat Biotechnol 2000; 18(6):630-634.
85. Margulies M, Egholm M, Altman WE et al. Genome sequencing in microfabricated high-density picolitre reactors. Nature 2005; 437(7057):376-380.
86. Dressman D, Yan H, Traverso G et al. Transforming single DNA molecules into fluorescent magnetic particles for detection and enumeration of genetic variations. Proc Natl Acad Sci USA 2003; 100(15):8817-8822.
87. Ronaghi M, Karamohamed S, Pettersson B et al. Real-time DNA sequencing using detection of pyrophosphate release. Anal Biochem 1996; 242(1):84-89.
88. Bainbridge MN, Warren RL, Hirst M et al. Analysis of the prostate cancer cell line LNCaP transcriptome using a sequencing-by-synthesis approach. BMC Genomics 2006; 7:246.
89. Dear PH. One by one: Single molecule tools for genomics. Brief Funct Genomic Proteomic 2003; 1(4):397-416.
90. Greulich KO. Single-molecule studies on DNA and RNA. Chemphyschem 2005; 6(12):2458-2471.
91. Wang P, Yan B, Guo JT et al. Structural genomics analysis of alternative splicing and application to isoform structure modeling. Proc Natl Acad Sci USA 2005; 102(52):18920-18925.

Chapter 6

Searching for Splicing Motifs

Lawrence A. Chasin*

Abstract

Intron removal during pre-mRNA splicing in higher eukaryotes requires the accurate identification of the two splice sites at the ends of the exons, or exon definition. The sequences constituting the splice sites provide insufficient information to distinguish true splice sites from the greater number of false splice sites that populate transcripts. Additional information used for exon recognition resides in a large number of positively or negatively acting elements that lie both within exons and in the adjacent introns. The identification of such sequence motifs has progressed rapidly in recent years, such that extensive lists are now available for exonic splicing enhancers and exonic splicing silencers. These motifs have been identified both by empirical experiments and by computational predictions, the validity of the latter being confirmed by experimental verification. Molecular searches have been carried out either by the selection of sequences that bind to splicing factors, or enhance or silence splicing in vitro or in vivo. Computational methods have focused on sequences of 6 or 8 nucleotides that are over- or under-represented in exons, compared to introns or transcripts that do not undergo splicing. These various methods have sought to provide global definitions of motifs, yet the motifs are distinctive to the method used for identification and display little overlap. Astonishingly, at least three-quarters of a typical mRNA would be comprised of these motifs. A present challenge lies in understanding how the cell integrates this surfeit of information to generate what is usually a binary splicing decision.

Splice Site Sequences Are Necessary but Not Sufficient

In the process of converting a pre-mRNA molecule to a mature mRNA, introns are removed by the spliceosome, a very large protein-RNA complex that contains five small nuclear RNA molecules and scores of proteins (refer to chapter by Matlin and Moore). During this reaction, the two bordering exons must be brought close together, much as two substrates in a synthetic reaction of intermediary metabolism. But in the latter case, each of the two substrates usually consists of a population of identical molecules, whereas the two ends of the intron have a varied composition. The enzyme that is the spliceosome must bring together the GU and AG (which are almost always identical) in the midst of some variety among the adjacent nucleotides.

The Splice Sites

The adjacent nucleotides at each splice site are far from random; they comprise two easily distinguished consensus sequences of nine bases for the 5' splice site and about 15 bases for the 3' splice site. Position specific scoring matrices (PSSM) compiled from thousands of introns reflect the relative contributions of each base at each position and allow any given sequence to be quantitatively evaluated for its degree of agreement to a consensus. One widely used such index is the consensus value (CV), which ranges from 100 (perfect consensus) to 0 (the worst consensus).[1,2] The median CVs of human 5' and 3' splice sites are 82 and 80, respectively, and the distribution of

*Lawrence A. Chasin—Department of Biological Sciences, Columbia University, New York, New York 10027, USA. Email: lac2@columbia.edu

Alternative Splicing in the Postgenomic Era, edited by Benjamin J. Blencowe and Brenton R. Graveley.

scores is wide: cutoffs of 78 for 5' splice sites and 75 for 3' splice sites capture only three-quarters of the sites. Interestingly, the consensus sequences themselves do not represent a majority among splice sites. For instance, in a set of 5000 randomly chosen constitutive exons, less than 5% contain 5' splice sites that perfectly match the consensus ((C or A)AG/GT(A or G)AGT) and the consensus sequences do not represent the four most common 5' splice site sequences (Fig. 1A). 5' Splice sites that contain only three of the seven variable bases are not uncommon; the data in Figure 1B suggest that about 20,000 such mismatched 5' splice site sequences are present in the human transcriptome.

The protein factors that recognize the 5' and 3' splice sites need to bind to many distinct sequences. As an example, U2AF,[65] which binds to the polypyrimidine tract of 3' splice sites, can accommodate a wide variety of pyrimidine rich sequences in its binding site.[3] Thus, this degree of diversity might be tolerable if introns and exons had evolved to lack sequences that resemble the splice site consensuses, so that the splice sites would be easily recognizable despite their degeneracy. But just the opposite is the case: pseudo splice site sequences (false splice site sequences that are not used) are about an order of magnitude more abundant than the real splice sites in large transcripts and are present at a frequency similar to or greater than that expected by chance (see Fig. 1C,D for the human *HPRT* gene). Moreover, many pseudo 5' splice sites exist that perfectly match the sequence of real splice sites; in these cases factors other than intrinsic strength[4] must play a role in distinguishing between the real and pseudo sites.

Different splice site sequences have different strengths and this strength generally correlates with the CV score. Thus the words "strong" and "weak" usually refer to the CV and not to a splicing measurement. Indeed, splicing regulation takes advantage of this strength—on average, alternative splice sites are slightly weaker than constitutive splice sites.[5-7] However, the correlation between splice site strength and splicing is far from perfect. For instance, Eperon and colleagues placed different 5' splice sites in competition with a constant globin 5' splice site and measured the proportion of splicing at the test site.[8] Correlation coefficients between splicing efficiency and agreement to the consensus were respectable (0.68 to 0.76) but far from perfect. Strength experiments have usually been set up as competitions between two nearby splice sites,[8] a situation that is not always the case for endogenous splice sites. That is, a weak splice site may be recognized efficiently if no nearby competitor is present. Inefficient splicing of a splice site in a heterologous context implies that in the natural context a splice site communicates with other nearby sequence elements. Splice site sequences may even have to be tailored to their context. For examples, mutation of a *DHFR* 5' splice site from AGA/GTAAGT (CV 79.6) to AGG/GTCAGT (CV 80.9) preserved the CV and the predicted ability to form a duplex with U1 snRNA, yet reduced splicing efficiency from 100% to 3%.[9] More sophisticated methods, such as treating the PSSM as probabilities,[10] using maximal dependence decomposition (MDD),[11] or a support vector machine (SVM),[12] may marginally improve splice site predictions. However, such enhancements do not change the conclusion that many real weak splice sites must be efficiently recognized, while many strong pseudo splice sites must be ignored in the course of splicing a typical pre-mRNA.

The Branch Point

A third element that plays a central role in pre-mRNA splicing is the branch point. The human branchpoint consensus is YNYCRAY, although this sequence was derived from the biochemical characterization of a small number of branchpoints.[13] The conserved adenosine attacks the 5' splice site and is usually located 18 to 40 nt upstream of the 3' splice site, although it can be more distant.[14] The variable distance and poor conservation of the branchpoint makes it a poor predictor of real 3' splice sites. For instance, including the branchpoint in a computational search for real 3' splice sites in the *HPRT* gene did not increase the accuracy of 3' splice site predictions.[15]

Exon Definition

The excision of an intron requires the pairing of splice sites at the ends of the intron, which can be considered "intron definition". However, the initial recognition of most splice sites probably involves

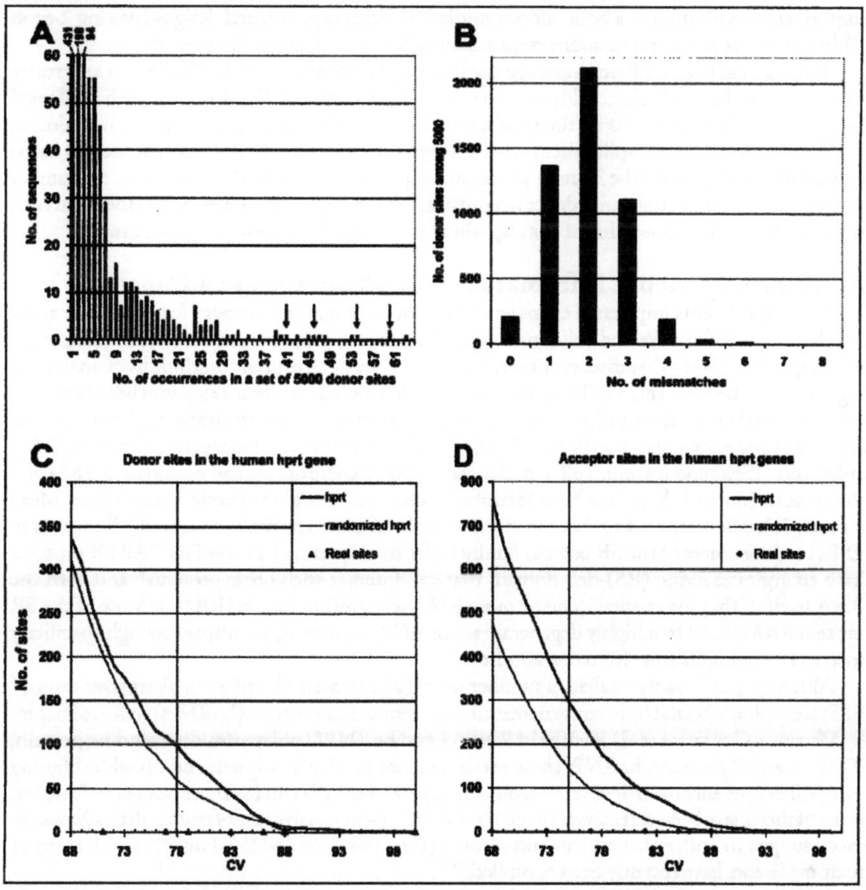

Figure 1. A) Histogram of the number of occurrences of each unique splice donor site sequence found among a set of 5000. The arrows show the points representing the four consensus sequences. For instance, there are 2 sequences that each appear 61 times in this set of 5000, one of which is a consensus sequence. B) Distribution of the number of mismatches to the 4 donor site consensus sequences among 5000 human donor sites. C) Frequency of pseudo donor splice site sequences in the 40,000 nt human *HPRT* transcript having the indicated minimum CV score. Also shown is the same analysis of randomized versions of the HPRT transcript (average of 10 randomizations). The symbols along the abscissa indicate the values for the eight real splice sites. The vertical line indicates the third quartile score for donor sites of real exons (i.e., one-quarter of real donor sites have CV scores below that value). D) As C, but for acceptor sites.

"exon definition", the identification of the two splice sites across the exon. There is plentiful genetic evidence supporting this idea, in that the usual consequence of mutating one splice site is skipping of the exon—the remaining wild type splice site on the other end of the exon is not used.[9,16] Similarly, a downstream 5' splice site greatly enhances splicing to an upstream 3' splice site in vitro.[17,18] Terminal exons are defined by interactions between factors that recognize the 5' cap and U1 snRNP for the first exon[19,20] and polyadenylation factors and splicing factors that recognize the 3' splice site for the last exon.[21] Despite the widespread acceptance of exon definition, the molecular basis for the implied communication has been seldom studied,[18,22] with experimental designs favoring 2-exon RNA molecules and interpretations emphasizing spliceosomal interactions across introns.

Internal human exons have an average size of about 120 nt and less than 5% of exons are greater than 250 nt in length.[23] If one adds the constraint that a potential 3' splice site must be followed by a 5' splice site within 250 nt, then the number of false 3' splice sites is substantially reduced, but the number of false 5' splice sites is not, as they become the limiting factor. The pseudo exons that are defined by these false 3' and 5' splice sites which are assumed to be never used, outnumber the real exons by more than an order of magnitude.[15] If we accept exon definition as the usual case, then the problem becomes that of distinguishing real exons from these pseudo exons.

Additional Sequence Information Lies within Exons and Introns

Early experiments implicated exons as a source of information necessary for alternative splicing.[24,25] In 1993, Shimura and colleagues defined an exonic enhancer sequence as a short purine rich sequence.[26] Since that discovery there has been a steady stream of descriptions of analogous regulatory sequences. These splicing regulatory sequences fall into four categories based on their location and their mode of action: exonic splicing enhancers and silencers (exonic splicing enhancers [ESEs] and exonic splicing silencers [ESSs]) and their intronic counterparts (intronic splicing enhancers [ISEs] and intronic splicing silencers [ISSs]). There are myriad examples of ESEs and the great majority of these have been identified from studies of alternatively spliced exons. More recently, it has become evident that constitutively spliced exons require ESEs as well.[27,28] In general, ESEs bind members of the SR protein family (refer to chapter by Lin and Fu).[29] All SR proteins have an arginine-serine (RS)-rich domain that can interact with other proteins[30] and with the RNA itself.[31] They also contain one or more RNA recognition motifs (RRMs). Most of the SR protein RRMs bind to a highly degenerate set of RNA sequences, yet display enough specificity so as to be distinguishable from one another.[32]

Although less broadly studied, a number of ESSs have been identified in alternative exons.[33] ESSs are typically bound by heterogeneous nuclear ribonuclear proteins (hnRNPs) (refer to chapter by Martinez-Contreras et al), such as hnRNPA1 and hnRNPI (polypyrimidine binding protein, PTB). Like SR proteins, hnRNPs show preferences for particular sequence motifs while binding many other sequences with less, yet notable, affinity.[34] Fewer ISEs and ISSs have been described, but some of these have been extensively characterized.[35] Comprehensive lists of mammalian alternative exons subject to both enhancement and silencing (both by ESSs and ISSs) together with many of their mediators have recently been compiled.[33,36]

Global Approaches for Defining Sequence Motifs for Splicing

A powerful approach to understanding how splice sites are recognized and regulated is to use bioinformatics or experimental approaches to define all the cis-elements that are implicated in splice site recognition. The hope here is that general rules will become evident as one uncovers the "splicing code".[37] The global approaches have been principally two: (1) statistical analysis of genomic sequences to find motifs associated with enhancement or silencing; and (2) molecular selection to define all the sequence motifs that will enhance or silence splicing in a particular context and/or in response to a particular splicing factor, or to find the sequences that bind best to a purified splicing factor. The remainder of this chapter will focus on such global approaches. Understanding the splicing code will allow for a more exhaustive identification of exons and therefore of genes and proteins, with wide implications for genomics and medicine. In addition, it will help us predict the patterns of alternative splicing and understand the mechanism and regulation of splicing.

Exonic Splicing Enhancers (ESEs) Predicted by Computation

ESEs lie by definition within exons and most exons differ from the rest of the genome in containing sequences that must code for proteins. Thus a search for motifs that are abundant in exons vs. other regions would be confounded by the inevitable emergence of common codon sequences. This problem has been dealt with in several different ways.

Fedorov et al[38] compared the frequencies of tetramers and pentamers in exons to those in intronless genes, reasoning that while both code for proteins, the former require splicing signals but the latter do not. Twenty-three sequences were identified that were significantly more abundant in exons, ranging from 17% to 42% overrepresentation. The number of intronless genes used in this study was relatively small,[63] perhaps limiting significance scores; and the biological activity of the sequences found was not tested.

Fairbrother et al[39] got around the protein coding problem by comparing exons to exons, thus neutralizing the effect of protein coding. They reasoned that ESEs should be more abundant in exons with weak splice sites than in exons with strong splice sites. Using all 4096 possible hexamers, they identified motifs for which this frequency difference was high, treating donor sites and acceptor sites separately. To sharpen the selection, they added another criterion: the motifs must also be more abundant in exons compared to flanking intronic regions. Using a cutoff of >2.5 standard deviations for both criteria yielded a combined set of 238 hexamers, or 5.8% of all possibilities. About a quarter of these were common to 5' and 3' splice sites. Many of these motifs were shown to be active in functional assays demonstrating the validity of this approach. Thus most of these hexamers are capable of acting as ESEs and are known as "RESCUE-ESEs". Since the success rate of the validity tests was high, one must conclude that this selection was stringent and that additional hexamers falling below the cutoffs may also act as ESEs. Even at a selection rate of 5.8%, about 23% of randomized exon sequences would correspond to RESCUE-ESEs (Table 1). Thus this study suggested that ESEs are abundant motifs. RESCUE-ESE sequences can be found at http://genes.mit.edu/burgelab/rescue-ese/ESE.txt.

In all motif selection experiments (computational or molecular) there are caveats due to biases inherent in each selection strategy. For example, in the RESCUE-ESE approach, by focusing on exons with weak splice sites, there may have been a biased selection for ESEs associated with alternatively spliced exons, since as a whole they exhibit weaker splice site sequences than constitutive exons.[5-7] A more subtle bias arises from the fact that the transcriptome has an intrinsically high A + T content of 57%.[2] When that content is reasonably used as a background to calculate splice site PSSM scores, G + C-rich splice site sequences will tend to stand out in information content as "strong" (more distinct) whereas A + T-rich sequences will appear "weaker" (less distinct from background). A search for weak splice sites using PSSM values will thus favor A + T-rich sequences for this reason alone and these sites will be associated with A + T rich genes (in A + T rich isochores) and consequently A + T-rich ESE candidates. This argument could explain why RECUE-ESEs have a relatively high (61%) A + T content (Table 2).

The validity of RESCUE-ESEs was subsequently tested by examining evolutionary conservation. SNP density at synonymous sites within these motifs is lower than expected, especially when located nearer to splice sites, supporting the idea that these motifs have been subject to purifying selection and thus are functional.[40]

A second study used the same general approach, but different criteria to search for ESE candidates. Our laboratory[41] circumvented the protein coding problem by limiting the analysis to nonprotein-coding exons. Forty percent of human genes contain noncoding first exons[42] and there are also a substantial number of genes with translation initiation sites located in the 3rd exon, or an exon that is further downstream. The latter represent a pool of about 2000 internal noncoding exons, of which about 500 were chosen that were less likely to have originated from sequencing errors. We searched for all possible octamers in this exon set, allowing a single mismatch per octamer in order to obtain a sufficient number of hits. Octamers were identified that were present at a much higher frequency in the noncoding exons compared to the sequences of two different negative control sets: (1) pseudo exons from the same genes; and (2) the 5' untranslated regions (UTRs) of intronless genes. Neither of these sequences code for proteins and the intronless UTR sequences may contain information for stability, transport and translation that should also be present in the noncoding exons and so would be filtered out. Motifs that fell above 2.8 standard deviations from the mean were considered putative ESEs (PESEs) and numbered 2069 of 65536 possibilities, or 3.2%. This comparison also allowed the identification of motifs that are

Table 1. Exon coverage by predicted splicing motifs

	Mean Hits per Exon[1]	Hits per nt per Motif	Mean Density[2]	nt per 127 Base Exon[3]	Mean Density in Randomized Exons[4]	Real / Randomized Density[1]	% of Exons with No Hits	Pseudo Exon Mean Density[5]
PESE	9.3	3.8×10^{-5}	0.305	39	0.169	1.80	10.6	0.167
RESCUE-ESE	13.6	4.7×10^{-4}	0.319	41	0.231	1.38	3.2	0.219
PESE + RESCUE-ESE	22.9		0.472	60	0.322	1.47	1.2	0.304
Goren ESR	14.9	4.3×10^{-4}	0.526	67	0.335	1.57	0.4	0.444
PESS	1.5	1.2×10^{-5}	0.067	9	0.084	0.80	36.3	0.174
FAS-hex3	2.0	1.6×10^{-4}	0.068	9	0.094	0.72	19.0	0.117
PESS + FAS-hex3	3.5		0.127	16	0.161	0.79	9.1	0.263
All 5	41.3		0.740	94	0.595	1.24	0.02	

[1]Based on 5000 randomly chosen constitutive exons, 50 to 250 nt in length. [2]Mean nt per exonic nt for each exon. [3]The average size of the 5000 exons tested. [4]All nts in each exon were shuffled randomly. [5]In a set of 1803 randomly chosen pseudo exons 50 to 250 in length with acceptor/donor site CVs > 75/78.

rare in real exons compared to pseudo exons and the 5' UTRs of intronless genes and these were considered putative ESSs (PESSs). Eight of eight PESEs enhanced splicing in a functional assay and single base mutations that reduced the PESE scores to near neutrality reduced this activity. Of 58 examples of mutations reported in the literature to reduce splicing, 33% could be explained by the disruption of a PESE (and 28% could be explained be the creation of a PESS). Again, because the success rate of the validation tests was high, it is likely that additional PESEs would be found among octamers with somewhat lower scores than those chosen. At the conservative threshold of 3.2%, about 17% of a randomized exon sequence would be represented by PESEs (Table 1); so like RESCUE-ESEs, these motifs are abundant. The average internal (coding) constitutive exon of 120 nt contains nine PESEs, often in overlapping clusters. PESEs are 2-fold more abundant in exons compared to introns. A full set of PESE sequences can be found at http://www.columbia.edu/cu/biology/faculty/chasin/xz3/pese262.txt. A list of the scores for each of the two criteria for all 65,536 octamers can be found at http://www.columbia.edu/cu/biology/faculty/chasin/xz3/octamers.txt.

Again, biases could have influenced the types of motifs that were selected. The 5' UTRs of intronless genes that were used as an ESE-under-represented data set are often situated within regions that are rich in CpG sequences, since the CpG islands that lie upstream of numerous genes often extend as much as 2 kb into the gene.[43] Moreover, noncoding exons used as the positive examples are low in CpG content relative to coding exons.[44] For both of these reasons, CpG-containing ESE motifs may have been under represented in this selection, since they may not be enriched over the relatively high background of CpG-containing octamers in the 5' UTR of intronless genes. In the earlier comparison of exons to intronless genes mentioned above,[38] most of the candidate ESS pentamers identified as being relatively scarce in exons contained CpG dinucleotides. Although they have a CpG content similar to that of exons as a whole (Table 2, compare columns 2 and 12), PESEs do not include many ESEs predicted from molecular selections and these tend to have very high CpG contents (Table 2, columns 5 to 8). Another possible weakness in the selection method described above stems from the assumption that noncoding genes do not contain protein coding information. In fact, such exons may have coded for proteins in the evolutionary past and maintained a vestige of this nonrandomness. In this case, PESE candidates that merely overlap with highly used dicodons could have been isolated as false positives. However, such sequences may tend to be ESEs nonetheless and the high validation rates of PESEs argues against this possibility.

PESEs were subsequently tested in a more rigorous fashion.[28] Six real mammalian exons (five constitutive and one alternative) were computationally scanned for PESEs. About four PESE clusters per 100 nt were found. By knocking out each individual PESE cluster with single base substitutions and assaying splicing in vivo, 18 of the 22 predicted PESEs were shown to be functional. In addition to functionality, this result showed that each exon required nearly all of its ESEs to work in concert to promote efficient splicing; i.e., most were not redundant. A similar test has been carried out using a minigene containing an alternatively spliced alpha-tropomyosin exon.[45] Eleven PESEs or PESSs were mutated to reduce their absolute scores and in 10 of the 11 trials the splicing levels responded accordingly (J. Coles and C.W. Smith, personal communication). As well as providing additional validation of PESEs, this study provides the first such experimental test of PESSs.

Although entirely different criteria were used to select RESCUE-ESEs and PESEs, they show considerable overlap, a fact that further supports the validity of both sets (Table 3). At the same time, the two sets contain distinct ESEs. As can be seen in Table 1 (column 4), each motif set covers about 30% of the nucleotides in a collection of 5000 human exons, but together they cover 47%, only slightly less than would be expected if they were randomly associated (~52%). Thus these two conservatively derived sets of ESEs already cover half of an average exon and there are several additional motif sets yet to be discussed.

Table 2. Base compositions of motif sets

1 Sequence Source:	2 RESCUE-ESEs	3 PESEs[4]	4 ESRs	5 Func. SELEX In Vivo ESEs	6 Func. SELEX (nuc. ext.)	8 Func. SELEX (nuc. ext.)	7 Func. SELEX ESEfinder (4 SRs)	9 Func. SELEX (ASF/SF2)[2]	10 PESSs[4]	11 Sironi ESSs[3]	12 Func. SELEX In Vivo ESSs	13 Real Exons[1]	14 Pseudo Exons[1]
Ref.:	Fairbrother, 2002[42]	Zhang, 2004[44]	Goren, 2006[51]	Coulter, 1997[83]	Tian, 1995[76], 2001[77]	Schaal, 1998[107]	Liu[78,79]; Cartegni, 2003[80]	Smith, 2006[49]	Zhang, 2004[44]	Sironi, 2004[53]	Wang, 2004[85]	1	1
No. of motifs	238	2060	285	30	101	46	104	60	1018	NA	103		
Sequence space	4096	65536	3721	4^{14}	4^{20}	4^{18}	4^{20}	4^{7}	65536	NA	4^{10}		
% of sequence space	5.8	3.1	7.7	10^{-5}	10^{-8}	10^{-7}	10^{-8}	0.4	1.6	NA	10^{-4}		
A%	48	31.2	28	25	34	12	22	20	29	15	17	26	26
C%	14	24.4	23	36	21	37	23	42	9	9	9	25	22
G%	25	28.5	25	30	31	35	36	30	15	53	36	27	23
T%	13	15.9	25	10	14	16	19	7	47	22	38	22	29
GC%	39	52.9	47	66	52	72	60	72	24	62	45	52	44
AT%	61	47.1	53	34	48	28	40	28	76	38	55	48	56
CpG%[5]	2.6	3.4	0.7	9.3	10.8	12.3	8.8	18.3	0.6	NA	2.6	3.2	1.3

NA: not available
[1] data from 5000 randomly chosen exons and 1803 randomly chosen pseudo exons with CV scores in the top 3 quartiles.
[2] 60 7mers from 7mers and 14mers
[3] Extracted from positional scoring matrix
[4] Using corrected data (http://www.columbia.edu/cu/biology/faculty/chasin/xz3/octamers.txt)
[5] Expectation by chance is about 6% (1/16)

Table 3. Overlaps between motif sets[1]

	Motif Set Size	PESEs[2]	Expected by Chance[3]	PESSs[2]	Expected by Chance[3]	Motif Reference
RESCUE-ESE[42]	238	74%	36%	11%	23%	45
FAS-hex3[85]	103	10%	37%	54%	23%	87
Goren ESR[51]	285	53%	44%	19%	21%	54

[1] Percentage of the indicated hexamer set members that can be found within PESE or PESS octamers.
[2] PESE, PESS: ref. 44, as amended at http://www.columbia.edu/cu/biology/faculty/chasin/xz3/octamers.txt
[3] calculated by simulation. Goren ESRs based on 3721 allowed hexamers.

Exonic Splicing Silencers (ESSs) Predicted by Computation

Global computational searches for ESS motifs have also been carried out. Sironi et al[146] collected a subset of pseudo exons that was rich in predicted ESEs and then searched for overrepresented hexamers as candidates for ESSs. A second criterion, overabundance in pseudo exons compared to sequences flanking the pseudo exons, was applied to normalize for possible base compositional differences between pseudo exon and exon regions. This second criterion also sharpened the search to make it test the hypothesis that ESSs function to prevent the splicing of pseudo exons, as opposed to simply being avoided in real exons. The winning motifs were clustered into families to generate three consensus sequences. One of the three ((T/G)G(T/A)GGGG) reduced exon inclusion about five-fold in a functional assay. This G-rich motif was overrepresented in a test set of pseudo exons compared to real exons.

A large set of putative ESSs emerged from our statistical analysis described above for PESEs.[41] By searching for octamers that were underrepresented in real exons compared to both pseudo exons and the 5' UTRs of intronless genes, the influence of codons was avoided and the influence of other nonsplicing signals residing in mRNA was minimized. A set of 974 PESSs was identified, grouped into families and a sampling tested in functional assays. Eleven of 12 PESSs increased exon skipping and single base mutations reversed this skipping. Sixteen of 58 exonic splicing mutations in the literature could be explained by PESS formation, a number comparable to those that could be explained by PESE disruption. The PESS set represents about 1.5% of all octamers. These sequences are very T-rich (47%) and C-poor (Table 2, column 9). PESSs are 3.5-fold more abundant in introns compared to exons overall and show an additional increase just downstream of real 5' splice sites, suggesting that they may function to facilitate accurate recognition of the real sites. They are also found at a higher frequency in the vicinity of pseudo exons, suggesting a use in repressing false splice sites. The combination of PESEs and PESSs increases the discrimination between real and pseudo exons: the ratio of PESEs to PESSs in real exons is 5.5 as opposed to 0.6 for pseudo exons. This difference has been used as a guide to suggest whether a given sequence is a exon or a pseudo exon (e.g., see Fig. 6 in reference 45). However, the frequency distribution of ESEs and ESSs in exons and pseudo exons overlap considerably, making them less than a reliable predictor of real exons. A list of PESSs is located at http://www.columbia.edu/cu/biology/faculty/chasin/xz3/pess262.txt .

Exonic Splicing Regulators (ESRs) Predicted by Computation

Yet another computational strategy to search for splicing regulatory motifs was devised by Goren et al.[47] Reasoning that splicing signals would be both conserved and abundant in exons, they ranked hexamers that stood out in these two respects. To get around the protein coding problem conservation was scored only at synonymous sites. Overabundant hexamers were chosen as dicodons that appeared more frequently than expected if codons were paired randomly; here again only codons differing at synonymous sites were compared so as to avoid the influence of protein coding. Hexamers with high scores for both criteria were collected, resulting in a set of 285 (7.7% of all hexamers considered) that represented the best combination of scores. Ten of these sequences were tested in functional assays and most of these were shown to either increase or decrease exon inclusion while none of nine control hexamers with low abundance and conservation scores significantly affected splicing. One might have thought that a selection based on abundance and conservation would favor ESEs, but both enhancer and silencer effects were observed, depending on the hexamer and on the host exon. The authors followed up on this dichotomy by placing each of two winning hexamers at 26 different positions within an 81 nt test exon. Here again, both enhancing and silencing results were obtained, this time dependent on position. Finally, when four winning hexamers resident in real exons were mutated so as to lose their high score, a mixture of positive and negative effects was observed while mutation of nonwinners had no effect. The authors thus called these motifs ESRs, for exonic splicing regulators, since their effects could be either positive or negative depending on the context. They further suggested that putative ESEs and ESSs identified by others but untested for a position effect should be similarly regarded. A list of all hexamers surveyed and their scores in terms of p-values for the significance of their deviation from mean frequencies can be found at http://www.tau.ac.il/~gilast/sup_mat7.htm.

The hexamers selected here will be biased toward sequences that harbor synonymous codons. Extreme examples are hexamers that contain a stop codon as either the first or last three positions; these 375 hexamers are removed from consideration. A second limitation is that the criteria used do not specifically target splicing motifs but apply to any function implicit in mRNA (transport, stability, etc.). However, the fact remains that greater splicing effects were seen with many of these sequences compared to controls. Some of the substitutions made to test for splicing phenotypes also caused sequence changes in overlapping endogenous PESE or RESCUE-ESE motifs (not shown), an outcome that is not surprising given the coverage figures shown in Table 1 and since any hexamer substitution changes 10 overlapping hexamers (or 13 octamers). Nevertheless, there were so many ESRs tested here at so many different positions, that it is likely that the ESR set does contain many novel motifs that affect splicing.

An interesting idea that this work gives rise to is the possibility that the same motif can act as an enhancer in one context and a silencer in another. The efficacy of some enhancers is known to be dependent on the distance from a target splice site. For example, the enhancement of splicing at the *dsx* 3' splice site has been shown to drop off when ESEs are placed more than 150 to 200 nt downstream.[48] On a chemical level, the ability of an RS domain to crosslink to a splice site also falls off at distances greater than 100 nt.[49] However, while decreasing efficiency, these far positionings do not reverse the effect of an enhancer. Individual natural ESEs have also been shown to be able to act negatively when placed within an intron[50] and there are examples in which a splicing factor acts positively at one splice site and negatively at another (e.g., hnRNP H/F and SRrp86[35]), or a single sequence element is a target for both positive and negative factors.[51-53] One could argue that there is a need for ESSs in exons in order to silence internal pseudo splice sites. But there are few of these in constitutive exons: summed pseudo 5' and 3' splice sites numbered less than 0.2 per 120 nt of exon in a set of 5000 examined (with a size limit of 250 nt) and over 80% of exons had neither such site. (Exons that contain alternative 5' or 3' splice sites obviously have more than one splice site per exon, but these should be considered real sites, not pseudo sites.)

The cautionary note sounded by Goren et al[54] presents a serious challenge, as there have been few systematic studies of the effect of position on motifs isolated by global searches. For the most part, however, experiments have not shown a context-dependent effect on activity. For example, eight PESSs that were originally found to be effective when inserted just downstream of a 3' splice site in a first test exon were equally effective when inserted just upstream of a 5' splice in a second test exon.[41] ESSs isolated by molecular selection similarly acted consistently as silencers in several different contexts.[55] Still, in these studies position was not systematically varied within a single context. Analysis of natural occurrences of splicing motifs may be more relevant and here too, the results have so far been consistent with prediction. For example, when we knocked out predicted ESEs in a beta-globin exon 2, five of the six disruptions decreased splicing and the exception did not significantly increase splicing. If some of the predicted ESEs were really ESSs, as predicted by the ESR idea, some of the knockouts should have increased splicing. Finally, almost all of a panel of human mutations affecting splicing can be explained by the disruption of predicted ESEs or the creation of predicted ESSs,[56] in accord with the expected behavior of the motif. It is possible that these motifs function as predicted when in their natural contexts but that their normal activity can be subverted when experimentally placed in ectopic positions. An analogous result, known as transcriptional interference, has been seen with experimentally manipulated promoters.[57]

Molecular Selections

Most molecular selections have targeted ESEs. Almost all of these types of experiments are based on the Systematic Evolution of Ligands by Exponential Enrichment (SELEX), originally designed to select for nucleic acid sequences that bind to a given protein or small molecule.[58] SELEX has been applied in two ways: (1) determining the sequences that can be recognized by a given RNA binding protein (binding SELEX); and (2) isolating sequences that functionally enhance splicing (functional SELEX).

Protein Binding SELEX

In protein-binding SELEX, a complex pool of RNA molecules containing a randomized region 8 to 20 nt long is incubated with a purified RNA-binding protein or domain. The RNAs that are bound by the protein are isolated, converted to cDNA, amplified by PCR and then transcribed into RNA for a subsequent round of selection. This process is repeated several times to enrich for RNA molecules with high affinity for the RNA binding protein. In this way, the binding specificities of several SR proteins, hnRNPs and other splicing factors have been determined.[32,34,59-67] The sequences are then analyzed to determine a consensus binding site(s).

The consensus motifs that have emerged from these binding SELEX experiments[68] (refer to chapters by Lin and Fu, Martinez-Contreras et al, and Ule and Darnell) illustrate that each protein binds to a distinct set of sequences but at the same time can recognize a diverse repertoire of sequences.[68] For instance, Cavaloc et al[63] sequenced over 90 oligonucleotides bound by the SR protein SC35 and found five distinct consensuses. However, such degeneracy is not always the case. Only a single long consensus was found by Tacke and Manley in binding SELEX experiments performed with SRp40 (ref. 66). Why more than one consensus sequence appears in many of these experiments is not clear. In cases of proteins with more than one RRM, it is possible that each binds a distinct sequence. Yet, when SELEX was applied to a single RRM derivative of ASF/SF2, the one resulting consensus sequence differed from the two consensus sequences yielded by the intact, wild type protein.[59] Alternatively, a single binding site may be endowed with some flexibility to accommodate a specific set of different sequences.[69] A certain dichotomy in binding behavior might also be related to the multiple roles SR proteins play in the splicing process:[70] initial recognition, exon definition,[71] spliceosome assembly[71,72] and perhaps the catalytic steps themselves.[49,73-75] (refer to chapter by Lin and Fu for a more detailed description of the functions of SR proteins in splicing.) It is also likely that the diversity of sequences recognized is related to the fact that many RNA binding proteins must bind to protein coding exons. Thus, the range of sequence motifs that can exist in a given exon is confined by the protein sequence encoded in that exon.

The advantage of binding SELEX experiments is that they probe and define the binding specificity of a purified protein, in the absence of possible interference by other factors. As such they provide a valuable starting point in the interpretation of the roles of SR proteins and the motifs dictating their action. On the other hand, SELEX may identify RNA sequences that bind to any surface of a purified protein, not necessarily the natural RNA binding site and could therefore include sequences that are not functionally relevant in a biological context.

Functional SELEX

In this strategy motifs that are able to influence splicing activity are selected. The iterative isolation and amplification steps of SELEX are used here to select for short sequences that, when inserted into an exon, enhance pre-mRNA splicing. In these studies, a pre-mRNA pool is first synthesized that contains a weak test exon containing a localized randomized region. This pre-mRNA pool is used in in vitro or in vivo splicing assays and the successfully spliced mRNAs are isolated and amplified by RT-PCR. As with the binding SELEX experiments, the winning RNA sequences from the first round are recycled through several additional rounds, enriching for RNA sequences that best enhance splicing.

The first experiments of this kind were carried out in vitro by Tian and Kole.[76] They found that the winning sequences were quite heterogeneous and could be divided into two classes: a majority that were purine-rich, typically consisting of short runs of 5-6 purine nucleotides and a significant minority (15% to 30%) that were not rich in purines. Almost all of the retested sequences produced efficient splicing, whereas only 1 in 10 of the unselected sequences promoted splicing. In a subsequent refinement of this procedure, shorter versions of the winning sequences were produced and these yielded a less heterogeneous group with a consensus GACGAC...CAGCAG (the core being of variable length) that was shown to bind SRp30.[77]

Two larger studies used random sequences inserted into the second exon of a 2-exon transcript spliced in vitro. Liu et al[78,79] selected sequences that responded to one of four different SR proteins by using S100 extracts for splicing. These extracts lack all SR proteins, but splicing activity can be restored upon supplementation with individual SR proteins.

Using this approach, the authors selected 20-mer sequences that promoted splicing in response to SRp40, ASF/SF2, SC35 and SRp55. Each of these selections yielded a consensus sequence that was distinct from the others. However, there was considerable heterogeneity within each group, the consensuses were often short (5 to 8 nucleotides) and contained many ambiguous positions. Some of the degeneracy might be explained by assigning a role to the sequences flanking each test motif, as not all the motifs could promote splicing when tested on their own. Nevertheless, it was possible to assemble a PSSM for each class of motifs and this information has been incorporated into an ESE searching program "ESEfinder" (http://rulai.cshl.edu/tools/ESE/) that scores sequences for their ability to match each of the SR protein-specific consensuses.[80] High-scoring motifs are found at a significantly higher frequency within exons as opposed to introns,[79,81] although the difference is modest (10%-20%). These motifs are also more strongly associated with weak compared to strong 3' splice sites.[82]

This experimental approach was refined in a later study that focused on motifs that mediate the activity of ASF/SF2.[56] Here, the random oligomer pools were restricted to either 7 or 14 nucleotides and were used to replace a 7-nucleotide natural ESE in *BRCA1* exon 18. In this construct, exon 18 was present as the central exon of a 3-exon transcript, an internal exon situation that is more commonly found in nature. Once again a degenerate though obvious sequence preference was evident from these experiments. Most of these sequences not only enhanced inclusion of BRCA1 exon 18, but also functioned in the heterologous context of exon 7 of the *SMN1* gene. A PSSM was derived using not only the relative prevalence of bases at each position, but also, in a novel approach, taking into account the degree to which each sequence enhanced splicing. The PSSM derived from the 7 nt oligomers (consensus of CCCCGCA) proved to be the better predictor of enhancement than the PSSM derived from the 14 nt oligomers. This consensus differed from that of the earlier derived ASF/SF2 consensus sequences (CACACGA) from the functional SELEX experiment employing a 2-exon pre-mRNA and both of these consensus sequences differ from the consensus sequences (AGGACAGAGC and RGAAGAAC) derived from binding SELEX experiments.[59] The authors combined the matrices from the two functional selections and showed that the resulting PSSM (with the consensus CGCACGA) was able to predict, with a statistically significant frequency, the outcome of a set of exonic mutations known to affect splicing.

Schaal and Maniatis[83] took a slightly different approach to defining consensus sequences that mediate SR protein function. As in the study by Liu et al described above, they inserted random oligomers into the second exon of a 2-exon transcript but assayed for splicing in vitro using nuclear extracts instead of S100 extracts. Sequences that enhanced splicing were identified after multiple rounds of selection using the same procedure as in Liu et al. The winning sequences were subsequently screened for their ability to respond to specific SR proteins in S100 extracts and the sequences were grouped according to their response. Here again heterogeneity and distinctness characterized the sequences identified. Most SR proteins displayed some sequence preferences but in general these sequences do not match the Liu et al consensuses mentioned above, nor do they match well with the results of binding SELEX experiments. That different motifs emerge from the functional SELEX experiments could be due to the effect of context, including the position of the insert or substitution, or related to the another aspect of the test exon used, e.g., small size, mutational weakening of splice sites, natural weakness of an alternative splice site, etc.

Functional SELEX for splicing was performed in vivo by Coulter et al,[84] who inserted random 14-mers into a poorly spliced second exon in a 2-exon construct. After several rounds of selection based on transient transfection, the winning clones were sequenced. These fell into three categories, purine-rich, adenine- plus cytosine-rich (ACE) and neither. Within the first two categories, the identifiable motifs were quite degenerate. An ACE motif in the human *CD44* gene was subsequently shown to act as an ESE and to be bound by the nonSR protein YB-1.[85] This last result represents a cautionary tale: by testing only for responsiveness or binding to SR proteins, other, possibly more significant, mediators targeting the isolated motifs may be overlooked.

In an interesting variation on the functional SELEX scheme, Woerful et al tested the activity of ~50 bp fragments of the *CD44* mRNA in an enhancer-dependent exon in vitro.[86] About half of the active sequences tested enhanced splicing and many of these mapped to a specific region within the *CD44* mRNA. Most of the sequences contained a short AC-rich motif whereas others contained purine-rich runs. This starting material was quite limited compared to random oligomers and the fact that a restricted subset of ESEs was overrepresented suggests that the test exon likely has a preference for particular ESEs.

ESSs

Wang et al[87] used an elegant genetic selection to isolate sequences that could cause exon skipping in vivo. When the central exon is included, it interrupts the reading frame of GFP; when skipped, functional GFP is expressed, allowing the positive cells to be isolated by FACS. After insertion of a set of random decamers, 133 unique sequences promoting GFP expression were isolated from this screen. Many of the most common hexamers caused skipping in a heterologous exon. This collection of 103 common ESS hexamers is known as FAS-hex3. By avoiding the iterative enrichment process of SELEX, any sequences that inhibit splicing were isolated, rather than only those that have the strongest ESS activity. FAS-hex3 sequences are 2-fold overrepresented in introns vs. exons and are especially overrepresented in exonic regions located between alternative 5' or 3' splice sites.[55] Like the computationally selected PESSs, they show a peak just upstream and downstream of 3' and 5' splice sites, perhaps to prevent neighboring pseudo sites from being used. Mutation of these sequences in natural exons increased the use of the proximal sites and resulted in increased exon inclusion, attesting to the silencing function of these sequences in natural alternative splicing. Alternative intron retention events were also inhibited by FAS-hex3 sequences, often in favor of skipping of the retained intron along with its flanking exons. Interestingly, FAS-hex3 sequences that could inhibit intron retention by increased skipping tended to be different from those that acted to decrease the use of an alternative splice site, suggesting that different ESSs act via different mechanisms.

ISEs and ISSs

Whereas there has been extensive investigation of the effect of intronic sequences on the alternative splicing of individual genes (e.g., 35, 88-90), there has been relatively little global searching for intronic splicing regulatory motifs. Early studies by Nussinov[91] identified G-runs as overrepresented in introns near both the 3'and 5' ends of exons. This work was extended by McCullough et al[92,93] to show that these sequences can enhance splicing at 5' splice sites by enhancing the binding of U1 snRNP. Statistical analyses and comparative genomics showed that intronic flanks of exons harbor short runs of G, C or T.[43,94] Evidence for a role of intronic flanks in splicing regulation comes from the finding that the flanks of alternatively spliced exons are more conserved than those of constitutive exons.[6,95]

Subsets of the genome have been searched for ISE motifs. Statistical analysis of sequences at the ends of short introns in several different organisms produced pentamers that could be used to enhance the accuracy of splice site prediction in such introns.[96] For humans, 8 of the 10 top pentamers were rich in G-triplets. Individual motifs have also been strongly associated with alternative splicing in the brain. Brain-specific ISE and ISS candidate motifs were identified statistically by analyzing 25 brain-specific alternatively spliced exons[97]; in contrast to introns overall, G-triplets were underrepresented downstream of these exons. One of the ISE sequences, UGCAUG, was subsequently shown to bind the Fox-1 and Fox-2 splicing factors and to be associated with regulated alternative splicing in different tissues, including brain and muscle in a number of species.[98-102] A downstream intronic G-tetramer motif was found to be associated with exon-skipping in alternatively spliced exons in the brain by Han et al and, interestingly, this element was shown to function in conjunction with an exonic UAGG to effect silencing.[103]

Our own laboratory has used machine learning to assess whether information for splice site recognition is present in sequences flanking constitutive exons.[12] A support vector machine (SVM) found that sequences residing within ~50 bases of the splice sites can help distinguish real exons from pseudo exons and identified overrepresented (ISE candidates) or underrepresented (ISS candidates) pentameric motifs that best aided the distinction. These included some novel motifs as wells as G-triplets mentioned above and, despite its degeneracy, branchpoint-like sequences, with a clear peak 24 nt upstream of the 3' splice sites. This work was followed up by a statistical test for pentamers overrepresented in human exon flanks compared to pseudo exons flanks.[104] A conservation filter was also applied here: only those pentamers that were also present within a 50 nuccleotide region flanking the orthologous mouse exons were retained. The resulting ISE candidates fell into two distinct groups based on G + C content. A survey of 100,000 constitutive exons showed that their 50-nt flanks in general fell into distinct GC-rich or GC-poor categories; remarkably, the extent of this dichotomy was much greater that that exhibited by the host genes overall (i.e., due to residence in a particular isochore). Thus the factors that recognize these putative ISEs are probably different for GC-rich genes and AT-rich genes, leaving open the possibility that distinct mechanisms operate for these two gene classes. The GC-rich exons differed from the AT-rich exons in other ways: the GC-rich ISEs tended to have a complementary ISE in the opposite flank whereas the AT-rich ISEs tended to have the same ISE in the opposite flank and predicted base pairing between the flanks and the exon tended to be avoided for GC-rich exons but not for AT-rich exons.

Although our predicted ISE/S motifs were not specifically tested, we did show that intronic sequences are often important for efficient splicing.[104] Specifically, we found that splicing of an exon is often inefficient when it is not flanked by the 50 nt intronic sequence beyond its splice site sequence (i.e., −63 to −14 and +7 to +56). In addition, two of three exons lacking their flanking introns exhibited decreased or aberrant splicing when moved to ectopic positions within the same intron. These studies show that intronic sequences proximal to exons contribute to splice site recognition in an exon-specific manner. The interplay between exonic elements and presumed ISEs could also be seen in our test of natural PESEs. In transcripts (from the *HBB-2* and *THBS4-13*) tested for ESEs by mutational analysis, the wild type exons exhibited 50% to 80% inclusion when their flanking intronic sequences were removed; in this situation, the removal of any one of several PESEs reduced inclusion several-fold. However, if the flanking intronic sequences were retained, splicing was refractory to such single PESE knockouts.[28] A similar interaction was seen with a PESS knockout, the mutated exon in this case shedding its intron flank requirement (X. Zhang, unpublished result).

Functional SELEX for Splice Site Sequences

Branchpoint sequences are active participants in catalysis rather than ISEs, but they do present an analogous tension between specificity and degeneracy. Functional SELEX has also been used for the selection of effective branchpoint sequences. After seven rounds of selection at a fixed position the consensus that emerged was TACTAAC,[105] which can optimally pair with U2 snRNA. However, when only a single round of selection was carried out, a wide variety of effective sequences were found, many with only 3 to 5 bases capable of pairing with U2 snRNA.[105,106] If the starting transcript had a weak polypyrimidine tract, or if an ESE was removed, a better match to U2 snRNA was selected, another indication that a balance of compensatory factors determines a splicing outcome.

Functional SELEX has also been used to define splice site sequences, including the polypyrimidine tract.[105-107] Here again, multiple rounds of selection converged on the consensus sequences and so no additional insight was provided into how exons with weak splice sites are recognized. Interestingly, the 5' splice site consensus was also selected in extracts lacking the U1 snRNA 5' end,[107] suggesting that this sequence may be recognized by protein components as well as by RNA-RNA hybridization.

Comparison of Computationally Predicted and Functional SELEX Selected Exonic Motifs

As shown in Table 3, RESCUE-ESEs show considerable overlap with PESEs and avoid overlap with PESSs. Similarly, fas3-hex3 silencers overlap with PESSs and avoid PESEs. Goren ESRs, which exhibit either enhancer or silencer activities, overlap less with PESEs and not significantly with PESSs (Table 3). Thus each set contains common and unique information. In contrast, ESEs defined by functional SELEX[56,76-79,83] appear quite distinct from their computationally-derived ESEs. These differences can be seen at the level of base composition and particularly in CpG content, which is remarkably high in SELEX winners obtained in four different laboratories (Table 2, columns 5 to 9). Binding SELEX motifs also usually differ from or show only weak similarities to those yielded by functional SELEX,[68] although certain of the former also display a high CpG content.[63,66] Given the methylated status of most CpGs in genomic DNA, one is tempted to speculate on connections to transcription: the slowing of transcriptional elongation at methylated CpGs[108] may provide time for the reloading of splicing factors onto the C-terminal domain (CTD) of RNA polymerase,[109] as well as increased time for the association of factors with weak splicing signals independent of any association with the CTD (refer to chapter by Kornblihtt).

One way to compare computationally derived motifs with those obtained from functional SELEX is to use ESEfinder, a Web-based program[80] that scores sequences using a PSSM derived from motifs selected for responsiveness to four different SR proteins.[78,79] About 40% of the PESE octamers contain sequences that fall above the threshold for at least one of the four SR proteins, a proportion that is not unreasonable given that PESEs presumably include binding sites for all SR proteins whereas ESEfinder covers only four. However, 28% of a random set of octamers also achieves this rather undemanding benchmark. ASF/SF2 motifs were the most common among PESEs at 15%. Looking more directly for overlaps, Wang et al also concluded that there was no significant overlap between ESEfinder motifs and RESCUE-ESES or PESEs, with the exception of ASF/SF2 motifs with PESEs.[81]

An Embarrassment of Riches?

The multi-pronged global attacks on defining splicing regulatory motifs summarized above have promoted optimism that the "splicing code" may soon be solved.[110,111] At this moment however, the number of effective motifs that have been experimentally defined or predicted may have reached the point of diminishing returns. The use of four computationally selected sets of ESEs and one genetically selected set of ESSs covers about three-quarters of the nucleotides in a typical exon (Table 1). Moreover, there is no reason to think that all ESEs and ESSs have been identified. While it is probably true that some of the predicted but untested ESE candidates will turn out to be inactive, it is even more certain that many more as yet unidentified motifs will have enhancer activity. These motifs are to be found in the sequence space below the conservative thresholds that have been used in selecting predicted ESEs. Hard evidence for such new sequences can be found in the tests of the predicted sequences. The specificity of both RESCUE-ESEs and PESEs was assessed by mutating the predicted enhancer to a sequence that did not score highly. In almost every case, the predicted decrease in splicing was indeed observed. But in half these tests the decrease was modest, with more than 50% of the enhancer activity remaining and a several fold splicing enhancement was still in evidence.[39,41] Thus the percentage of sequence space assigned to ESEs in these two studies must undoubtedly be increased by at least 50% and probably more. Stadler et al[112] have now formalized a search for additional ESEs and ESSs by developing an algorithm, called Neighborhood Inference, that searches for new ESEs on the basis of sequence similarity to known ESEs and dissimilarity from ESSs and vice versa. A sample of high-scoring hexamers identified by this algorithm proved to act as ESEs or ESSs as predicted. The authors conclude that the list of splicing regulatory motifs is much greater than previously thought.

An unreported reservoir of ESEs is also evident in most SELEX experiments. For example of the 28 ASF/SF2-responsive heptamer motifs isolated in a functional selection, no two were identical.[56] A repeat of this experiment would therefore be expected to turn up many additional unique heptamers.

The extensiveness of the ESE list results in a ubiquity of these elements. Although present at lower densities in introns than in exons, computationally defined ESEs nevertheless heavily populate pseudo exons (Table 1, column 9), and ESEs predicted by functional SELEX occur in introns at about 80% to 90% of the level found in exons.[81] While from a strictly bioinformatics point of view one can take solace in the high statistical significance of these differences, the fact remains that the overall differences are modest. Thus we are once again challenged with figuring out how, in the face of this richness of signals, the cell distinguishes real splice sites from pseudo sites and real exons from pseudo exons. This present situation has engendered the less optimistic view that we have indeed reached a point "right on the edge of chaos".[113]

One consequence of the prevalence of splicing motifs is technical: one must be careful in drawing conclusions from mutational perturbations of pre-mRNA sequences. A single base change can impinge on many possible resident motifs; and insertions, deletions and new junctions increase the number of collaterally emergent sequences considerably. It may be necessary to turn to very simple exons or to aim at easily characterized or isolated regions to minimize ambiguity in the results. It will also be prudent to take this accumulated list of ESEs into account when interpreting the result of any selection experiment (computational or SELEX based). An examination of the results of three types of ESEs (sequences underlying ESEfinder, binding SR proteins or discovered in individual genes) showed three-quarters to be populated by at least one RESCUE-ESE or PESE (data not shown). The second consequence is conceptual. How can we explain how so many sequences can act together to produce the binary decision that is made for the great majority of exons? How can the perceived modest yet real differences in ESE and ESS densities between exons and pseudo exons be leveraged to produce that binary decision? To say that it involves a "balance of combinatorial factors" is not much better than saying that it depends on "context", in that it describes the situation we see without really explaining it.

One easy way out is to invoke secondary structure acting to "present" some of these motifs but not others (e.g., as loops; see ref. 114) or to mask some motifs and not others (e.g., see ref. 115). One technical problem in evaluating the role of secondary structure on the splicing of natural exons is that we do not really know how transcripts fold in vivo. In particular, many conformations that would be considered too unstable to contribute to predicted equilibrium structures could be kinetically trapped during transcription and last long enough to influence splicing outcomes. It is clear that secondary structures do play role in many splicing decisions (reviewed in ref. 116) (refer to chapter by Park and Graveley) but whether this influence is pervasive is not yet clear. If secondary structure is not invoked then we need models that seek to explain why so many ESEs are present.

It is reasonable to think that the majority of these splicing motifs are playing a role in exon definition.[17,18] The main feature of this model is that both ends of an exon must be recognized before splicing at either end can occur. The implied communication between the two ends of the exon could be realized via a bridge of proteins, each a necessary link in a flow of information mediated by a series of allosteric transitions. This is a rather complicated model but does explain the necessity of multiple ESEs to construct this bridge. Moreover, a fitting combination of proteins may be needed to ensure proper signal propagation, although this need not be a unique assemblage

(Fig. 2A). Consistent with this model is the finding that different pre-mRNA molecules associate with different sets of nuclear proteins.[117] Arguments against such a model are that random sequences can be inserted into exons often without dire consequences. Thus insertion of bacterial sequences from 100[118] to 1000[119] nucleotides long into the central exon of a 3-exon construct did not impair splicing despite the probability that they do not have high densities of ESEs. Although we found that one-third of a randomly chosen set of human genomic sequences of about 100 nt decreased splicing when inserted into an exon, the other two-thirds had little effect.[118] An argument against a requirement for specific proteins lies in the fact that insertion of any of a wide variety of predicted ESE sequences can enhance the splicing of a crippled exon.[39,41,112] Indeed it should not be necessary to invoke a continuous bridge for the two ends of an exon to communicate in exon definition: an ESE at each end could suffice to recruit splicing factors that could then interact by simply forming a loop (Fig. 2B).

An alternative model that necessitates large numbers of ESEs puts the emphasis on ESSs. If ESSs are fairly common and if the binding of an inhibitory factor to a single ESS is sufficient to inhibit splicing[120,121] then it may be necessary to have enough ESEs per exon to prevent even a single silencing protein from binding (Fig. 2C-E). If there are no ESSs this is not a problem as may be the case of the bacterial inserts mentioned above. But the protein coding requirements of the exon may not allow the complete exclusion of ESSs, and these sequence elements may have additional downstream roles even in constitutively spliced exons, such as in translation.[122,123] Furthermore, splicing inhibitors such as hnRNP A1 can undergo multimerization leading to cooperative binding[124] with the consequent displacement of many more distant ESE binding factors (Fig. 2F-H). Extensive coverage of the exon with ESEs and their binding factors may prevent proteins like hnRNP A1 from gaining a foothold. The ESE in the immunoglobulin M2 exon acts in this way by disrupting the association of PTB with an ESS.[120] Of course neither of these anti-ESS models excludes a positive role for other ESEs in promoting splicing at the same time.

It is possible that ESSs provide more of the information for discriminating real exons from pseudo exons than ESEs (for a review see ref. 46). As can be seen in Table 1, RESCUE-ESEs plus PESEs are about 60% more abundant in real exons than pseudo exons, but PESSs and FAS-hex3 silencers are 2-fold more abundant in pseudo exons than real exons. There are now several examples

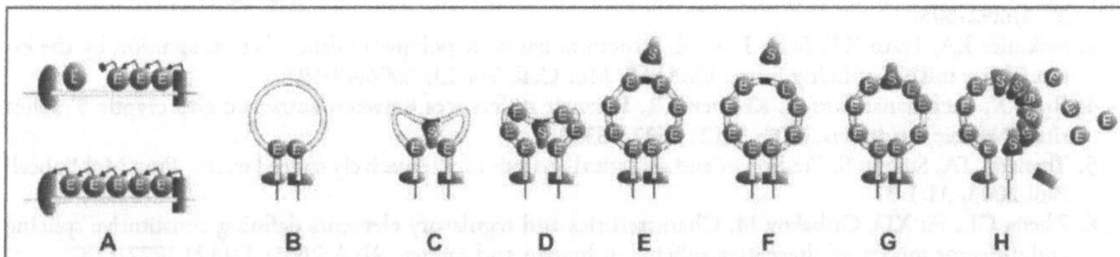

Figure 2. Models incorporating a role for extensive exon coverage in exon definition. Thick lines, exons; thin lines, introns; E, enhancer binding proteins; S, silencer binding proteins, rectangle and oval, spliceosomal or prespliceosomal complexes. A) A bridge of proteins is required in order to sense that both ends of the exon have been recognized as splice sites. Complete exon coverage is required for efficient transmission of this information via allosteric transitions. B) The opposite case, in which enhancers help recruit splicing factors to the splice sites and then interact with each other directly to convey the information that both sites have been recognized. C) A single silencer disrupting the interaction in B, so that splicing does not occur and the exon is skipped. D) Even if many splicing activators are bound to many enhancers, the binding of a single inhibitory protein to an available silencer can still inhibit splicing. E) Coverage of the exon is extensive enough for steric hindrance to prevent the binding of even a single silencer protein. F-H) Despite many enhancers, leaving a single silencer unobstructed allows binding by an inhibitory protein (e.g., hnRNP A1). Once bound, the inhibitory protein multimerizes, leading to the displacement of splicing factors.

of intronic mutations that create new exons apparently by the inactivation of silencers elements (e.g., see ref. 125). Such events have led to the exonization of Alu sequences[126] and could underlie the evolution of new genes in general.[127] There is little doubt that ESSs can play a role in the silencing of pseudo exons, but whether this mechanism represents a global role has not been established.

The Future

Structure determines function in biology and the structure of RNA within an RNP is ultimately dictated by its sequence. What proteins bind to what RNAs,[128] how the position and order of motifs influence factor binding and splicing and how positive and negative intron sequences factor into the equation are all questions that are approachable experimentally. Answers to these questions will help move us from lists to mechanisms. From bioinformatics we can expect yet more global information defining ISEs and ISSs as well as refining ESEs and ESSs. Relationships between motifs (e.g., see ref. 103) and between motifs and expressed splicing factors[129-131] are beginning to be revealed using combined computational and molecular approaches. The problem of how the cell integrates information from a large set of overlapping signals is not unique to splicing. In transcriptional regulation the choice of true promoters from among pseudo signals has parallels in the identification of true splice sites, and even the cellular decisions that are made during embryonic development are somewhat analogous in that slight differences in a morphogenetic gradient are amplified to produce a binary response. For splicing we now know many of the players, a necessary step in order to decipher the rules of the game.

Acknowledgements

I thank Mauricio Arias, Shengdong Ke and Xiang Zhang for useful discussions and Xiang Zhang for the databases used in calculations performed here. Adrian Krainer generously provided the 20-mer sequences underlying ESEfinder. Splicing research in our laboratory is supported by a grant from the NIH.

References

1. Senapathy P, Shapiro MB, Harris NL. Splice junctions, branch point sites and exons: sequence statistics, identification and applications to genome project. Methods Enzymol 1990; 183:252-278.
2. Zhang XH, Leslie CS, Chasin LA. Computational searches for splicing signals. Methods 2005; 37(4):292-305.
3. Sickmier EA, Frato KE, Shen H et al. Structural basis for polypyrimidine tract recognition by the essential pre-mRNA splicing factor U2AF65. Mol Cell. Vol 23; 2006:49-59.
4. Roca X, Sachidanandam R, Krainer AR. Intrinsic differences between authentic and cryptic 5' splice sites. Nucleic Acids Res 2003; 31(21):6321-6333.
5. Thanaraj TA, Stamm S. Prediction and statistical analysis of alternatively spliced exons. Prog Mol Subcell Biol 2003; 31:1-31.
6. Zheng CL, Fu XD, Gribskov M. Characteristics and regulatory elements defining constitutive splicing and different modes of alternative splicing in human and mouse. RNA 2005; 11(12):1777-1787.
7. Itoh H, Washio T, Tomita M. Computational comparative analyses of alternative splicing regulation using full-length cDNA of various eukaryotes. RNA 2004; 10(7):1005-1018.
8. Lear AL, Eperon LP, Wheatley IM et al. Hierarchy for 5' splice site preference determined in vivo. J Mol Biol. Vol 211; 1990:103-115.
9. Carothers AM, Urlaub G, Grunberger D et al. Splicing mutants and their second-site suppressors at the dihydrofolate reductase locus in Chinese hamster ovary cells. Mol Cell Biol 1993; 13(8):5085-5098.
10. Schneider TD. Information content of individual genetic sequences. J Theor Biol 1997; 189(4):427-441.
11. Burge C, Karlin S. Prediction of complete gene structures in human genomic DNA. J Mol Biol 1997; 268(1):78-94.
12. Zhang XH, Heller KA, Hefter I et al. Sequence information for the splicing of human pre-mRNA identified by support vector machine classification. Genome Res 2003; 13(12):2637-2650.
13. Green MR. Biochemical mechanisms of constitutive and regulated pre-mRNA splicing. Annu Rev Cell Biol 1991; 7:559-599.
14. Smith CW, Nadal-Ginard B. Mutually exclusive splicing of alpha-tropomyosin exons enforced by an unusual lariat branch point location: implications for constitutive splicing. Cell 1989; 56(5):749-758.

15. Sun H, Chasin LA. Multiple splicing defects in an intronic false exon. Mol Cell Biol 2000; 20(17):6414-6425.
16. Krawczak M, Reiss J, Cooper DN. The mutational spectrum of single base-pair substitutions in mRNA splice junctions of human genes: causes and consequences. Hum Genet 1992; 90(1-2):41-54.
17. Berget SM. Exon recognition in vertebrate splicing. J Biol Chem 1995; 270(6):2411-2414.
18. Robberson BL, Cote GJ, Berget SM. Exon definition may facilitate splice site selection in RNAs with multiple exons. Mol Cell Biol 1990; 10(1):84-94.
19. Lewis JD, Izaurralde E, Jarmolowski A et al. A nuclear cap-binding complex facilitates association of U1 snRNP with the cap-proximal 5' splice site. Genes Dev 1996; 10(13):1683-1698.
20. Zeng C, Berget SM. Participation of the C-terminal domain of RNA polymerase II in exon definition during pre-mRNA splicing. Mol Cell Biol 2000; 20(21):8290-8301.
21. Cooke C, Hans H, Alwine JC. Utilization of splicing elements and polyadenylation signal elements in the coupling of polyadenylation and last-intron removal. Mol Cell Biol 1999; 19(7):4971-4979.
22. Query CC, McCaw PS, Sharp PA. A minimal spliceosomal complex A recognizes the branch site and polypyrimidine tract. Mol Cell Biol 1997; 17(5):2944-2953.
23. Zhang MQ. Statistical features of human exons and their flanking regions. Hum Mol Genet 1998; 7(5):919-932.
24. Cooper TA, Ordahl CP. Nucleotide substitutions within the cardiac troponin T alternative exon disrupt pre-mRNA alternative splicing. Nucleic Acids Res 1989; 17(19):7905-7921.
25. Streuli M, Saito H. Regulation of tissue-specific alternative splicing: exon-specific cis-elements govern the splicing of leukocyte common antigen pre-mRNA. EMBO J 1989; 8(3):787-796.
26. Watakabe A, Tanaka K, Shimura Y. The role of exon sequences in splice site selection. Genes Dev 1993; 7(3):407-418.
27. Schaal TD, Maniatis T. Multiple distinct splicing enhancers in the protein-coding sequences of a constitutively spliced pre-mRNA. Mol Cell Biol 1999; 19(1):261-273.
28. Zhang XH, Kangsamaksin T, Chao MS et al. Exon inclusion is dependent on predictable exonic splicing enhancers. Mol Cell Biol 2005; 25(16):7323-7332.
29. Graveley BR. Sorting out the complexity of SR protein functions. RNA 2000; 6(9):1197-1211.
30. Kohtz JD, Jamison SF, Will CL, et al. Protein-protein interactions and 5'-splice-site recognition in mammalian mRNA precursors. Nature 1994; 368(6467):119-124.
31. Shen H, Green MR. RS domains contact splicing signals and promote splicing by a common mechanism in yeast through humans. Genes Dev. Vol 20; 2006:1755-1765.
32. Tacke R, Manley JL. Determinants of SR protein specificity. Curr Opin Cell Biol 1999; 11(3):358-362.
33. Pozzoli U, Sironi M. Silencers regulate both constitutive and alternative splicing events in mammals. Cell Mol Life Sci 2005; 62(14):1579-1604.
34. Abdul-Manan N, Williams KR. hnRNP A1 binds promiscuously to oligoribonucleotides: utilization of random and homo-oligonucleotides to discriminate sequence from base-specific binding. Nucleic Acids Res 1996; 24(20):4063-4070.
35. Black DL. Mechanisms of alternative pre-messenger RNA splicing. Annu Rev Biochem. 2003; 72:291-336.
36. Zheng ZM. Regulation of alternative RNA splicing by exon definition and exon sequences in viral and mammalian gene expression. J Biomed Sci 2004; 11(3):278-294.
37. Fu XD. Towards a splicing code. Cell 2004; 119:736-738.
38. Fedorov A, Saxonov S, Fedorova L et al. Comparison of intron-containing and intron-lacking human genes elucidates putative exonic splicing enhancers. Nucleic Acids Res 2001; 29(7):1464-1469.
39. Fairbrother WG, Yeh RF, Sharp PA et al. Predictive identification of exonic splicing enhancers in human genes. Science 2002; 297(5583):1007-1013.
40. Fairbrother WG, Holste D, Burge CB et al. Single nucleotide polymorphism-based validation of exonic splicing enhancers. PLoS Biol 2004; 2(9):E268.
41. Zhang XH, Chasin LA. Computational definition of sequence motifs governing constitutive exon splicing. Genes Dev 2004; 18(11):1241-1250.
42. Davuluri RV, Grosse I, Zhang MQ. Computational identification of promoters and first exons in the human genome. Nat Genet 2001; 29(4):412-417.
43. Majewski J, Ott J. Distribution and characterization of regulatory elements in the human genome. Genome Res 2002; 12(12):1827-1836.
44. Saxonov S, Berg P, Brutlag DL. A genome-wide analysis of CpG dinucleotides in the human genome distinguishes two distinct classes of promoters. Proc Natl Acad Sci USA 2006; 103(5):1412-1417.
45. Grellscheid SN, Smith CW. An apparent pseudo-exon acts both as an alternative exon that leads to nonsense-mediated decay and as a zero-length exon. Mol Cell Biol 2006; 26(6):2237-2246.

46. Sironi M, Menozzi G, Riva L et al. Silencer elements as possible inhibitors of pseudoexon splicing. Nucleic Acids Res 2004; 32(5):1783-1791.
47. Goren A, Ram O, Amit M et al. Comparative analysis identifies exonic splicing regulatory sequences—The complex definition of enhancers and silencers. Mol Cell. Vol 22; 2006:769-781.
48. Graveley BR, Hertel KJ, Maniatis T. A systematic analysis of the factors that determine the strength of pre-mRNA splicing enhancers. EMBO J 1998; 17(22):6747-6756.
49. Shen H, Green MR. RS domains contact splicing signals and promote splicing by a common mechanism in yeast through humans. Genes Dev 2006; 20(13):1755-1765.
50. Kanopka A, Muhlemann O, Akusjarvi G. Inhibition by SR proteins of splicing of a regulated adenovirus pre-mRNA. Nature 1996; 381(6582):535-538.
51. Wu JY, Kar A, Kuo D et al. SRp54 (SFRS11), a Regulator for tau Exon 10 Alternative Splicing Identified by an Expression Cloning Strategy. Mol Cell Biol 2006; 26(18):6739-6747.
52. Cartegni L, Hastings ML, Calarco JA et al. Determinants of exon 7 splicing in the spinal muscular atrophy genes, SMN1 and SMN2. Am J Hum Genet 2006; 78(1):63-77.
53. Kashima T, Manley JL. A negative element in SMN2 exon 7 inhibits splicing in spinal muscular atrophy. Nat Genet 2003; 34(4):460-463.
54. Goren A, Ram O, Amit M et al. Comparative analysis identifies exonic splicing regulatory sequences—The complex definition of enhancers and silencers. Mol Cell 2006; 22(6):769-781.
55. Wang Z, Xiao X, Van Nostrand E et al. General and specific functions of exonic splicing silencers in splicing control. Mol Cell 2006; 23(1):61-70.
56. Smith PJ, Zhang C, Wang J et al. An increased specificity score matrix for the prediction of SF2/ASF-specific exonic splicing enhancers. Hum Mol Genet 2006; 15(16):2490-2508.
57. Eszterhas SK, Bouhassira EE, Martin DI et al. Transcriptional interference by independently regulated genes occurs in any relative arrangement of the genes and is influenced by chromosomal integration position. Mol Cell Biol 2002; 22(2):469-479.
58. Tuerk C, Gold L. Systematic evolution of ligands by exponential enrichment: RNA ligands to bacteriophage T4 DNA polymerase. Science 1990; 249(4968):505-510.
59. Tacke R, Manley JL. The human splicing factors ASF/SF2 and SC35 possess distinct, functionally significant RNA binding specificities. EMBO J 1995; 14(14):3540-3551.
60. Kim S, Shi H, Lee DK et al. Specific SR protein-dependent splicing substrates identified through genomic SELEX. Nucleic Acids Res 2003; 31(7):1955-1961.
61. Hui J, Hung LH, Heiner M et al. Intronic CA-repeat and CA-rich elements: a new class of regulators of mammalian alternative splicing. EMBO J 2005; 24(11):1988-1998.
62. Amarasinghe AK, MacDiarmid R, Adams MD et al. An in vitro-selected RNA-binding site for the KH domain protein PSI acts as a splicing inhibitor element. RNA 2001; 7(9):1239-1253.
63. Cavaloc Y, Bourgeois CF, Kister L et al. The splicing factors 9G8 and SRp20 transactivate splicing through different and specific enhancers. RNA 1999; 5(3):468-483.
64. Wang J, Dong Z, Bell LR. Sex-lethal interactions with protein and RNA. Roles of glycine-rich and RNA binding domains. J Biol Chem 1997; 272(35):22227-22235.
65. Buckanovich RJ, Darnell RB. The neuronal RNA binding protein Nova-1 recognizes specific RNA targets in vitro and in vivo. Mol Cell Biol 1997; 17(6):3194-3201.
66. Tacke R, Chen Y, Manley JL. Sequence-specific RNA binding by an SR protein requires RS domain phosphorylation: creation of an SRp40-specific splicing enhancer. Proc Natl Acad Sci USA 1997; 94(4):1148-1153.
67. Faustino NA, Cooper TA. Identification of putative new splicing targets for ETR-3 using sequences identified by systematic evolution of ligands by exponential enrichment. Mol Cell Biol 2005; 25(3):879-887.
68. Bourgeois CF, Lejeune F, Stevenin J. Broad specificity of SR (serine/arginine) proteins in the regulation of alternative splicing of pre-messenger RNA. Prog Nucleic Acid Res Mol Biol 2004; 78:37-88.
69. Sickmier EA, Frato KE, Shen H et al. Structural basis for polypyrimidine tract recognition by the essential pre-mRNA splicing factor U2AF65. Mol Cell 2006; 23(1):49-59.
70. Sanford JR, Ellis J, Caceres JF. Multiple roles of arginine/serine-rich splicing factors in RNA processing. Biochem Soc Trans 2005; 33(Pt 3):443-446.
71. Boukis LA, Liu N, Furuyama S et al. Ser/Arg-rich protein-mediated communication between U1 and U2 small nuclear ribonucleoprotein particles. J Biol Chem 2004; 279(28):29647-29653.
72. MacMillan AM, McCaw PS, Crispino JD et al. SC35-mediated reconstitution of splicing in U2AF-depleted nuclear extract. Proc Natl Acad Sci USA 1997; 94(1):133-136.
73. Shen H, Green MR. A pathway of sequential arginine-serine-rich domain-splicing signal interactions during mammalian spliceosome assembly. Mol Cell 2004; 16(3):363-373.
74. Shen H, Kan JL, Green MR. Arginine-serine-rich domains bound at splicing enhancers contact the branchpoint to promote prespliceosome assembly. Mol Cell 2004; 13(3):367-376.

75. Chew SL, Liu HX, Mayeda A et al. Evidence for the function of an exonic splicing enhancer after the first catalytic step of pre-mRNA splicing. Proc Natl Acad Sci USA 1999; 96(19):10655-10660.
76. Tian H, Kole R. Selection of novel exon recognition elements from a pool of random sequences. Mol Cell Biol 1995; 15(11):6291-6298.
77. Tian H, Kole R. Strong RNA splicing enhancers identified by a modified method of cycled selection interact with SR protein. J Biol Chem 2001; 276(36):33833-33839.
78. Liu HX, Chew SL, Cartegni L et al. Exonic splicing enhancer motif recognized by human SC35 under splicing conditions. Mol Cell Biol 2000; 20(3):1063-1071.
79. Liu HX, Zhang M, Krainer AR. Identification of functional exonic splicing enhancer motifs recognized by individual SR proteins. Genes Dev 1998; 12(13):1998-2012.
80. Cartegni L, Wang J, Zhu Z et al. ESEfinder: A web resource to identify exonic splicing enhancers. Nucleic Acids Res 2003; 31(13):3568-3571.
81. Wang J, Smith PJ, Krainer AR et al. Distribution of SR protein exonic splicing enhancer motifs in human protein-coding genes. Nucleic Acids Res 2005; 33(16):5053-5062.
82. Wu Y, Zhang Y, Zhang J. Distribution of exonic splicing enhancer elements in human genes. Genomics 2005; 86(3):329-336.
83. Schaal TD, Maniatis T. Selection and characterization of pre-mRNA splicing enhancers: identification of novel SR protein-specific enhancer sequences. Mol Cell Biol 1999; 19(3):1705-1719.
84. Coulter LR, Landree MA, Cooper TA. Identification of a new class of exonic splicing enhancers by in vivo selection. Mol Cell Biol 1997; 17(4):2143-2150.
85. Stickeler E, Fraser SD, Honig A et al. The RNA binding protein YB-1 binds A/C-rich exon enhancers and stimulates splicing of the CD44 alternative exon v.4 EMBO J 2001; 20(14):3821-3830.
86. Woerfel G, Bindereif A. In vitro selection of exonic splicing enhancer sequences: identification of novel CD44 enhancers. Nucleic Acids Res 2001; 29(15):3204-3211.
87. Wang Z, Rolish ME, Yeo G et al. Systematic identification and analysis of exonic splicing silencers. Cell 2004; 119(6):831-845.
88. Wagner EJ, Garcia-Blanco MA. Polypyrimidine tract binding protein antagonizes exon definition. Mol Cell Biol 2001; 21(10):3281-3288.
89. Dredge BK, Darnell RB. Nova regulates GABA(A) receptor gamma2 alternative splicing via a distal downstream UCAU-rich intronic splicing enhancer. Mol Cell Biol 2003; 23(13):4687-4700.
90. Singh NN, Androphy EJ, Singh RN. In vivo selection reveals combinatorial controls that define a critical exon in the spinal muscular atrophy genes. RNA 2004; 10(8):1291-1305.
91. Nussinov R. Conserved signals around the 5' splice sites in eukaryotic nuclear precursor mRNAs: G-runs are frequent in the introns and C in the exons near both 5' and 3' splice sites. J Biomol Struct Dyn 1989; 6(5):985-1000.
92. McCullough AJ, Berget SM. G triplets located throughout a class of small vertebrate introns enforce intron borders and regulate splice site selection. Mol Cell Biol 1997; 17(8):4562-4571.
93. McCullough AJ, Berget SM. An intronic splicing enhancer binds U1 snRNPs to enhance splicing and select 5' splice sites. Mol Cell Biol 2000; 20(24):9225-9235.
94. Louie E, Ott J, Majewski J. Nucleotide frequency variation across human genes. Genome Res 2003; 13(12):2594-2601.
95. Sorek R, Ast G. Intronic sequences flanking alternatively spliced exons are conserved between human and mouse. Genome Res 2003; 13(7):1631-1637.
96. Lim LP, Burge CB. A computational analysis of sequence features involved in recognition of short introns. Proc Natl Acad Sci USA 2003; 98(20):11193-11198.
97. Brudno M, Gelfand MS, Spengler S et al. Computational analysis of candidate intron regulatory elements for tissue-specific alternative pre-mRNA splicing. Nucleic Acids Res 2001; 29(11):2338-2348.
98. Minovitsky S, Gee SL, Schokrpur S et al. The splicing regulatory element, UGCAUG, is phylogenetically and spatially conserved in introns that flank tissue-specific alternative exons. Nucleic Acids Res 2005; 33(2):714-724.
99. Auweter SD, Fasan R, Reymond L et al. Molecular basis of RNA recognition by the human alternative splicing factor Fox-1. EMBO J 2006; 25(1):163-173.
100. Zhou HL, Baraniak AP, Lou H. A role for Fox-1/Fox-2 in mediating the neuronal pathway of calcitonin/CGRP alternative RNA processing. Mol Cell Biol 2006.
101. Nakahata S, Kawamoto S. Tissue-dependent isoforms of mammalian Fox-1 homologs are associated with tissue-specific splicing activities. Nucleic Acids Res 2005; 33(7):2078-2089.
102. Jin Y, Suzuki H, Maegawa S et al. A vertebrate RNA-binding protein Fox-1 regulates tissue-specific splicing via the pentanucleotide GCAUG. EMBO J 2003; 22(4):905-912.
103. Han K, Yeo G, An P et al. A combinatorial code for splicing silencing: UAGG and GGGG motifs. PLoS Biol 2005; 3(5):e158.

104. Zhang XH, Leslie CS, Chasin LA. Dichotomous splicing signals in exon flanks. Genome Res 2005; 15(6):768-779.
105. Lund M, Tange TO, Dyhr-Mikkelsen H et al. Characterization of human RNA splice signals by iterative functional selection of splice sites. RNA 2000; 6(4):528-544.
106. Buvoli M, Mayer SA, Patton JG. Functional crosstalk between exon enhancers, polypyrimidine tracts and branchpoint sequences. EMBO J 1997; 16(23):7174-7183.
107. Lund M, Kjems J. Defining a 5' splice site by functional selection in the presence and absence of U1 snRNA 5' end. RNA 2002; 8(2):166-179.
108. Lorincz MC, Dickerson DR, Schmitt M et al. Intragenic DNA methylation alters chromatin structure and elongation efficiency in mammalian cells. Nat Struct Mol Biol 2004; 11(11):1068-1075.
109. Listerman I, Sapra AK, Neugebauer KM. Cotranscriptional coupling of splicing factor recruitment and precursor messenger RNA splicing in mammalian cells. Nat Struct Mol Biol 2006; 13(9):815-822.
110. Matlin AJ, Clark F, Smith CW. Understanding alternative splicing: towards a cellular code. Nat Rev Mol Cell Biol 2005; 6(5):386-398.
111. Fu XD. Towards a splicing code. Cell 2004; 119(6):736-738.
112. Stadler MB, Shomron N, Yeo GW et al. Inference of splicing regulatory activities by sequence neighborhood analysis. PLoS Genet 2006; 2(11):e191.
113. Buratti E, Baralle M, Baralle FE. Defective splicing, disease and therapy: searching for master checkpoints in exon definition. Nucleic Acids Res 2006; 34(12):3494-3510.
114. Varani G, Nagai K. RNA recognition by RNP proteins during RNA processing. Annu Rev Biophys Biomol Struct 1998; 27:407-445.
115. Buratti E, Muro AF, Giombi M et al. RNA folding affects the recruitment of SR proteins by mouse and human polypurinic enhancer elements in the fibronectin EDA exon. Mol Cell Biol 2004; 24(3):1387-1400.
116. Buratti E, Baralle FE. Influence of RNA secondary structure on the pre-mRNA splicing process. Mol Cell Biol 2004; 24(24):10505-10514.
117. Bennett M, Pinol-Roma S, Staknis D et al. Differential binding of heterogeneous nuclear ribonucleoproteins to mRNA precursors prior to spliceosome assembly in vitro. Mol Cell Biol 1992; 12(7):3165-3175.
118. Fairbrother WG, Chasin LA. Human genomic sequences that inhibit splicing. Mol Cell Biol 2000; 20(18):6816-6825.
119. Chen IT, Chasin LA. Large exon size does not limit splicing in vivo. Mol Cell Biol 1994; 14(3):2140-2146.
120. Shen H, Kan JL, Ghigna C et al. A single polypyrimidine tract binding protein (PTB) binding site mediates splicing inhibition at mouse IgM exons M1 and M2. RNA 2004; 10(5):787-794.
121. Zahler AM, Damgaard CK, Kjems J et al. SC35 and heterogeneous nuclear ribonucleoprotein A/B proteins bind to a juxtaposed exonic splicing enhancer/exonic splicing silencer element to regulate HIV-1 tat exon 2 splicing. J Biol Chem 2004; 279(11):10077-10084.
122. Bonnal S, Pileur F, Orsini C et al. Heterogeneous nuclear ribonucleoprotein A1 is a novel internal ribosome entry site trans-acting factor that modulates alternative initiation of translation of the fibroblast growth factor 2 mRNA. J Biol Chem 2005; 280(6):4144-4153.
123. Valcarcel J, Gebauer F. Posttranscriptional regulation: the dawn of PTB. Curr Biol 1997; 7(11):R705-708.
124. Zhu J, Mayeda A, Krainer AR. Exon identity established through differential antagonism between exonic splicing silencer-bound hnRNP A1 and enhancer-bound SR proteins. Mol Cell 2001; 8(6):1351-1361.
125. Pagani F, Buratti E, Stuani C et al. A new type of mutation causes a splicing defect in ATM. Nat Genet 2002; 30(4):426-429.
126. Sorek R, Lev-Maor G, Reznik M et al. Minimal conditions for exonization of intronic sequences: 5' splice site formation in alu exons. Mol Cell 2004; 14(2):221-231.
127. Zhang XH, Chasin LA. Comparison of multiple vertebrate genomes reveals the birth and evolution of human exons. Proc Natl Acad Sci USA 2006; 103(36):13427-13432.
128. Jurica MS, Licklider LJ, Gygi SR et al. Purification and characterization of native spliceosomes suitable for three-dimensional structural analysis. RNA 2002; 8(4):426-439.
129. Blanchette M, Green RE, Brenner SE et al. Global analysis of positive and negative pre-mRNA splicing regulators in Drosophila. Genes Dev 2005; 19(11):1306-1314.
130. Ule J, Jensen KB, Ruggiu M et al. CLIP identifies Nova-regulated RNA networks in the brain. Science 2003; 302(5648):1212-1215.
131. Ule J, Stefani G, Mele A et al. An RNA map predicting Nova-dependent splicing regulation. Nature 2006; 444(7119):580-586.

CHAPTER 7

SR Proteins and Related Factors in Alternative Splicing

Shengrong Lin and Xiang-Dong Fu*

Abstract

SR proteins are a family of RNA binding proteins that contain a signature RS domain enriched with serine/arginine repeats. The RS domain is also found in many other proteins, which are collectively referred to as SR-related proteins. Several prototypical SR proteins are essential splicing factors, but the majority of RS domain-containing factors are characterized by their ability to alter splice site selection in vitro or in transfected cells. SR proteins and SR-related proteins are generally believed to modulate splice site selection via RNA recognition motif-mediated binding to exonic splicing enhancers and RS domain-mediated protein-protein and protein-RNA interactions during spliceosome assembly. However, the biological function of individual RS domain-containing splicing regulators is complex because of redundant as well as competitive functions, context-dependent effects and regulation by cotranscriptional and post-translational events. This chapter will focus on our current mechanistic understanding of alternative splicing regulation by SR proteins and SR-related proteins and will discuss some of the questions that remain to be addressed in future research.

Introduction

SR proteins were discovered in the early 1990s by the identification of factors associated with purified spliceosomes,[1,2] by the purification of critical non-snRNP splicing activities in constitutive and alternative splicing,[3-6] and by the analysis of components of a nuclear body that could be selectively precipitated with Mg^{++}.[7] By virtue of its ability to complement splicing-deficient S100 cytoplasmic extracts from HeLa cells and to stimulate splice site switching in HeLa nuclear extracts, SF2/ASF was the first SR protein shown to have dual roles in constitutive and alternative splicing.[3,4,6,8] This observation was quickly extended to other SR proteins.[9-11] The S100 complementation and splice site switch assays have since become standard functional tests for SR proteins isolated from higher eukaryotic organisms.

Sequence analysis has revealed that SR protein family members consist of one or two RNA recognition motifs and a signature RS domain enriched with serine/arginine repeats.[12,13] These structural features have been commonly used to classify SR proteins. Clearly, not all SR proteins behave like prototypical SR proteins. For example, a subset have different fractionation properties and/or are not sufficient to complement S100 extracts. In addition, several new SR protein family members exhibit activities in both constitutive and alternative splicing that are opposite to those possessed by prototypical SR proteins. Because of the functional diversity among SR proteins, we propose to define SR proteins based on their common structural features including at least one RNA recognition motif and an RS domain. Using this classification, several RS

*Corresponding Author: Xiang-Dong Fu—Department of Cellular and Molecular Medicine, University of California, San Diego, La Jolla, California, USA. Email: xdfu@ucsd.edu

Alternative Splicing in the Postgenomic Era, edited by Benjamin J. Blencowe and Brenton R. Graveley. ©2007 Landes Bioscience and Springer Science+Business Media.

domain-containing RNA binding proteins, including human TRA2β and RNPS1, can now be classified as SR proteins (Table 1).

In addition to SR proteins, many other splicing factors contain an RS domain. These proteins are collectively referred to as SR-related proteins.[14] In mammalian cells, SR-related proteins include other RNA binding proteins, such as both subunits of the U2AF heterodimer, the U1 snRNP specific protein U1-70K and various enzymes, including several ATPases involved in RNA rearrangement within the spliceosome[15-18] (Table 1). It is generally thought that the RS domains in SR proteins and SR-related splicing factors facilitate spliceosome assembly by mediating protein-protein interactions.[19] However, recent studies have revealed direct binding of the RS domain to critical splicing signals in pre-mRNA transcripts.[20,21]

Interestingly, budding yeast express a few RNA binding proteins that structurally resemble SR proteins.[22] However, there is no direct evidence that these proteins are essential pre-mRNA processing factors in this organism and it is interesting to note in this context that ~5% of the genes in budding yeast contain a single intron and alternative splicing is rare. Therefore, splicing can take place in the absence of SR proteins, which begs the question as to why SR proteins are essential splicing factors in higher eukaryotic cells. The differential requirement for SR proteins in yeast and higher eukaryotic cells probably reflects the fact that the splicing signals in yeast pre-mRNAs are essentially invariant, whereas those in mammals are diverse. Thus, the RS domain in SR proteins may function to strengthen the recognition of weak splicing signals, as has been

Table 1. SR proteins and SR-related splicing regulators

Classification	Factors	Key Domains	Functions
Classic SR Proteins	SRp20,[1] SF2/ASF,[2] SC35,[3] 9G8,[4] SRp40,[5] SRp55/B52,[6] SRp75[7]	One or two RRMs plus an RS domain	Constitutive and alternative splicing
Additional SR proteins	hTRA2α,[8] hTRA2β,[9] RNPS1,[10] SRp38,[11] SRp30c,[12] p54,[13] SRrp35,[14] SRrp53,[15] SRp86[16]	One or two RRMs plus an RS domain	Positive and negative regulation of alternative splicing
RNA binding SR related factors	U2AF65,[17] U2AF35,[18] Urp,[19] HCC1/CAPER,[20] U1-70K,[21] hSWAP,[22] Pinin,[23] Sip1,[24] SR-A1,[25] ZNF265,[26] SRm160,[27] SRm300,[28]	RRM, PWI domain, Zn finger plus an RS domain	Splicing factors or co-activators
Enzymes and regulators carrying an RS domain	hPRP5,[29] hPRP16,[30] Prp22/HRH1,[31] U5-100K/ hPRP28,[32] ClkSty-1,[33] 2,[34] 3,[35] CLASP,[36] Prp4K,[37] CrkRS/CRK7/CDK12,[38] CDC2L5,39 CCNL1,[40] CCNL2,[41] SR-cyp,[42]	DEAH box, kinase domains, peptidyl-prolyl isomerase domain	Spliceosome rearrangement and modification of splicing factors

Key literature information and protein sequence for each gene can be found by individual NCBI accession number: [1]NP_003008 [2]NP_008855 [3]NP_003007 [4]NP_001026854 [5]NP_008856 [6]NP_006266 [7]NP_005617 [8]NP_037425 [9]NP_004584 [10]NP_542161 [11]NP_473357 [12]NP_003760 [13]NP_004759 [14]NP_542781 [15]NP_057709 [16]NP_631907 [17]NP_009210 [18]NP_006749 [19]NP_005080 [20]NP_909122 [21]NP_003080 [22]NP_008987 [23]NP_002678 [24]NP_004710 [25]NP_067051 [26]NP_976225 [27]NP_005830 [28]NP_057417 [29]NP_055644 [30]NP_054722 [31]NP_004932 [32]NP_004809 [33]NP_004062 [34]NP_003984 [35]NP_003983 [36]NP_008987 [37]NP_003904 [38]NP_057591 [39]NP_003709 [40]NP_064703 [41]NP_112199 [42]NP_004783 recently documented.[23] In addition, SR proteins are critical for pairing complexes assembled on the 5' and 3' splice sites. This functional requirement may not be critical for splicing in yeast, where introns are relatively short and the communication between splice sites may not require RS domain-mediated interactions during splicing assembly.

The Role of SR Proteins in Splice Site Selection

Prototypical SR proteins, such as SC35, SF2/ASF and 9G8, are required to initiate spliceosome assembly in nuclear extracts. This early function of SR proteins is mediated by their sequence-specific binding to cis-acting elements, which are mostly located in exons and functionally characterized as exonic splicing enhancers (ESEs). The binding specificity of individual SR proteins has been experimentally defined using a technique called SELEX, based either on in vitro binding[24,25] or on the functional consequence of in vitro splicing.[26-28] The ESEs characterized to date have been used to develop an ESE-finder program[29] to assist with the identification of potential cis-acting regulatory elements in pre-mRNAs. While the program is a useful guide for searching for cis-acting regulatory elements in various pre-mRNAs, the information derived is preliminary for several reasons. First, similar analyses have not been extended to other SR proteins. Second, many ESEs may be recognized by non-SR proteins. Third, some complex ESEs may require the action of more than one RS domain-containing splicing factor, as observed in the *Drosophila doublesex* pre-mRNA.[30,31] Consequently, the vast majority of computationally deduced and/or experimentally verified ESEs remain to be characterized with regards to the specific trans-acting factors involved.[32-36] Furthermore, it is unclear as to why SR proteins generally do not bind to intronic sequences that resemble ESEs. An interesting possibility is that SR proteins may bind to all potential sites in an initial scanning mode before stabilization at specific functional ESEs via their interactions with other splicing factors that promote spliceosome assembly.

Two non-exclusive models have been proposed to explain the functional consequence of initial SR protein binding to an ESE (Fig. 1). One model emphasizes the effect of ESE-bound SR proteins on the recruitment and stabilization of additional splicing factors, such as U1 at the 5' splice site[37-39] and the U2AF complex at the 3' splice site.[40-43] Both SR proteins and RS domain-containing splicing co-activators have been implicated in promoting communication between the 5' and 3' splice sites.[44-49] The second model stresses the role of ESE-bound SR proteins in preventing or displacing other RNA binding proteins, such as hnRNP A1, from binding at exonic splicing silencers (ESSs).[50,51] These two mechanisms are likely operating in a synergistic fashion to favor spliceosome assembly on functional splice sites.

The early function of SR proteins in splice site recognition is probably similar in both constitutive and alternative splicing. Based on in vitro analysis of several prototypical SR proteins in alternative splicing, binding of SR proteins promotes the selection of proximal sites over distal ones in alternative 5' or 3' splice site choices.[8,9,52,53] In such processes, splice site selection may be dictated by the intrinsic strength of the competing splice sites and/or the frequency of competing exonic splicing silencer (ESS) sequences.[54] SR protein binding may enhance complex assembly on both strong and weak splice sites to make them equally competitive.[55] The proximal site is then selected because of the insulating function of SR proteins, allowing the closest pair of splice sites to be linked in later spliceosome assembly events[56] (Fig. 1). This insulating function may play a critical role in preventing exon skipping during the removal of multiple introns in a pre-mRNA transcript.

The ability of SR proteins to bind RNA is essential for the activity of SR proteins in both constitutive and alternative splicing.[57-59] In contrast, the RS domain seems to be important for constitutive splicing, but dispensable in alternative splicing, at least for the small number of pre-mRNA substrates analyzed.[58,60] The reason why the RS domain is not required for alternative splicing is not completely understood. It is possible that SR proteins lacking the RS domain may be sufficient to compete with the binding of negative splicing factors to adjacent splicing silencer sequences.[50,51] Given the fact that the dispensability of the RS domain in alternative splicing has only been tested with a limited number of alternative splicing substrates, it remains possible that certain alternative splicing events may require the domain to promote the selection of weak splice sites.

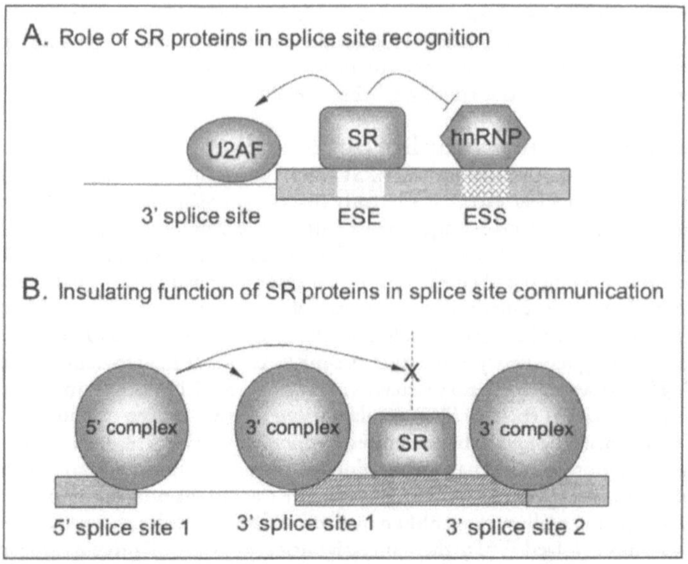

Figure 1. Role of SR proteins in splice site selection. A) An ESE-bound SR protein may stimulate complex assembly at a nearby functional splice site and/or antagonize the negative effect of an hnRNP protein on spliceosome assembly. B) An insulating function of SR proteins may promote the selection of the proximal splice site and prevent the use of the distal splice site.

SR Proteins Modulate Alternative Splicing in Both Ways

As described above, SR proteins seem to promote exon inclusion and the selection of intron-proximal splice sites over distal ones. However, further studies indicate that different SR proteins may influence splice site selection in both positive and negative fashions. Three distinct mechanisms by which SR proteins negatively modulate splice site selection have been reported in the literature (Fig. 2). SR proteins may recognize some intronic sequences that resemble ESEs, therefore resulting in the activation of an intronic cryptic splice site at the expense of a native splice site[61] (Fig. 2A). Mechanistically, this mode of negative regulation is similar to the activity of SR proteins in promoting the selection of a proximal, weak splice site in competition with a strong, distal one.

SR proteins may be actively involved in suppressing splice sites in a substrate-dependent manner. This was observed in SR knockout cardiomyocytes, where loss of SF2/ASF induced exon inclusion in the alternatively spliced CaMKIIδ gene.[62] While the direct effect of SR proteins in CaMKIIδ exon skipping event remains to be confirmed by in vitro analysis, a more recent study demonstrated that SF2/ASF acted on an ESE to promote exon skipping in the *Ron* proto-oncogene.[63] Similarly, SRp30c was found to suppress splice site selection of an alternative exon in the hnRNP A1 gene.[64] While the mechanism for these SR protein-dependent exon skipping events remains elusive, the phenomenon may be related to a number of earlier observations that different SR proteins appear to have opposite effects on regulated splicing.[65-69] In these cases, different SR proteins may act on their respective cis-acting elements to antagonize each other, thereby influencing the final choice of alternative splice sites. The opposite effects observed with different SR proteins may be due to the possibility that some SR proteins are more productive in promoting splice site selection than others, such that less productive SR proteins may interfere with productive ones in a competitive manner (Fig. 2B). Furthermore, it was recently shown that the positive and negative effects may be also related to the location of SR protein binding site with respect to splice sites.[70]

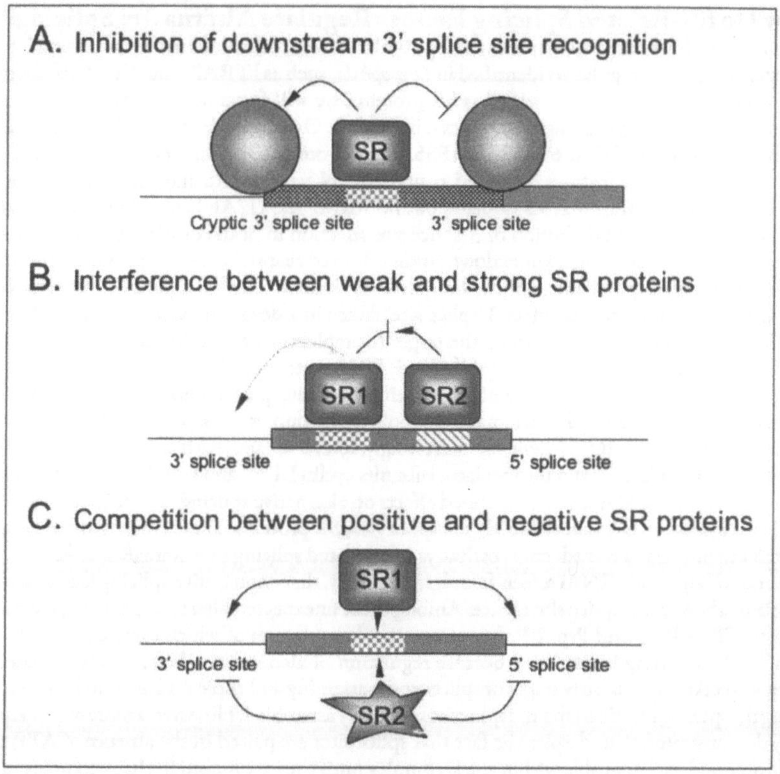

Figure 2. Positive and negative effects of SR proteins on splice site selection. A) An SR protein may bind to an intronic sequence resembling an ESE, thereby activating an upstream cryptic 3' splice site and inhibiting the use of the normal, downstream 3' splice site. B) The function of an ESE-bound SR protein (SR2) may be blocked by another ESE-bound SR protein with a weaker activity in splicing activation. C) The same cis-acting ESE may be recognized by both positive and negative SR proteins.

Aside from the substrate-dependent effects of typical SR proteins, some SR proteins appear to only function in splicing in a negative fashion (Fig. 2C). The best characterized example is SRp86, which appears to antagonize typical SR proteins in splice site selection.[71-73] Likewise, the SR protein p54, which was initially identified as a U2AF65-interacting protein, promotes the selection of an intron-distal splice site in the E1A pre-mRNA.[74] In a recent functional screen using a tau-based alternative splicing reporter, p54 was found to compete with hTRA2β for binding to an ESE and to promote exon skipping.[75] Joining this list of "negative" SR proteins are two new SR-related RNA binding proteins, SRrp35 and SRrp40 (also known as NSSR, TASR or SRp38), which should be classified as SR proteins.[76] SRp38 was isolated as an alternative splicing regulator in several independent studies.[76-79] Interestingly, SRp38 normally seems to have little activity in splicing. However, following heat shock and during cell mitosis, dephosphorylation of the RS domain of SRp38 results in a strong inhibitory effect on splicing.[80,81] However, when the RS domain of SRp38 was linked to an MS2 binding site or to the RNA recognition motif (RRM) of a typical SR protein, the hybrid protein appeared to act as a typical splicing activator, like other SR proteins.[79,82] Thus, both the RNA binding activity and the phosphorylation state of its RS domain contribute to the inhibitory effect of SRp38 on splicing.

How Do SR-Related Splicing Factors Regulate Alternative Splicing?

In the past, SR-related alternative splicing regulators were often referred to as mammalian homologues of splicing regulators identified in *Drosophila*, such as hTRA2α and hTRA2β[83]. Because these splicing factors can be classified as SR proteins, we will focus our discussion on the other RS domain-containing splicing factors listed in Table 1. One example is the U2AF heterodimer, which is comprised of U2AF65 and U2AF35. These proteins are structurally related to SR proteins, but have distinct features: U2AF65 contains an N-terminal RS and three RRMs, whereas U2AF35 carries a C-terminal RS domain, but no RRM. The U2AF heterodimer is believed to play a critical role in the definition of 3' splice site selection in both constitutive and alternative splicing. Indeed, recent RNAi knockdown studies showed that the U2AF heterodimer is directly involved in regulating splicing in both *Drosophila* and human cells.[84,85] Unlike SR proteins, however, U2AF does not seem to affect 3' splice site choice in a dosage dependent manner. Instead, the U2AF heterodimer appears to be the target for replacement by other polypyrimidine tract binding proteins, such as Sxl in *Drosophila*[86] or PTB in vertebrates.[87-89]

Besides U2AF, a growing number of RS domain-containing proteins have been implicated in alternative splicing, including the mammalian homologue of suppressor-of-white-apricot[90] and a large Zn-finger protein ZNF265[91] (Table 1). Interestingly, several kinases, such as Clk/Sty,[92,93] CrkRS,[94] Prp4K,[95] and CDC2L5,[96] and the regulator subunits cyclin L1[97,98] and L2,[99,100] also contain an RS domain. While these kinases have exhibited effects on alternative splicing in transfected cells, only Clk/Sty is known to target and directly phosphorylate SR proteins. These kinase systems have the potential to link signal transduction pathways to regulated splicing in mammalian cells.

A recent large-scale RNAi screen found, surprisingly, that constitutive splicing factors are also capable of altering the splice site choice. Among these unexpected alternative splicing regulators are the ATPase Prp5 and Prp22,[85] the mammalian homologues of which carry an extra RS domain.[15,18] This finding is surprising because regulation of alternative splicing has been generally thought to take place in early stages of spliceosome assembly and these ATPases are known to act during the splicing reaction after the spliceosome is fully assembled. However, a more recent kinetic study demonstrated that, despite the fact that splice sites are paired in the absence of ATP, they are flexible and exchangeable within the E complex until they are locked in the A complex in the presence of ATP.[101] Thus, many factors that act after spliceosome assembly may still be capable of functioning as regulators in alternative splicing. This finding is consistent with the role of Prp5, Prp22 and other "late" splice factors in regulated splicing. The recent recognition of the dynamic nature of the spliceosome provides a conceptual framework for understanding how many known factors for constitutive splicing show an ability to modulate alternative splicing.[102]

Functional Requirement of SR Proteins In Vivo

While regulated splicing was initially recognized and extensively studied by genetics in the *Drosophila* system, most concepts and mechanistic insights into the regulation of alternative splicing by SR proteins and SR-related proteins have been based on biochemical analysis in vitro or in transfected cells. It is therefore important to test and extend the biochemical studies to in vivo systems. To this end, the RNAi approach has been used to determine the role of SR proteins in *C. elegans*.[103] Strikingly, most SR protein knockdowns resulted in no detectable phenotype, except for a late embryonic lethal phenotype induced by RNAi against SF2/ASF. These findings suggest an extensive functional overlap among the SR family of splicing factors in this model organism. A more extensive RNAi screen performed in *Drosophila* S2 cells revealed the role of several SR proteins and SR-related splicing factors in alternative splicing.[83] Although the RNAi approach has been applied to mammalian cells to demonstrate specific requirements of SR proteins in alternative splicing,[104,105] a similar systematic undertaking remains to be extended to the mammalian system where regulated splicing may be more dynamic and thus more complex.

Complementary to the RNAi approach, gene targeting in chicken DT40 cells and in mice has permitted the analysis of SR proteins in vivo. A study performed on SF2/ASF knockout DT40 cells revealed that SF2/ASF is required for cell viability,[106] has an unexpected role in maintaining genomic stability,[107] and has a regulatory function in DNA fragmentation during apoptosis.[108] At least one of these in vivo functions (DNA fragmentation) was linked to SF2/ASF-regulated alternative splicing.[108] These studies have significantly extended our understanding of SR proteins in vivo.

So far, all SR protein knockout mice studied to date have shown an early embryonic lethal phenotype, thus demonstrating the fundamental function of SR proteins in vivo.[62,109,110] Surprisingly however, SC35 seems to be dispensable in nondividing mature cardiomyocytes, indicating that SR proteins are not universally required for cell viability in vivo.[111] This observation is in agreement with an RNAi result in *C. elegans*.[103] Importantly, specific alternative splicing events have been directly linked to some defined phenotypes in SC35 and SF2/ASF knockout mice, showing that SR proteins are indeed regulators of alternative splicing in mammalian cells.

Interestingly, an SF2/ASF mutant lacking the RS domain could rescue cell viability in SF2/ASF-depleted mouse embryonic fibroblasts.[112] Because the RS domain in SF2/ASF is required for constitutive splicing but dispensable in alternative splicing in most cases, this observation suggests that most cellular malfunctions might result from defects in alternative splicing. This possibility is consistent with the studies of the SF2/ASF orthologue in *Drosophila*, in which dASF appeared to lack any activity in constitutive splicing, but functioned as a regulator in alternative splicing.[113] Furthermore, the global pattern of gene expression was not dramatically altered in SR protein-depleted cells, indicating that inactivation of individual SR proteins may not cause widespread defects in constitutive splicing.[111,114]

SR Proteins as Splicing Regulators In Vivo: Why So Few Targets?

Members of the SR family of splicing factors are among the bestcharacterized splicing regulators and have been extensively studied by biochemical analysis. One surprising finding from the study of SR protein knockout cells was that most splicing events (both constitutive and alternative) remained unaltered in response to depletion of individual SR proteins in vivo. This result has been assumed to be due to functional redundancy among SR proteins, which may be explained by two potential mechanisms (Fig. 3). First, more than one SR protein may be able to recognize

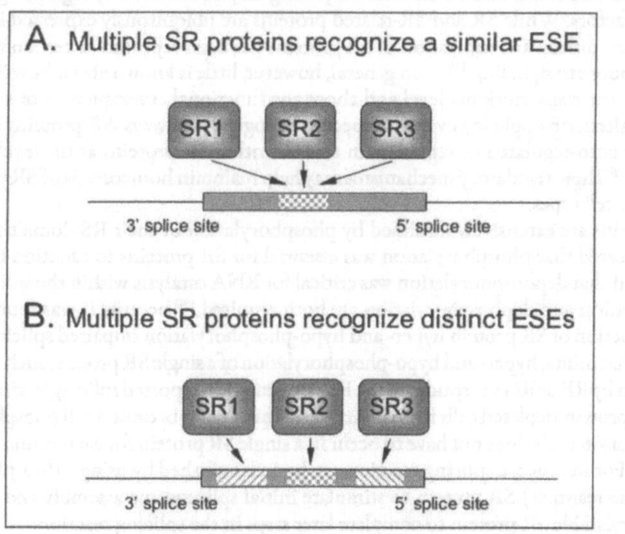

Figure 3. Potential functional redundancy of SR proteins. A) Multiple SR proteins may recognize the same ESE in a pre-mRNA. B) Multiple SR proteins may interact with several distinct ESEs in a pre-mRNA. As a result, deficiency of a single SR protein may have little effect on most constitutive splicing events.

a similar set of ESEs present in most exons; this has been observed in vitro with SF2/ASF and hTRA2β, which are both are capable of recognizing purine-rich ESEs.[25,41,83,115] Second, most exonic sequences appear to harbor multiple ESEs that are responsive to distinct SR proteins,[116] which may act independently or in a synergistic manner.[31,117] As a result, many splicing events may be responsive to SR protein overexpression, but relatively insensitive to down regulation or depletion of a single SR protein. Overexpression of SR proteins may exert a dominant effect on exons containing related ESEs. Therefore, caution must be taken in interpreting overexpression results in transfected cells, in which an affected splicing event may not be the natural substrate for the SR protein under study. This problem can be addressed by comparing results from both overexpression and RNAi knockdown studies.

According to the theory of functional redundancy, one might expect a more prevalent effect of SR protein depletion on alternative splicing versus constitutive splicing in vivo, since alternative splicing is often coupled with weak splice sites in conjunction with specific ESEs.[33] In this regard, alternative splicing would be more dependent on individual ESEs and thus more sensitive to variations in SR protein expression. As a result, SR proteins may be collectively essential, but individually dispensable for constitutive splicing in most cases. On the other hand, individual SR proteins may each control a defined spectrum of substrates via weak splice sites coupled with ESEs and these substrates may be limited in type or in number. Therefore, SR proteins may function as alternative splicing regulators in vivo more extensively than previously appreciated. The challenge is in identifying key alternative splicing events involving specific SR proteins and to link these molecular alterations to defined biological phenotypes.

Regulation of SR Splicing Regulators

SR proteins and SR-related splicing factors are direct effectors in alternative splicing and are likely subject to regulation at the transcriptional and post-translational levels. Additional regulation likely takes place in response to cell signaling events. Regulation of SR proteins and other splicing regulators by signaling is reviewed in the chapter by Lynch in this book. Accordingly, we will focus our discussion on how alternative splicing may be achieved by regulating the SR family of splicing factors. While SR and SR-related proteins are ubiquitously expressed in most tissues and cell types, differential expression of SR proteins has been reported in certain tissues and cell types in response to signaling.[118-122] In general, however, little is known about how SR proteins are regulated at the transcriptional level and about the functional consequences of such regulation on specific alternative splicing events in specific biological pathways. SR proteins have also been found to be auto-regulated or regulated in trans by other SR proteins at the level of alternative splicing.[123-126] These regulatory mechanisms may help maintain homeostasis of SR protein expression in most cell types.

SR proteins are extensively modified by phosphorylation in their RS domains. Several early studies indicated that phosphorylation was essential for SR proteins to function in spliceosome assembly and that dephosphorylation was critical for RNA catalysis within the spliceosome.[127-129] Phosphorylation and dephosphorylation are both required,[130] because it was found that experimental induction of SR protein hyper- and hypo-phosphorylation impaired splicing.[92] However, mutations that mimic hyper- and hypo-phosphorylation of a single SR protein, such as substitution of RS repeats by RE or RG dipeptides in the RS domain, still supported splicing in vitro and complemented SR protein-depleted cells for viability.[112,131] This is likely because a full phosphorylation/dephosphorylation cycle does not have to occur in a single SR protein for each round of the splicing reaction.[129] For instance, a splicing reaction can be accomplished by using a thio-phosphorylated (phosphatase-resistant) SR protein to stimulate initial spliceosome assembly and using another dephosphorylatable SR protein to complete later steps in the splicing reaction.

Because the activity of SR proteins in constitutive splicing is clearly modulated by phosphorylation, it is conceivable that regulated phosphorylation may have a profound influence on alternative splicing. Indeed, overexpression or inhibition of an SR protein-specific kinase has been shown to modulate splice site selection.[92,132-135] The activation of various signal transduction pathways has also been shown to affect alternative splicing via, at least in part, differential phosphorylation of SR proteins.[136,137] However, we are far from understanding how SR protein phosphorylation might affect the activity of SR proteins in constitutive and regulated splicing. While phosphorylation of the RS domain is generally believed to prevent SR proteins from non-specific binding to RNA, the impact varies with respect to RS domain-mediated protein-protein interactions that enhance the interaction in certain cases and suppress the interaction in others.[138,139] Importantly, it is essentially unknown as to which proteins are actually engaging in the interaction with the RS domain of an SR protein within the spliceosome and how such interactions might be influenced by phosphorylation. Moreover, SR proteins are phosphorylated at multiple sites in their RS domains.[140] It is currently unclear whether the activity of SR proteins might be affected by phosphorylation in a context or site-specific manner. Finally, phosphorylation has been shown to regulate the localization of SR proteins[93,141-143] and their recruitment to the transcriptional machinery has been shown to facilitate cotranscriptional splicing in the nucleus.[139,144,145] Because SR proteins are known to affect alternative splicing in a dosage-dependent manner, the impact of phosphorylation on the availability (localization) and targeting efficiency (recruitment) of SR proteins may contribute to the complex pattern of alternative splicing in mammalian cells.

One approach to investigate the regulation of splicing by phosphorylation is to identify and characterize specific kinases and phosphatases involved in the process. To date, several protein kinases have been implicated as SR protein kinases, including SRPKs,[141,143] Clk/Sty,[93,132,134] and Akt.[136,137] The family of SRPK and Clk/Sty kinases catalyzed phosphorylation of SR proteins in multiple sites in the RS domain, but with different substrate specificity.[140,146] It is important to emphasize the fact that these kinases were mostly identified by in vitro kinase assays and their effect on splicing, if any, was only tested in transfected cells. Genetic evidence will be required to firmly establish the enzyme-substrate relationship for all of the reported SR protein kinases. In *Drosophila*, a Clk/Sty-related kinase has been shown to phosphorylate endogenous SR proteins and more importantly, mutations in the kinase altered the sex determination pathway.[147] The SRPK family of kinases was initially identified based on their ability to alter the localization of SR proteins in interphase cells as well as during cell mitosis.[141,143,148] A recent RNAi study showed a major impact of SRPK1 depletion on SR protein phosphorylation in vivo.[149] These observations provide genetic evidence for the involvement of these kinases in SR protein phosphorylation in vivo; how these kinases are involved in the regulation of alternative splicing is an important subject for future studies.

The action of kinases is often counteracted by phosphatases. Unfortunately, phosphatases specifically involved in SR protein dephosphorylation are largely unknown. In vitro, both PP1 and PP2A were able to act on SR proteins and activated splicing.[127,128,150,151] Several PP2A family members have been copurified with spliceosomal components.[152] Intriguingly, a recent study demonstrates the essential role of both PP1 and PP2A phosphatases in the second step of splicing, but their main substrates are U2 and U5 snRNP components, instead of SR proteins, indicating that multiple phosphatases are involved in the splicing regulators and those specific for SR proteins remain to be identified and functionally characterized.[153] In particular, because SRp38 is particularly sensitive to dephosphorylation in response to mitotic transitions and heat shock,[81] it will be of great interest to identify the phosphatase(s) responsible and the potential role of these enzymes in regulated splicing. Interestingly, although alternative splicing is not common in budding yeast, a member of the SRPK family of kinases is conserved in the organism and is responsible for phosphorylation of the SR-related RNA binding protein Npl3p.[154] This action is counteracted by the yeast PP1 family phosphatase Glc7p, suggesting that the mammalian counterpart of Glc7c may function as an SR protein-specific phosphatase.[155]

SR Protein-Regulated Splicing in Development and Disease

As splicing is an essential component of gene expression and a key point in expression regulation, splicing defects have been linked to various diseases in humans.[156,157] Given the role of SR proteins and related splicing factors in alternative splicing and cell growth control, they are primary candidates for causing specific disease phenotypes. Available evidence indicates that SR proteins may be involved in development and disease in several ways. First, they may function as critical regulators of disease-causing genes, such as oncogenes or tumor suppressor genes.[63,158,159] Consistent with this possibility, a recent study showed that the alternative splicing of the *Ron* proto-oncogene was subject to regulation by SF2/ASF and the protein product from an alternatively spliced isoform appeared to contribute directly to the invasive behavior of tumor cells.[63] In knockout mice, SF2/ASF was found to play a critical role in the developmental control of CaMKIIδ alternative splicing in the heart, resulting in differential cellular targeting of the kinase and malfunction in Ca^{++} signaling in cardiomyocytes.[62] Because SR proteins affect alternative splicing in a dosage-dependent manner, it is conceivable that altered expression of SR proteins may manifest the effect by changing the alternative splicing pattern of their target genes, thereby causing specific defects in the regulation of cell proliferation and differentiation. Indeed, altered expression of SR proteins and SR protein-specific kinases has been detected in multiple types of cancer.[160-165]

The second way for SR proteins to act in disease pathways lies in their ability to recognize specific point mutations and small deletions directly in disease-causing genes, thereby manifesting the disease phenotype via the mutation-triggered alternative splicing events.[157] One of the best such examples is the disease gene *SMN* in spinal muscular atrophy (SMA). The *SMN* gene is duplicated in the human genome, but the disease phenotype is only associated with molecular defects in the *SMN1* gene.[166] The reason why *SMN2* is insufficient to complement the defective *SMN1* gene in SMA is because of a point mutation in exon 7 in the *SMN2* gene which converts an ESE to an ESS, thereby causing exon skipping to result in a partly defective gene product.[166-169] These findings illustrate how some silent mutations may be linked to specific diseases because of their impact on the regulatory information embedded in the sequence. Therefore, although SR proteins and SR-related splicing factors have not yet been directly mapped as disease genes, they may play a larger role in the expression of specific disease phenotypes than previously anticipated. This may be one of the major tumor selection mechanisms resulting from an unstable genome due to defects, for example, in the DNA repair pathway.

Concluding Remarks

Despite the significant progress that has improved our understanding of alternative splicing mechanisms and the functional consequences of regulated splicing in development and disease, we are still confronted with a large array of challenges, which may be expressed in the following questions: (1) Why do SR proteins generally recognize exonic splicing enhancers, but not similar sequences in introns? (2) Which protein(s) interact with the RS domains of SR proteins during spliceosome assembly? (3) Why is the RS domain differentially required for constitutive and alternative splicing? (4) What is the molecular basis by which some SR proteins act positively and others act negatively on splicing? (5) To what extent do SR proteins share redundant functions in splicing? (6) How do SR proteins cooperate with other splicing RNA binding proteins to regulate alternative splicing? (7) How are SR proteins regulated in vivo and in response to signals? (8) To what extent does the activity of SR proteins in alternative splicing contribute to their functional requirement in development and cell proliferation control? In this chapter, we have speculated on some of these questions based on the available evidence. Additional experiments that address these biological and mechanistic questions are clearly needed to understand the function and regulation of this important class of splicing regulators in development and disease.

Acknowledgements

We thank Jonathan Hagopian for critical reading of the manuscript. Work in the authors' lab was supported by grants from NIH.

References

1. Fu XD, Maniatis T. Factor required for mammalian spliceosome assembly is localized to discrete regions in the nucleus. Nature 1990; 343:437-441.
2. Fu XD, Maniatis T. Isolation of a complementary DNA that encodes the mammalian splicing factor SC35. Science 1992; 256:535-538.
3. Ge H, Zuo P, Manley JL. Primary structure of the human splicing factor ASF reveals similarities with Drosophila regulators. Cell 1991; 66:373-382.
4. Ge H, Manley JL. A protein factor, ASF, controls cell-specific alternative splicing of SV40 early pre-mRNA in vitro. Cell 1990; 62:25-34.
5. Krainer AR, Conway GC, Kozak D. Purification and characterization of pre-mRNA splicing factor SF2 from HeLa cells. Genes Dev 1990; 4:1158-1171.
6. Krainer AR, Mayeda A, Kozak D et al. Functional expression of cloned human splicing factor SF2: homology to RNA-binding proteins, U1 70K and Drosophila splicing regulators. Cell 1991; 66:383-394.
7. Zahler AM, Lane WS, Stolk JA et al. SR proteins: a conserved family of pre-mRNA splicing factors. Genes Dev 1992; 6:837-847.
8. Krainer AR, Conway GC, Kozak D. The essential pre-mRNA splicing factor SF2 influences 5' splice site selection by activating proximal sites. Cell 1990; 62:35-42.
9. Fu XD, Mayeda A, Maniatis T et al. General splicing factors SF2 and SC35 have equivalent activities in vitro and both affect alternative 5' and 3' splice site selection. Proc Natl Acad Sci USA 1992; 89:11224-11228.
10. Zahler AM, Neugebauer KM, Lane WS et al. Distinct functions of SR proteins in alternative pre-mRNA splicing. Science 1993; 260:219-222.
11. Cavaloc Y, Popielarz M, Fuchs JP et al. Characterization and cloning of the human splicing factor 9G8: a novel 35 kDa factor of the serine/arginine protein family. EMBO J 1994; 13:2639-2649.
12. Fu XD. The superfamily of arginine/serine-rich splicing factors. RNA 1995; 1:663-680.
13. Graveley BR. Sorting out the complexity of SR protein functions. RNA 2000; 6:1197-1211.
14. Boucher L, Ouzounis CA, Enright AJ et al. A genome-wide survey of RS domain proteins. RNA 2001; 7:1693-1701.
15. Ono Y, Ohno M, Shimura Y. Identification of a putative RNA helicase (HRH1), a human homolog of yeast Prp22. Mol Cell Biol 1994; 14:7611-7620.
16. Teigelkamp S, Mundt C, Achsel T et al. The human U5 snRNP-specific 100-kD protein is an RS domain-containing, putative RNA helicase with significant homology to the yeast splicing factor Prp28p. RNA 1997; 3:1313-1326.
17. Ortlepp D, Laggerbauer B, Mullner S et al. The mammalian homologue of Prp16p is overexpressed in a cell line tolerant to Leflunomide, a new immunoregulatory drug effective against rheumatoid arthritis. RNA 1998; 4:1007-1018.
18. Sukegawa J, Blobel G. A putative mammalian RNA helicase with an arginine-serine-rich domain colocalizes with a splicing factor. J Biol Chem 1995; 270:15702-15706.
19. Wu JY, Maniatis T. Specific interactions between proteins implicated in splice site selection and regulated alternative splicing. Cell 1993; 75:1061-1070.
20. Shen H, Green MR. A pathway of sequential arginine-serine-rich domain-splicing signal interactions during mammalian spliceosome assembly. Mol Cell 2004; 16:363-373.
21. Shen H, Kan JL, Green MR. Arginine-serine-rich domains bound at splicing enhancers contact the branchpoint to promote prespliceosome assembly. Mol Cell 2004; 13:367-376.
22. Birney E, Kumar S, Krainer AR. Analysis of the RNA-recognition motif and RS and RGG domains: conservation in metazoan pre-mRNA splicing factors. Nucleic Acids Res 1993; 21:5803-5816.
23. Shen H, Green MR. RS domains contact splicing signals and promote splicing by a common mechanism in yeast through humans. Genes Dev 2006; 20:1755-1765.
24. Cavaloc Y, Bourgeois CF, Kister L et al. The splicing factors 9G8 and SRp20 transactivate splicing through different and specific enhancers. RNA 1999; 5:468-483.
25. Tacke R, Manley JL. The human splicing factors ASF/SF2 and SC35 possess distinct, functionally significant RNA binding specificities. EMBO J 1995; 14:3540-3551.
26. Liu HX, Zhang M, Krainer AR. Identification of functional exonic splicing enhancer motifs recognized by individual SR proteins. Genes Dev 1998; 12:1998-2012.
27. Liu HX, Chew SL, Cartegni L et al. Exonic splicing enhancer motif recognized by human SC35 under splicing conditions. Mol Cell Biol 2000; 20:1063-1071.
28. Schaal TD, Maniatis T. Selection and characterization of pre-mRNA splicing enhancers: identification of novel SR protein-specific enhancer sequences. Mol Cell Biol 1999; 19:1705-1719.
29. Cartegni L, Wang J, Zhu Z et al. ESEfinder: A web resource to identify exonic splicing enhancers. Nucleic Acids Res 2003; 31:3568-3571.

30. Lynch KW, Maniatis T. Assembly of specific SR protein complexes on distinct regulatory elements of the Drosophila doublesex splicing enhancer. Genes Dev 1996; 10:2089-2101.
31. Lynch KW, Maniatis T. Synergistic interactions between two distinct elements of a regulated splicing enhancer. Genes Dev 1995; 9:284-293.
32. Wang Z, Rolish ME, Yeo G et al. Systematic identification and analysis of exonic splicing silencers. Cell 2004; 119:831-845.
33. Fairbrother WG, Yeh RF, Sharp PA et al. Predictive identification of exonic splicing enhancers in human genes. Science 2002; 297:1007-1013.
34. Fairbrother WG, Yeo GW, Yeh R et al. RESCUE-ESE identifies candidate exonic splicing enhancers in vertebrate exons. Nucleic Acids Res 2004; 32:W187-190.
35. Zhang XH, Leslie CS, Chasin LA. Computational searches for splicing signals. Methods 2005; 37:292-305.
36. Zhang XH, Chasin LA. Computational definition of sequence motifs governing constitutive exon splicing. Genes Dev 2004; 18:1241-1250.
37. Zahler AM, Roth MB. Distinct functions of SR proteins in recruitment of U1 small nuclear ribonucleoprotein to alternative 5' splice sites. Proc Natl Acad Sci USA 1995; 92:2642-2646.
38. Kohtz JD, Jamison SF, Will CL et al. Protein-protein interactions and 5'-splice-site recognition in mammalian mRNA precursors. Nature 1994; 368:119-124.
39. Zuo P, Manley JL. The human splicing factor ASF/SF2 can specifically recognize pre-mRNA 5' splice sites. Proc Natl Acad Sci USA 1994; 91:3363-3367.
40. Wang Z, Hoffmann HM, Grabowski PJ. Intrinsic U2AF binding is modulated by exon enhancer signals in parallel with changes in splicing activity. RNA 1995; 1:21-35.
41. Li Y, Blencowe BJ. Distinct factor requirements for exonic splicing enhancer function and binding of U2AF to the polypyrimidine tract. J Biol Chem 1999; 274:35074-35079.
42. Graveley BR, Hertel KJ, Maniatis T. The role of U2AF35 and U2AF65 in enhancer-dependent splicing. RNA 2001; 7:806-818.
43. Zuo P, Maniatis T. The splicing factor U2AF35 mediates critical protein-protein interactions in constitutive and enhancer-dependent splicing. Genes Dev 1996; 10:1356-1368.
44. Hertel KJ, Maniatis T. Serine-arginine (SR)-rich splicing factors have an exon-independent function in pre-mRNA splicing. Proc Natl Acad Sci USA 1999; 96:2651-2655.
45. Boukis LA, Liu N, Furuyama S et al. Ser/Arg-rich protein-mediated communication between U1 and U2 small nuclear ribonucleoprotein particles. J Biol Chem 2004; 279:29647-29653.
46. Fu XD, Maniatis T. The 35-kDa mammalian splicing factor SC35 mediates specific interactions between U1 and U2 small nuclear ribonucleoprotein particles at the 3' splice site. Proc Natl Acad Sci USA 1992; 89:1725-1729.
47. Stark JM, Bazett-Jones DP, Herfort M et al. SR proteins are sufficient for exon bridging across an intron. Proc Natl Acad Sci U S A 1998; 95:2163-2168.
48. Blencowe BJ, Issner R, Nickerson JA et al. A coactvator of pre-mRNA splicing. Gene & Dev. 1998; 12:996-1009.
49. Blencowe BJ, Bauren G, Eldridge G et al. The SRm160/300 splicing coactivator subunites. RNA 200; 6:111-120.
50. Zhu J, Mayeda A, Krainer AR. Exon identity established through differential antagonism between exonic splicing silencer-bound hnRNP A1 and enhancer-bound SR proteins. Mol Cell 2001; 8:1351-1361.
51. Kan JL, Green MR. Pre-mRNA splicing of IgM exons M1 and M2 is directed by a juxtaposed splicing enhancer and inhibitor. Genes Dev 1999; 13:462-471.
52. Reed R, Maniatis T. A role for exon sequences and splice-site proximity in splice-site selection. Cell 1986; 46:681-690.
53. Mayeda A, Krainer AR. Regulation of alternative pre-mRNA splicing by hnRNP A1 and splicing factor SF2. Cell 1992; 68:365-375.
54. Wang Z, Xiao X, Van Nostrand E et al. General and specific functions of exonic splicing silencers in splicing control. Mol Cell 2006; 23:61-70.
55. Eperon IC, Makarova OV, Mayeda A et al. Selection of alternative 5' splice sites: role of U1 snRNP and models for the antagonistic effects of SF2/ASF and hnRNP A1. Mol Cell Biol 2000; 20:8303-8318.
56. Ibrahim el C, Schaal TD, Hertel KJ et al. Serine/arginine-rich protein-dependent suppression of exon skipping by exonic splicing enhancers. Proc Natl Acad Sci USA 2005; 102:5002-5007.
57. Chandler SD, Mayeda A, Yeakley JM et al. RNA splicing specificity determined by the coordinated action of RNA recognition motifs in SR proteins. Proc Natl Acad Sci USA 1997; 94:3596-3601.
58. Caceres JF, Krainer AR. Functional analysis of pre-mRNA splicing factor SF2/ASF structural domains. EMBO J 1993; 12:4715-4726.
59. Zuo P, Manley JL. Functional domains of the human splicing factor ASF/SF2. EMBO J 1993; 12:4727-4737.

60. Zhu J, Krainer AR. Pre-mRNA splicing in the absence of an SR protein RS domain. Genes Dev 2000; 14:3166-3178.
61. Kanopka A, Muhlemann O, Akusjarvi G. Inhibition by SR proteins of splicing of a regulated adenovirus pre-mRNA. Nature 1996; 381:535-538.
62. Xu X, Yang D, Ding JH et al. ASF/SF2-regulated CaMKIIdelta alternative splicing temporally reprograms excitation-contraction coupling in cardiac muscle. Cell 2005; 120:59-72.
63. Ghigna C, Giordano S, Shen H et al. Cell motility is controlled by SF2/ASF through alternative splicing of the Ron protooncogene. Mol Cell 2005; 20:881-890.
64. Simard MJ, Chabot B. SRp30c is a repressor of 3' splice site utilization. Mol Cell Biol 2002; 22:4001-4010.
65. Gallego ME, Gattoni R, Stevenin J et al. The SR splicing factors ASF/SF2 and SC35 have antagonistic effects on intronic enhancer-dependent splicing of the beta-tropomyosin alternative exon 6A. EMBO J 1997; 16:1772-1784.
66. Jumaa H, Nielsen PJ. The splicing factor SRp20 modifies splicing of its own mRNA and ASF/SF2 antagonizes this regulation. EMBO J 1997; 16:5077-5085.
67. Lemaire R, Winne A, Sarkissian M et al. SF2 and SRp55 regulation of CD45 exon 4 skipping during T-cell activation. Eur J Immunol 1999; 29:823-837.
68. ten Dam GB, Zilch CF, Wallace D et al. Regulation of alternative splicing of CD45 by antagonistic effects of SR protein splicing factors. J Immunol 2000; 164:5287-5295.
69. Watermann DO, Tang Y, Zur Hausen A et al. Splicing factor Tra2-beta1 is specifically induced in breast cancer and regulates alternative splicing of the CD44 gene. Cancer Res 2006; 66:4774-4780.
70. Goren A, Ram O, Amit M et al. Comparative analysis identifies exonic splicing regulatory sequences—The complex definition of enhancers and silencers. Mol Cell 2006; 22:769-781.
71. Barnard DC, Li J, Peng R et al. Regulation of alternative splicing by SRrp86 through coactivation and repression of specific SR proteins. RNA 2002; 8:526-533.
72. Li J, Barnard DC, Patton JG. A unique glutamic acid-lysine (EK) domain acts as a splicing inhibitor. J Biol Chem 2002; 277:39485-39492.
73. Li J, Hawkins IC, Harvey CD et al. Regulation of alternative splicing by SRrp86 and its interacting proteins. Mol Cell Biol 2003; 23:7437-7447.
74. Zhang WJ, Wu JY. Functional properties of p54 a novel SR protein active in constitutive and alternative splicing. Mol Cell Biol 1996; 16:5400-5408.
75. Wu JY, Kar A, Kuo D et al. SRp54 (SFRS11), a regualtor for tau exon 10 alternative splicing identified by an expression cloning strategy. Mol Cell Biol 2006; 26:6739-6747.
76. Cowper AE, Caceres JF, Mayeda A et al. Serine-arginine (SR) protein-like factors that antagonize authentic SR proteins and regulate alternative splicing. J Biol Chem 2001; 276:48908-48914.
77. Yang L, Embree LJ, Hickstein DD. TLS-ERG leukemia fusion protein inhibits RNA splicing mediated by serine-arginine proteins. Mol Cell Biol 2000; 20:3345-3354.
78. Komatsu M, Kominami E, Arahata K et al. Cloning and characterization of two neural-salient serine/arginine-rich (NSSR) proteins involved in the regulation of alternative splicing in neurones. Genes Cells 1999; 4:593-606.
79. Fushimi K, Osumi N, Tsukahara T. NSSRs/TASRs/SRp38s function as splicing modulators via binding to pre-mRNAs. Genes Cells 2005; 10:531-541.
80. Shin C, Manley JL. The SR protein SRp38 represses splicing in M phase cells. Cell 2002; 111:407-417.
81. Shin C, Feng Y, Manley JL. Dephosphorylated SRp38 acts as a splicing repressor in response to heat shock. Nature 2004; 427:553-558.
82. Shin C, Kleiman FE, Manley JL. Multiple properties of the splicing repressor SRp38 distinguish it from typical SR proteins. Mol Cell Biol 2005; 25:8334-8343.
83. Tacke R, Tohyama M, Ogawa S et al. Human Tra2 proteins are sequence-specific activators of pre-mRNA splicing. Cell 1998; 93:139-148.
84. Pacheco TR, Moita LF, Gomes AQ et al. RNAi Knockdown of hU2AF35 Impairs Cell Cycle Progression and Modulates Alternative Splicing of Cdc25 Transcripts. Mol Biol Cell 2006.
85. Park JW, Parisky K, Celotto AM et al. Identification of alternative splicing regulators by RNA interference in Drosophila. Proc Natl Acad Sci USA 2004; 101:15974-15979.
86. Valcarcel J, Singh R, Zamore PD et al. The protein Sex-lethal antagonizes the splicing factor U2AF to regulate alternative splicing of transformer pre-mRNA. Nature 1993; 362:171-175.
87. Lou H, Helfman DM, Gagel RF et al. Polypyrimidine tract-binding protein positively regulates inclusion of an alternative 3'-terminal exon. Mol Cell Biol 1999; 19:78-85.
88. Izquierdo JM, Majos N, Bonnal S et al. Regulation of Fas alternative splicing by antagonistic effects of TIA-1 and PTB on exon definition. Mol Cell 2005; 19:475-484.

89. Sharma S, Falick AM, Black DL. Polypyrimidine tract binding protein blocks the 5' splice site-dependent assembly of U2AF and the prespliceosomal E complex. Mol Cell 2005; 19:485-496.
90. Sarkissian M, Winne A, Lafyatis R. The mammalian homolog of suppressor-of-white-apricot regulates alternative mRNA splicing of CD45 exon 4 and fibronectin IIICS. J Biol Chem 1996; 271:31106-31114.
91. Adams DJ, van der Weyden L, Mayeda A et al. ZNF265—a novel spliceosomal protein able to induce alternative splicing. J Cell Biol 2001; 154:25-32.
92. Prasad J, Colwill K, Pawson T et al. The protein kinase Clk/Sty directly modulates SR protein activity: both hyper- and hypophosphorylation inhibit splicing. Mol Cell Biol 1999; 19:6991-7000.
93. Colwill K, Pawson T andrews B et al. The Clk/Sty protein kinase phosphorylates SR splicing factors and regulates their intranuclear distribution. EMBO J 1996; 15:265-275.
94. Ko TK, Kelly E, Pines J. CrkRS: a novel conserved Cdc2-related protein kinase that colocalises with SC35 speckles. J Cell Sci 2001; 114:2591-2603.
95. Dellaire G, Makarov EM, Cowger JJ et al. Mammalian PRP4 kinase copurifies and interacts with components of both the U5 snRNP and the N-CoR deacetylase complexes. Mol Cell Biol 2002; 22:5141-5156.
96. Even Y, Durieux S, Escande ML et al. CDC2L5, a Cdk-like kinase with RS domain, interacts with the ASF/SF2-associated protein p32 and affects splicing in vivo. J Cell Biochem 2006.
97. Dickinson LA, Edgar AJ, Ehley J et al. Cyclin L is an RS domain protein involved in pre-mRNA splicing. J Biol Chem 2002; 277:25465-25473.
98. Chen HH, Wang YC, Fann MJ. Identification and characterization of the CDK12/cyclin L1 complex involved in alternative splicing regulation. Mol Cell Biol 2006; 26:2736-2745.
99. Yang L, Li N, Wang C et al. Cyclin L2, a novel RNA polymerase II-associated cyclin, is involved in pre-mRNA splicing and induces apoptosis of human hepatocellular carcinoma cells. J Biol Chem 2004; 279:11639-11648.
100. de Graaf K, Hekerman P, Spelten O et al. Characterization of cyclin L2, a novel cyclin with an arginine/serine-rich domain: phosphorylation by DYRK1A and colocalization with splicing factors. J Biol Chem 2004; 279:4612-4624.
101. Lim SR, Hertel KJ. Commitment to splice site pairing coincides with A complex formation. Mol Cell 2004; 15:477-483.
102. Query CC, Konarska MM. Suppression of multiple substrate mutations by spliceosomal prp8 alleles suggests functional correlations with ribosomal ambiguity mutants. Mol Cell 2004; 14:343-354.
103. Longman D, Johnstone IL, Caceres JF. Functional characterization of SR and SR-related genes in Caenorhabditis elegans. EMBO J 2000; 19:1625-1637.
104. Gabut M, Mine M, Marsac C et al. The SR protein SC35 is responsible for aberrant splicing of the E1alpha pyruvate dehydrogenase mRNA in a case of mental retardation with lactic acidosis. Mol Cell Biol 2005; 25:3286-3294.
105. Massiello A, Chalfant CE. SRp30a (ASF/SF2) regulates the alternative splicing of caspase-9 pre-mRNA and is required for ceramide-responsiveness. J Lipid Res 2006; 47:892-897.
106. Wang J, Xiao SH, Manley JL. Genetic analysis of the SR protein ASF/SF2: interchangeability of RS domains and negative control of splicing. Genes Dev 1998; 12:2222-2233.
107. Li X, Manley JL. Inactivation of the SR protein splicing factor ASF/SF2 results in genomic instability. Cell 2005; 122:365-378.
108. Li X, Wang J, Manley JL. Loss of splicing factor ASF/SF2 induces G2 cell cycle arrest and apoptosis, but inhibits internucleosomal DNA fragmentation. Genes Dev 2005; 19:2705-2714.
109. Wang HY, Xu X, Ding JH et al. SC35 plays a role in T-cell development and alternative splicing of CD45. Mol Cell 2001; 7:331-342.
110. Jumaa H, Wei G, Nielsen PJ. Blastocyst formation is blocked in mouse embryos lacking the splicing factor SRp20. Curr Biol 1999; 9:899-902.
111. Ding JH, Xu X, Yang D et al. Dilated cardiomyopathy caused by tissue-specific ablation of SC35 in the heart. EMBO J 2004; 23:885-896.
112. Lin S, Xiao R, Sun P et al. Dephosphorylation-dependent sorting of SR splicing factors during mRNP maturation. Mol Cell 2005; 20:413-425.
113. Allemand E, Gattoni R, Bourbon HM et al. Distinctive features of Drosophila alternative splicing factor RS domain: implication for specific phosphorylation, shuttling and splicing activation. Mol Cell Biol 2001; 21:1345-1359.
114. Lemaire R, Prasad J, Kashima T et al. Stability of a PKCI-1-related mRNA is controlled by the splicing factor ASF/SF2: a novel function for SR proteins. Genes Dev 2002; 16:594-607.
115. Venables JP, Bourgeois CF, Dalgliesh C et al. Up-regulation of the ubiquitous alternative splicing factor Tra2beta causes inclusion of a germ cell-specific exon. Hum Mol Genet 2005; 14:2289-2303.

116. Schaal TD, Maniatis T. Multiple distinct splicing enhancers in the protein-coding sequences of a constitutively spliced pre-mRNA. Mol Cell Biol 1999; 19:261-273.
117. Hertel KJ, Maniatis T. The function of multisite splicing enhancers. Mol Cell 1998; 1:449-455.
118. Hanamura A, Caceres JF, Mayeda A et al. Regulated tissue-specific expression of antagonistic pre-mRNA splicing factors. RNA 1998; 4:430-444.
119. Du K, Leu JI, Peng Y et al. Transcriptional up-regulation of the delayed early gene HRS/SRp40 during liver regeneration. Interactions among YY1, GA-binding proteins and mitogenic signals. J Biol Chem 1998; 273:35208-35215.
120. Shinozaki A, Arahata K, Tsukahara T. Changes in pre-mRNA splicing factors during neural differentiation in P19 embryonal carcinoma cells. Int J Biochem Cell Biol 1999; 31:1279-1287.
121. Jumaa H, Guenet JL, Nielsen PJ. Regulated expression and RNA processing of transcripts from the Srp20 splicing factor gene during the cell cycle. Mol Cell Biol 1997; 17:3116-3124.
122. Chiu Y, Ouyang P. Loss of Pnn expression attenuates expression levels of SR family splicing factors and modulates alternative pre-mRNA splicing in vivo. Biochem Biophys Res Commun 2006; 341:663-671.
123. Jumaa H, Nielsen PJ. Regulation of SRp20 exon 4 splicing. Biochim Biophys Acta 2000; 1494:137-143.
124. Lejeune F, Cavaloc Y, Stevenin J. Alternative splicing of intron 3 of the serine/arginine-rich protein 9G8 gene. Identification of flanking exonic splicing enhancers and involvement of 9G8 as a trans-acting factor. J Biol Chem 2001; 276:7850-7858.
125. Sureau A, Gattoni R, Dooghe Y et al. SC35 autoregulates its expression by promoting splicing events that destabilize its mRNAs. EMBO J 2001; 20:1785-1796.
126. Stoilov P, Daoud R, Nayler O et al. Human tra2-beta1 autoregulates its protein concentration by influencing alternative splicing of its pre-mRNA. Hum Mol Genet 2004; 13:509-524.
127. Mermoud JE, Cohen P, Lamond AI. Ser/Thr-specific protein phosphatases are required for both catalytic steps of pre-mRNA splicing. Nucleic Acids Res 1992; 20:5263-5269.
128. Mermoud JE, Cohen PT, Lamond AI. Regulation of mammalian spliceosome assembly by a protein phosphorylation mechanism. EMBO J 1994; 13:5679-5688.
129. Xiao SH, Manley JL. Phosphorylation-dephosphorylation differentially affects activities of splicing factor ASF/SF2. EMBO J 1998; 17:6359-6367.
130. Cao W, Jamison SF, Garcia-Blanco MA. Both phosphorylation and dephosphorylation of ASF/SF2 are required for pre-mRNA splicing in vitro. RNA 1997; 3:1456-1467.
131. Cazalla D, Zhu J, Manche L et al. Nuclear export and retention signals in the RS domain of SR proteins. Mol Cell Biol 2002; 22:6871-6882.
132. Hartmann AM, Rujescu D, Giannakouros T et al. Regulation of alternative splicing of human tau exon 10 by phosphorylation of splicing factors. Mol Cell Neurosci 2001; 18:80-90.
133. Muraki M, Ohkawara B, Hosoya T et al. Manipulation of alternative splicing by a newly developed inhibitor of Clks. J Biol Chem 2004; 279:24246-24254.
134. Duncan PI, Stojdl DF, Marius RM et al. In vivo regulation of alternative pre-mRNA splicing by the Clk1 protein kinase. Mol Cell Biol 1997; 17:5996-6001.
135. Cardinali B, Cohen PT, Lamond AI. Protein phosphatase 1 can modulate alternative 5' splice site selection in a HeLa splicing extract. FEBS Lett 1994; 352:276-280.
136. Blaustein M, Pelisch F, Tanos T et al. Concerted regulation of nuclear and cytoplasmic activities of SR proteins by AKT. Nat Struct Mol Biol 2005; 12:1037-1044.
137. Patel NA, Kaneko S, Apostolatos HS et al. Molecular and genetic studies imply Akt-mediated signaling promotes protein kinase CbetaII alternative splicing via phosphorylation of serine/arginine-rich splicing factor SRp.40 J Biol Chem 2005; 280:14302-14309.
138. Xiao SH, Manley JL. Phosphorylation of the ASF/SF2 RS domain affects both protein-protein and protein-RNA interactions and is necessary for splicing. Genes Dev 1997; 11:334-344.
139. Yeakley JM, Tronchere H, Olesen J et al. Phosphorylation regulates in vivo interaction and molecular targeting of serine/arginine-rich pre-mRNA splicing factors. J Cell Biol 1999; 145:447-455.
140. Velazquez-Dones A, Hagopian JC, Ma CT et al. Mass spectrometric and kinetic analysis of ASF/SF2 phosphorylation by SRPK1 and Clk/Sty. J Biol Chem 2005; 280:41761-41768.
141. Gui JF, Lane WS, Fu XD. A serine kinase regulates intracellular localization of splicing factors in the cell cycle. Nature 1994; 369:678-682.
142. Koizumi J, Okamoto Y, Onogi H et al. The subcellular localization of SF2/ASF is regulated by direct interaction with SR protein kinases (SRPKs). J Biol Chem 1999; 274:11125-11131.
143. Wang HY, Lin W, Dyck JA et al. SRPK2: a differentially expressed SR protein-specific kinase involved in mediating the interaction and localization of pre-mRNA splicing factors in mammalian cells. J Cell Biol 1998; 140:737-750.

144. Misteli T, Caceres JF, Clement JQ et al. Serine phosphorylation of SR proteins is required for their recruitment to sites of transcription in vivo. J Cell Biol 1998; 143:297-307.
145. Misteli T, Spector DL. RNA polymerase II targets pre-mRNA splicing factors to transcription sites in vivo. Mol Cell 1999; 3:697-705.
146. Colwill K, Feng LL, Yeakley JM et al. SRPK1 and Clk/Sty protein kinases show distinct substrate specificities for serine/arginine-rich splicing factors. J Biol Chem 1996; 271:24569-24575.
147. Du C, McGuffin ME, Dauwalder B et al. Protein phosphorylation plays an essential role in the regulation of alternative splicing and sex determination in Drosophila. Mol Cell 1998; 2:741-750.
148. Gui JF, Tronchere H, Chandler SD et al. Purification and characterization of a kinase specific for the serine- and arginine-rich pre-mRNA splicing factors. Proc Natl Acad Sci USA 1994; 91:10824-10828.
149. Hayes GM, Carrigan PE, Beck AM et al. Targeting the RNA splicing machinery as a novel treatment strategy for pancreatic carcinoma. Cancer Res 2006; 66:3819-3827.
150. Kanopka A, Muhlemann O, Petersen-Mahrt S et al. Regulation of adenovirus alternative RNA splicing by dephosphorylation of SR proteins. Nature 1998; 393:185-187.
151. Misteli T, Spector DL. Serine/threonine phosphatase 1 modulates the subnuclear distribution of pre-mRNA splicing factors. Mol Biol Cell 1996; 7:1559-1572.
152. Tran HT, Ulke A, Morrice N et al. Proteomic characterization of protein phosphatase complexes of the mammalian nucleus. Mol Cell Proteomics 2004; 3:257-265.
153. Shi Y, Reddy B, Manley JL. PP1/PP2A phosphatases are required for the second step of Pre-mRNA splicing and target specific snRNP proteins. Mol Cell 2006; 23:819-829.
154. Siebel CW, Feng L, Guthrie C et al. Conservation in budding yeast of a kinase specific for SR splicing factors. Proc Natl Acad Sci USA 1999; 96:5440-5445.
155. Gilbert W, Guthrie C. The Glc7p nuclear phosphatase promotes mRNA export by facilitating association of Mex67p with mRNA. Mol Cell 2004; 13:201-212.
156. Faustino NA, Cooper TA. Pre-mRNA splicing and human disease. Genes Dev 2003; 17:419-437.
157. Cartegni L, Chew SL, Krainer AR. Listening to silence and understanding nonsense: exonic mutations that affect splicing. Nat Rev Genet 2002; 3:285-298.
158. Colapietro P, Gervasini C, Natacci F et al. NF1 exon 7 skipping and sequence alterations in exonic splice enhancers (ESEs) in a neurofibromatosis 1 patient. Hum Genet 2003; 113:551-554.
159. Guil S, Gattoni R, Carrascal M et al. Roles of hnRNP A1, SR proteins and p68 helicase in c-H-ras alternative splicing regulation. Mol Cell Biol 2003; 23:2927-2941.
160. Ghigna C, Moroni M, Porta C et al. Altered expression of heterogenous nuclear ribonucleoproteins and SR factors in human colon adenocarcinomas. Cancer Res 1998; 58:5818-5824.
161. Stickeler E, Kittrell F, Medina D et al. Stage-specific changes in SR splicing factors and alternative splicing in mammary tumorigenesis. Oncogene 1999; 18:3574-3582.
162. Maeda T, Furukawa S. Transformation-associated changes in gene expression of alternative splicing regulatory factors in mouse fibroblast cells. Oncol Rep 2001; 8:563-566.
163. Fischer DC, Noack K, Runnebaum IB et al. Expression of splicing factors in human ovarian cancer. Oncol Rep 2004; 11:1085-1090.
164. Zerbe LK, Pino I, Pio R et al. Relative amounts of antagonistic splicing factors, hnRNP A1 and ASF/SF2, change during neoplastic lung growth: implications for pre-mRNA processing. Mol Carcinog 2004; 41:187-196.
165. Pind MT, Watson PH. SR protein expression and CD44 splicing pattern in human breast tumours. Breast Cancer Res Treat 2003; 79:75-82.
166. Lorson CL, Hahnen E androphy EJ et al. A single nucleotide in the SMN gene regulates splicing and is responsible for spinal muscular atrophy. Proc Natl Acad Sci USA 1999; 96:6307-6311.
167. Cartegni L, Hastings ML, Calarco JA et al. Determinants of exon 7 splicing in the spinal muscular atrophy genes, SMN1 and SMN2. Am J Hum Genet 2006; 78:63-77.
168. Kashima T, Manley JL. A negative element in SMN2 exon 7 inhibits splicing in spinal muscular atrophy. Nat Genet 2003; 34:460-463.
169. Cartegni L, Krainer AR. Disruption of an SF2/ASF-dependent exonic splicing enhancer in SMN2 causes spinal muscular atrophy in the absence of SMN1. Nat Genet 2002; 30:377-384.

Chapter 8

hnRNP Proteins and Splicing Control

Rebeca Martinez-Contreras, Philippe Cloutier, Lulzim Shkreta, Jean-François Fisette, Timothée Revil and Benoit Chabot*

Abstract

Proteins of the heterogeneous nuclear ribonucleoparticles (hnRNP) family form a structurally diverse group of RNA binding proteins implicated in various functions in metazoans. Here we discuss recent advances supporting a role for these proteins in precursor-messenger RNA (pre-mRNA) splicing. Heterogeneous nuclear RNP proteins can repress splicing by directly antagonizing the recognition of splice sites, or can interfere with the binding of proteins bound to enhancers. Recently, hnRNP proteins have been shown to hinder communication between factors bound to different splice sites. Conversely, several reports have described a positive role for some hnRNP proteins in pre-mRNA splicing. Moreover, cooperative interactions between bound hnRNP proteins may encourage splicing between specific pairs of splice sites while simultaneously hampering other combinations. Thus, hnRNP proteins utilize a variety of strategies to control splice site selection in a manner that is important for both alternative and constitutive pre-mRNA splicing.

Introduction

Pre-RNA splicing is the process by which introns are removed to produce functional mRNAs. Alternative splicing occurs when exons, introns, or portions thereof, are differentially included to produce distinct mRNAs from an identical pre-mRNA. Alternative splicing is part of the expression program of more than 70% of all human genes,[1] and is a major source of protein diversity in plants and metazoans. The extraordinary contribution that alternative splicing makes to the biology of complex organs like the mammalian brain is only beginning to be appreciated (reviewed in ref. 2). A full description of the impact of alternative splicing on biology will require characterizing the function of a huge collection of splice isoforms produced in different tissues. Moreover, splicing mutations often cause human diseases and alternative splicing profiles are frequently altered in cancer and other pathological conditions (reviewed in ref. 3,4; refer to chapter by Orengo and Cooper). Consequently, there is much hope that monitoring global splicing profiles can provide rich and informative signatures for disease identity, progression and prognosis.[5,6]

An equally important task is to understand how authentic splice sites are selected and how these decisions are modulated in different cell types under a variety of environmental cues. Considerable progress has been accomplished towards understanding the basic principles of the general splicing reaction (reviewed in ref. 7; refer to chapter by Matlin and Moore). The removal of introns requires the recognition of four splicing signals: the 5' splice site, the branch site, the polypyrimidine tract (PPT) and the AG at the 3' splice site. The U1 snRNP uses RNA complementarity to interact with the 5' splice site. The U2AF heterodimer binds to the downstream 3' splice site: U2AF65 binds to the PPT and U2AF35 recognizes the AG. U2AF binding can be stimulated by U1 snRNP bound to a downstream 5' splice site, a process termed exon definition (Fig. 1A). U2AF stabilizes the interaction of factors bound at the branch site (i.e., initially the protein SF1 and later

*Corresponding Author: Benoit Chabot—Université de Sherbrooke, Sherbrooke, Québec, Canada. Email: benoit.chabot@usherbrooke.ca

Alternative Splicing in the Postgenomic Era, edited by Benjamin J. Blencowe and Brenton R. Graveley. ©2007 Landes Bioscience and Springer Science+Business Media.

U2 snRNP). Once the initial recognition of splicing signals has occurred, the ends of the intron are brought in close proximity to achieve intron definition (Fig. 1B), an event that will lead to complete spliceosome assembly and, ultimately, intron removal.

Many splicing signals are not strong enough to insure their efficient recognition by the basal splicing machinery. Moreover, alternative splicing often requires the activation of specific splice sites by enhancer elements (for a review see ref. 8). These elements are generally located in the vicinity of the splicing signals and their activity has been associated mostly with a group of RNA binding proteins named the SR proteins (refer to chapter by Lin and Fu). The binding of SR proteins to exonic enhancer elements can stimulate the binding of U2AF, U2 snRNP or U1 snRNP.[9-13]

The control of alternative splicing also entails the specific repression of splice sites. This involves silencer elements that are typically bound by hnRNP proteins, but can also be recognized by other factors including SR proteins (for a review see ref. 8).

In the following sections, we will examine the contributions that the major hnRNP proteins make to both general and alternative splicing, with an emphasis on their mechanisms of action. This compilation may underscore common strategies used by different groups of hnRNP proteins to control splicing decisions.

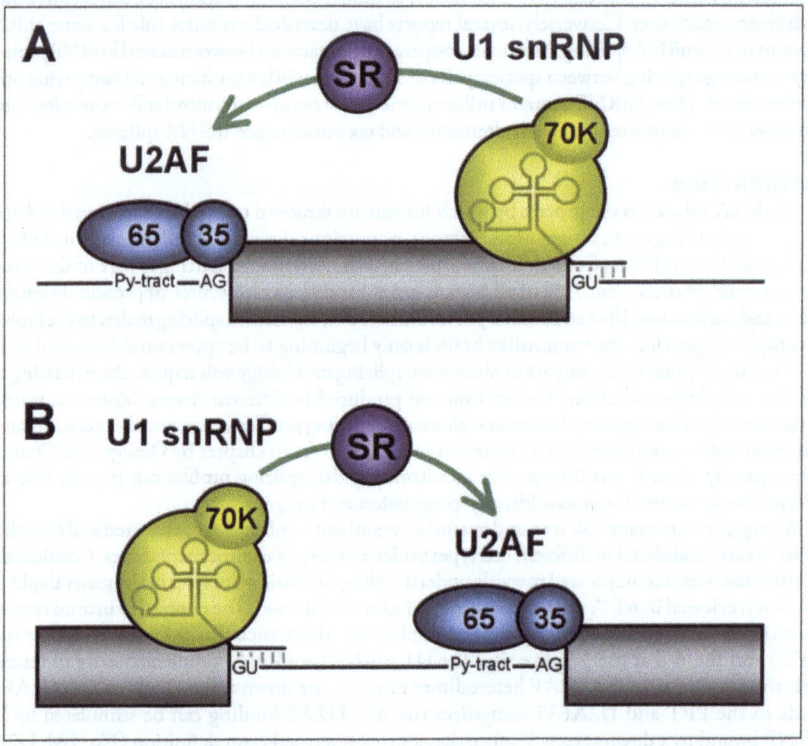

Figure 1. Exon and intron definition. A) Exon definition. The binding of U1 snRNP to a 5' splice site can stimulate the binding of U2AF at the 3' splice site in a process that requires SR proteins to interact with the U1 snRNP protein 70 kDa and the U2AF35-U2AF65 heterodimer. B) Intron definition. When the 5' splice site and the 3' splice site have been recognized by the U1 snRNP and U2AF, respectively, SR proteins are presumed to mediate communication between both splice sites to define the intron.[223]

A Brief History of hnRNP Proteins

Nascent pre-mRNAs emerging from RNA polymerase II transcription elongation complexes interact with a variety of proteins and snRNPs. The term hnRNP proteins was initially coined to indicate the group of proteins associated with high molecular weight nuclear RNA (heterogeneous nuclear RNA or pre-mRNA). hnRNP proteins were then operationally defined as the most abundant proteins associated with 40S particles that are generated from mild ribonuclease digestion of hnRNA-containing material sedimenting between 40S and 250S (reviewed in ref. 14). In rapidly dividing cells, the major core hnRNP proteins are present at high intranuclear concentrations (low millimolar levels, nearly as abundant as histones).[15] This definition later evolved to specify proteins that could be crosslinked by ultraviolet (UV) irradiation to hnRNA in vivo. The subsequent use of monoclonal antibodies and two-dimensional gel electrophoresis established our current definition of mammalian hnRNP proteins as a set of approximately 24 distinct polypeptides that stably associate with hnRNA.[16] The fruit fly *Drosophila melanogaster* contains ten to fifteen major hnRNP proteins.[17-19] Because the definition of an hnRNP protein does not reflect a precise biochemical activity or function other than being associated with newly synthesized precursor mRNAs, it may not be too surprising that hnRNP proteins display considerable structural diversity (for a detailed description, see Fig. 2). Antibodies specific to any one hnRNP protein recover the complete set of major hnRNP proteins,[20] an observation which could be interpreted as evidence that hnRNP proteins bind RNA nonspecifically. However, more recent experiments have revealed a nonrandom deposition of hnRNPs proteins on nascent pre-mRNAs,[21,22] suggesting sequence-specific binding.

Early on, it became clear that some hnRNP proteins associate with specific components of the splicing machinery. However, a functional role for hnRNP proteins in the general splicing reaction has remained controversial because the incubation of model pre-mRNAs in nuclear extracts leads to the formation of H complexes that are not substrates for the assembly of a functional spliceosome (refer to chapter by Matlin and Moore).[23,24] These complexes contain the same set of hnRNP proteins that bind to hnRNA in vivo, although their protein composition varies depending on the identity of the pre-mRNA. These reports were followed by the landmark demonstration of the capacity of hnRNP A1 to modulate 5' splice site selection in vitro.[25] As we will see below, many hnRNP proteins have now been shown to modulate splice site selection.

Although hnRNP and hnRNP-like proteins have been described in yeast and plants, this chapter will focus on metazoan hnRNP proteins. Moreover, although several of the major hnRNP proteins have been implicated in a variety of biological processes including telomere biogenesis, polyadenylation, translation, RNA editing and mRNA stability (reviewed in ref. 26), we will limit our discussion to their role in splicing.

hnRNP Proteins, Splicing and Alternative Splicing

A function in splicing has been documented or proposed for more than half of the major hnRNP proteins. Other hnRNP proteins like N, S and T remain poorly characterized and there is no evidence that the hnRNP proteins A0, A3, A/B, D, DL and U play a role in splicing.

hnRNP A1 and A2—hnRNP F, H, H' and 2H9

Although the A and H groups of hnRNP proteins display distinctive structural features (see Fig. 2), we have combined them here because they can bind to RNA sequences with GGG-containing motifs (see Table 1) and can control splicing through similar mechanisms (see below). The hnRNP A1 and A2 proteins contain two RNA recognition motifs (RRMs) followed by a glycine-rich domain (GRD) characterized by repeats of an arginine-glycine-glycine motif termed the RGG box (Fig. 2). The hnRNP proteins F, H, H' and 2H9 contain RRM-related domains that are loosely similar to the consensus RRM (quasi-RRMs or qRRMs) and regions rich in glycine, tyrosine and arginine (GYR) or glycine- and tyrosine-rich (GY) (Fig. 2).

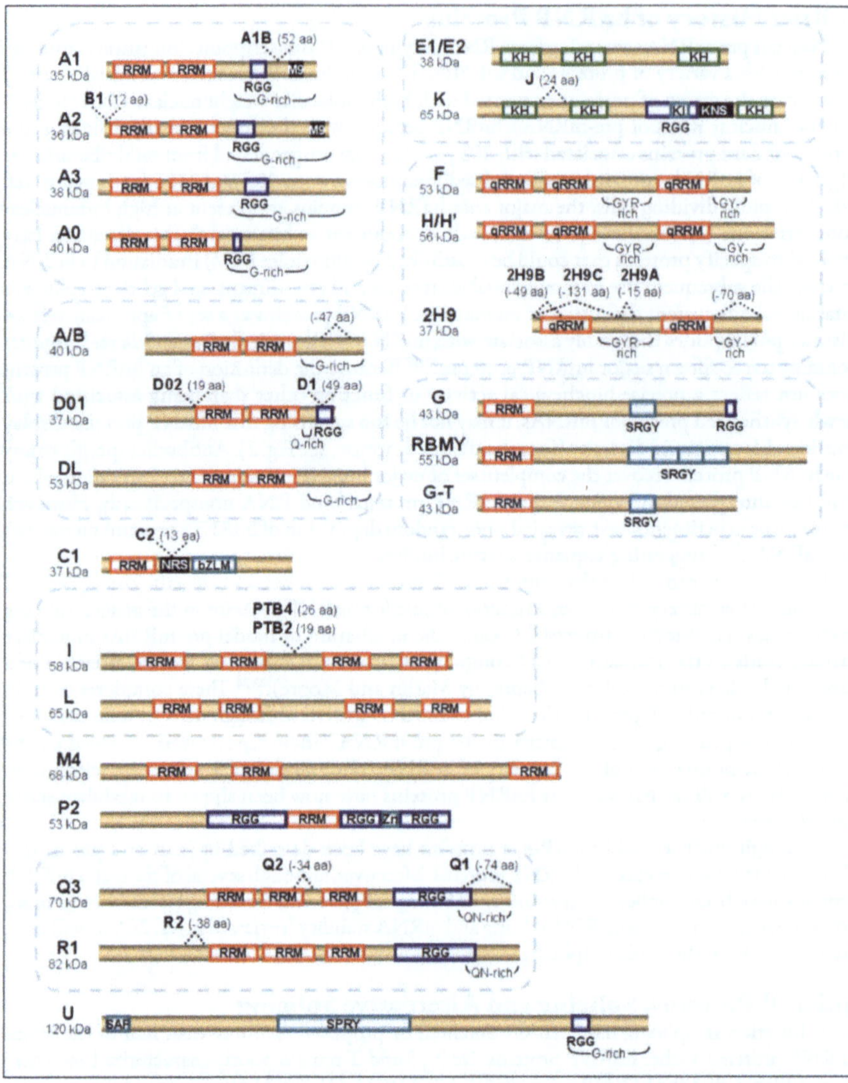

Figure 2. Structure of the major hnRNP proteins and splice isoforms. The molecular weights of each protein are indicated below the name. The type of RNA Recognition Motifs (RRM, quasi (q)RRM or KH) is indicated as well as the presence of auxiliary domains, such as the glycine-rich domain (G) and RGG motifs. The hnRNP proteins are grouped (dashed rectangles) according to global protein identity. hnRNP A1, A2, A3, A0, A/B and D proteins have been grouped according to the similarity between RRMs.[224] hnRNP A1 and A2 share 68% amino acid identity (80% for the RRMs). Isoforms of hnRNP A1 and A2 proteins are generated by alternative splicing to produce the A1B and B1 isoforms. A1B (initially designated B2) contains an additional 52 aa (53 aa in mouse) in the C-terminal glycine-rich region).[225,226] B1 contains an additional 12 aa near the amino terminus.[227] The A1 and A2 proteins contain a nucleo-cytoplasmic export signal (M9).[228] The hnRNP D group is composed of four main isoforms. *Figure legend continued on next page.*

Figure 2, continued. D01 (also known as AUF1 or p37) is the shortest while D02 (p40) and D1 (p42) result from the inclusion of exon 2 and exon 7, respectively. D2 (p45) is not shown, but is produced from the inclusion of both exons. hnRNP C1 and C2 are generated by alternative splicing and the C2 isoform contains an additional 13 aa.[227,229] The C1 and C2 proteins contain an N-terminal RRM,[209,230,231] and a basic leucine zipper (bZIP)-like motif (bZLM).[232-234] The RRM plays a minimal role in the overall affinity of hnRNP C for RNA. A highly basic 40 aa domain preceding the leucine zipper motif provides the high-affinity for RNA.[232,233] hnRNP C does not shuttle between the nucleus and cytoplasm due to a nuclear retention signal (NRS).[235] The family of hnRNP E proteins (also called CPs or PCBPs for poly rC binding proteins) includes at least two principal proteins (E1 and E2) that contain, along with hnRNP K, three KH domains.[228,236-239] The hnRNP E1 gene is believed to have arisen from the retrotransposition of a fully processed E2 mRNA. α-CP3 (PCBP3) and α-CP4 (PCBP4) are paralogs derived from gene duplication events.[240] There are four known alternatively spliced isoforms of hnRNP K,[241] and hnRNP J is believed to represent another isoform created by the exclusion of a central exon.[242] hnRNP K possesses a nuclear shuttling domain (KNS) and a KI (K Interactive) domain that is responsible for interactions with various proteins.[243] hnRNP F, H, H' and 2H9 are encoded by different genes. An amino acid alignment reveals 96% similarity between H and H', 78% between H and F and 75% between H' and F. hnRNP H, H' and F contain three quasi RNA recognition motifs (qRRMs). A region rich in glycine, tyrosine and arginine (GYR) is located between qRRM2 and qRRM3. These proteins also contain a C-terminal glycine- and tyrosine-rich (GY) domain.[244] The hnRNP 2H9 gene is alternatively spliced to produce 6 different isoforms (2H9, 2H9A, 2H9B, 2H9C, 2H9D and 2H9E) that all contain qRRM3 but varying flanking domains.[245] 2H9A, 2H9B and 2H9C are truncated as shown. 2H9D and 2H9E lack a 70 aa region at the C-terminus but differ in their N-termini. 2H9D is truncated like 2H9C, but 2H9E is further shortened by 6 aa. GRSF-1 is a GGG binding protein that has extensive similarity to members of the hnRNP H family of proteins.[246] The hnRNP G group includes three closely related proteins: hnRNP G (RBMX), RBMY and hnRNP G-T. hnRNP G contains a single N-terminal RRM domain.[247] The human genome contains several additional retrotransposed copies of *RBMX* derived from processed cDNAs.[248,249] Two of these human *RBMX* retrogenes are almost full length and their conceptual proteins display 92-96% identity to RBMX and are expressed in testis (RBMXL1) and brain (RBMXL9).[249] The human RBMY protein has four tandemly repeated copies of a 37 aa motif enriched in serine, arginine, glycine and tyrosine (SRGY), which is also found in some SR proteins and that functions as a protein interaction domain.[166] The SRGY motif is present only once in hnRNP G and G-T and is less well conserved.[250,251] Although the *HNRNPG-T* gene is retroposon-derived and does not contain any introns, it is highly conserved (73% identity with hnRNP G).[252] RBMY and hnRNP G share 76% similarity and 60% identity.[247,253] hnRNP I is also known as PTB or PTBP1. It is composed of four RRMs that diverge from the canonical motif but that are similar to the RRMs of hnRNP L.[254] PTB is alternatively spliced to produce a variety of isoforms that can affect the length of the interdomain linker. PTBP2 and PTBP4 differ from the most studied isoform PTBP1 by the insertion of 19 and 26 aa, respectively, between RRM2 and RRM3. These isoforms induce varying levels of repression on *α-TM* exon 3, with PTBP4 being the strongest repressor, followed by PTBP2 and PTBP1.[140] Neural-specific (n)PTB is a paralogue of PTB displaying 73% identity with PTB (80% for the RRM domains).[255] hnRNP L displays 55% similarity to hnRNP I/PTB (and 32% identity between the RRMs of the two proteins).[254,256] There are four isoforms of hnRNP M (M1-M4) and possibly two more that are present in LH-hnRNP complexes.[217,257] The exact splicing events that are responsible for these differences are unknown. hnRNP P2, also named TLS/FUS, possesses one RRM and several RGG motifs. However, RNA binding seems to be mediated by the zinc finger domain (Zn).[258] Three isoforms of hnRNP Q have been identified,[158] with Q3 displaying 83% identity to hnRNP R. Q3 is composed of three RRMs, two NLSs and an RGG box. Q1 lacks part of the RGG box and the second NLS. Q2 lacks 34 aa from the second RRM. The shorter isoforms of hnRNP Q lack the glutamine- and asparagine-rich (QN) C-terminus present in full-length hnRNP R and hnRNP Q. The two known splice isoforms for hnRNP R are shown. hnRNP U,[221] also known as Scaffold Attachment Factor A (SAF-A) can bind to RNA through an RGG domain located at the C-terminus. It also possesses an SP1a and ryanodine receptor (SPRY) homology domain of unknown function and a scaffold-associated region (SAR)-specific bipartite DNA binding domain (SAF) capable of binding specific DNA sequences.[259]

Table 1. Binding specificity of hnRNP proteins

hnRNP Protein	RNA Binding Site	References
A1/A2	UAGRG$^{A}/_{U}$	48,207
	UAGG	208
C	U-rich	209,210
	AGUAUUUUUGUGGA	211
D	AU-rich	212
	AUUUA	213
	UUAG	214
E	poly C	215
F/H/H'/2H9	GGGA	72
	GGGC (H and H')	
G	AAGU	103
I or PTB	UCUU(C)	114
	CUCUCU	121
	Y_6CUUCUCUCUY_6	120
K	poly C	216
	AUC$_{3/4}$($^{U}/_{A}$)($^{A}/_{U}$)	
L	CA-rich	142
M	poly G & poly U	217
	purine-rich (Hrp59)	152
P2	GGUG	218
Q	U- and AU-rich	219
R	poly U	220
U	poly G	221
	poly A	222

R, purine; Y, pyrimidine.

hnRNP A1 and H Proteins Are Splicing Repressors

hnRNP A1 can alter 5' splice site choice,[27,29] and promote exon skipping.[25,30-46] hnRNP A2 displays a similar function in alternative splicing.[31,32,47,48] Members of the hnRNP H group also promote alternative 5' splice site selection, splicing inhibition and exon skipping.[49,55] A high-throughput search for exon motifs that repress exon inclusion has uncovered groups of sequences that likely represent binding sites for hnRNP A1 and H proteins.[56] G-tracts, potential binding sites for hnRNP A1, A2 and H, are enriched in pseudo-exons and may be responsible for their repression.[57]

The mechanisms by which hnRNP A1 and H repress a splice site have been examined in a few cases. In *tat* exon 2 of HIV-1, the binding of hnRNP H near the 3' splice site interferes with the binding of U2AF35.[51] Likewise, the binding of hnRNP H to two intronic G-runs inactivates a 3' splice site in the *cystathionine β-synthase* pre-mRNA.[58] hnRNP H binding sites located near or overlapping a 5' splice site can also repress splicing by directly inhibiting U1 snRNP binding.[59] Thus, the steric occlusion of splicing signals by hnRNP H proteins can explain splicing inhibition in these cases. In a related fashion, the *Drosophila* hnRNP A1 ortholog hrp48 associates with the P-element-specific inhibitor (PSI) and U1 snRNP to form a complex that hinders spliceosome assembly at the authentic downstream 5' splice site.[60]

As shown for the HIV *tat* exon 2,[61] the binding of hnRNP A1 to exonic sequences can antagonize the interaction of the SR protein SC35 to an overlapping site (Fig. 3A). The ability of hnRNP A1 to interfere with the binding of SR or SR-related proteins has been described in many pre-mRNAs.[36,38,41,62-65] Likewise, the binding of hnRNP F and H to a silencer element in exon 2

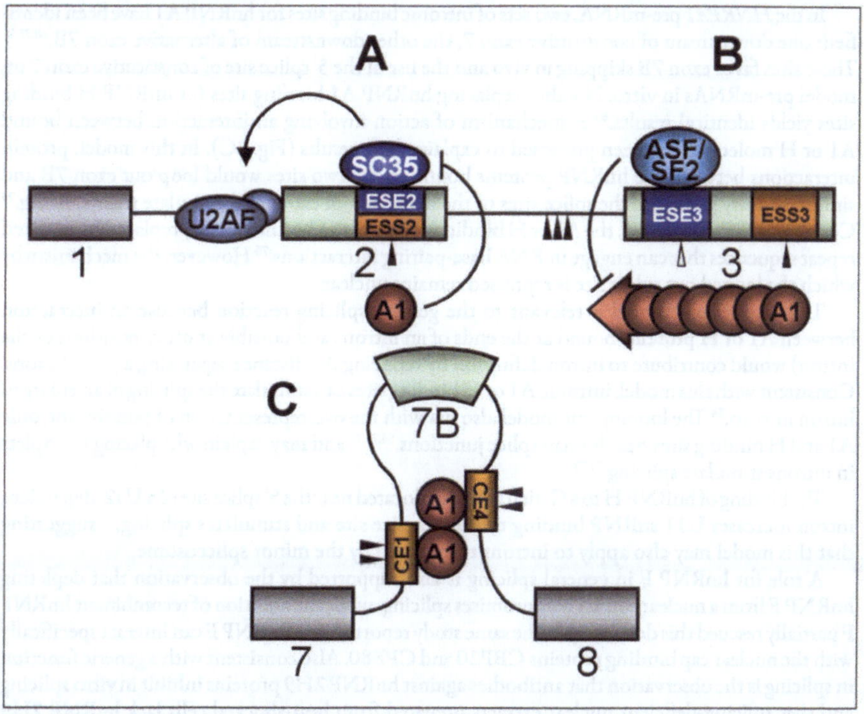

Figure 3. Different mechanisms of splicing control used by the hnRNP A1 protein. A) Occlusion. hnRNP A1 binds to a splicing silencer element in the HIV-1 *tat* exon 2 (ESS2) that overlaps with an enhancer element (ESE2) bound by SC35. The enhancer element stimulates U2AF binding. A1 antagonizes the binding of SC35 to inhibit the use of the 3' splice site.[54] B) Nucleation. hnRNP A1 binds to a high-affinity site in HIV exon 3 (ESS3). Propagation of A1 binding has been proposed to occlude 3' splice site usage. However, the binding of ASF/SF2 to ESE3 would block the nucleation process.[27] C) Looping out. hnRNP A1 molecules bound in the introns flanking alternative exon 7B in the *HNRP* A1 pre-mRNA are interacting to loop out exon 7B and approximate exons 7 and 8.[47,67] Black arrows indicate hnRNP A1 binding sites, the white arrow indicates the binding site for ASF/SF2.

of the rat α-*tropomyosin* antagonizes the positive effect of the SR protein 9G8.[66] When binding sites for hnRNP A1 and a SR protein are overlapping, a single A1 molecule may be sufficient to abrogate SR protein binding, which in turn should prevent the efficient recruitment of factors at splicing signals. When these sites are not overlapping however, cooperative binding of hnRNP A1 molecules along the exon has been proposed as the mechanism that elicit splicing repression.[27] This nucleation, or propagation, model (Fig. 3B) was suggested to explain how A1 interferes with U2AF binding at the 3' splice site of HIV-1 exons.[34,67]

Positive Roles for hnRNP A1 and H Proteins

The hnRNP F and H proteins play dual roles in the alternative splicing of *c-src*. On the one hand, they are part of an intronic splicing enhancer complex with KSRP and nPTB (see below) that promotes the neuro-specific inclusion of exon N1. In nonneuronal cells, hnRNP F and H associate with KSRP and PTB to form a complex that represses the inclusion of exon N1.[68,69] In several pre-mRNAs, the binding of hnRNP F or H in an intron activates a nearby 5' splice site.[52,70,71] The binding of hnRNP H to an exon can stabilize the interaction of U1 snRNP at a 5' splice site.[72]

In the *HNRPA1* pre-mRNA, two sets of intronic binding sites for hnRNP A1 have been identified: one downstream of constitutive exon 7, the other downstream of alternative exon 7B.[48,73,74] These sites favor exon 7B skipping in vivo and the use of the 5' splice site of constitutive exon 7 on model pre-mRNAs in vitro. Notably, replacing hnRNP A1 binding sites for hnRNP H binding sites yields identical results.[54] A mechanism of action involving an interaction between bound A1 or H molecules has been proposed to explain these results (Fig. 3C). In this model, protein interactions between the hnRNP proteins bound to the two sites would loop out exon 7B and simultaneously juxtapose the splice sites of the distal pair of exons and stimulate their splicing.[75] Consistent with the model, the A1 or H binding elements can be functionally replaced by inverted repeat sequences that can engage in RNA base-pairing interactions.[76] However, the mechanism by which the looped out splice site is repressed remains unclear.

The looping out model is relevant to the general splicing reaction because an interaction between A1 or H proteins bound at the ends of an intron (and possibly at other positions of the intron) would contribute to intron definition by reducing the distance separating a pair of exons. Consistent with this model, intronic A1 or H binding sites can stimulate the splicing of an enlarged intron in vitro.[54] The looping out model also fits with the overrepresentation of putative intronic A1 and H binding sites near human splice junctions,[54,74,77] and may explain why placing G-triplets in introns stimulate splicing.[77,78]

The binding of hnRNP H to a G-rich sequence located near the 5' splice site of a U12-dependent intron increases U11 snRNP binding to the 5' splice site and stimulates splicing,[79] suggesting that this model may also apply to introns recognized by the minor spliceosome.

A role for hnRNP F in general splicing is also supported by the observation that depleting hnRNP F from a nuclear extract compromises splicing, while the addition of recombinant hnRNP F partially rescued this deficiency.[80] The same study reported that hnRNP F can interact specifically with the nuclear cap binding proteins CBP20 and CPB80. Also consistent with a generic function in splicing is the observation that antibodies against hnRNP 2H9 proteins inhibit in vitro splicing and that splicing-deficient nuclear extracts prepared from heat-shocked cells lack hnRNP 2H9 proteins,[81] as seen previously for hnRNP M.[82]

hnRNP proteins have also been studied in *Drosophila*. Overexpression of hrp36, which is similar to hnRNP A1, has limited effects on alternative splicing.[83] Likewise, RNAi depletion of hrp36 affects the alternative splicing of *Dscam* exon 4 but not of exon 17, or of alternative exons in the *para* and *dAdar* pre-mRNAs.[84] Moreover, a recent microarray study indicates that RNAi depletion of hrp48, another hnRNP protein related to hnRNP A1, affects approximately 3% of alternative splicing events.[85] In contrast to what might have been expected based on the described antagonism between mammalian hnRNP A1 and SR proteins,[8] only a small fraction of the splicing events regulated by hrp36 are also regulated by SR proteins dASF/SF2 and B52.[85] Although these results cannot rule out indirect effects, they suggest that the impact of the *Drosophila* hnRNP A1-related proteins on regulated splicing is limited to a subset of pre-mRNAs.

hnRNP C

The mammalian *HNRPC* gene produces the splice isoforms C1 and C2. Monoclonal antibodies against hnRNP C or immunodepletion of hnRNP C inhibit splicing in vitro.[86,87] The exact function of hnRNP C in splicing is not yet clear. First, while hnRNP C binds to pre-mRNA in early splicing complexes, this interaction decreases at later steps of spliceosome assembly.[88] Second, although hnRNP C proteins bind to the U-rich PPT of pre-mRNAs,[89,90] mutations that prevent hnRNP C binding do not affect splicing.[91] On the other hand, mutations in the β-globin PPT that improve hnRNP C binding also decreased splicing efficiency.[92]

Although hnRNP C proteins could play a role in alternative splicing, the fact that these proteins are not required for cell viability,[93] makes it unlikely that they play an essential, nonredundant role in constitutive splicing.

The KH-Domain Containing hnRNP K and hnRNP E Proteins

Although mammalian hnRNP K can interact with splicing factors such as hnRNP A1, A2, G, L, D and U proteins, Sam68, TAF15, SRp20, YB-1 and 9G8,[94,95] only two studies have associated hnRNP K with splicing activity.[96,97] In *Drosophila*, the splicing regulatory protein PSI can be considered an hnRNP K-like protein because it contains KH domains. PSI is expressed only in somatic tissues and functions to repress splicing of the P-element *transposase* pre-mRNA, restricting P-element transposition to the germline.[60] PSI functions by binding to a pseudo-5' splice site sequence and forming a complex with hrp48 and U1 snRNP that inhibits splicing at the downstream authentic 5' splice site.[98]

For the hnRNP E proteins (E1 and E2), only hnRNP E2 has been shown to function in pre-mRNA splicing. hnRNP E binds to an exonic splicing silencer element located in variable exon 4 in the human transmembrane tyrosine phosphatase *CD45* pre-mRNA (see hnRNP L).[99] Although high levels of E2 repress splicing of this exon, this activity was not strictly dependent on the exonic splicing silencer and E2 continued to bind to an inactive version of the element. Thus, the precise role of hnRNP E2 in the splicing control of *CD45* remains to be understood.

Members of the hnRNP G Group

Three proteins make up the human hnRNP G group: hnRNP G or RBMX, RBMY and hnRNP G-T. RBMY and hnRNP G-T are expressed in germ cells while hnRNP G is more broadly distributed in somatic cells. hnRNP G and RBMY stimulate the in vivo inclusion of exon 7 of the *SMN2* gene.[100] This stimulation is most likely mediated by an interaction of the SRGY repeats in these hnRNP proteins with the SR protein Tra2β bound to a splicing enhancer in exon 7.[100-102] In contrast, Tra2β and hnRNP G can antagonize one another and this was first uncovered in the SK exon of the human α_s*-tropomyosin* pre-mRNA. Whereas Tra2β binds to a splicing enhancer in the SK exon to stimulate its inclusion, hnRNP G binds to a different sequence in the same exon to antagonize the Tra2β interaction.[103] A similar antagonism between hnRNP G and Tra2β was observed with a *dystrophin* pseudo-exon.[103] hnRNP G can also antagonize the activity of SRrp86, a SR-related protein that modulates the activity of other SR proteins.[104,105] While SRrp86 stimulates the inclusion of *CD44* exon v5, hnRNP G inhibits its inclusion and antagonizes the effect of SRrp86 in a dose-dependent manner.[106]

hnRNP I or PTB

hnRNP I or PTB functions as a splicing repressor in essentially all systems that have been studied so far (reviewed in refs. 107-109). The only case of apparent positive regulation by PTB concerns an alternative exon in the *CGRP* pre-mRNA, but this may be an indirect consequence of PTB negatively regulating a nearby exon.[109,110]

In the α*-tropomyosin* (α*-TM*) pre-mRNA, exons 2 and 3 are mutually exclusive due to the extreme proximity of the branch point of exon 3 to the 5' splice site of exon 2.[111] Stronger splice sites and a longer PPT lead to inclusion of exon 3 as the default pathway in most cells.[112] However, in smooth muscle cells, splicing of exon 3 is repressed by elements located both upstream (URE) and downstream (DRE) of this exon.[113] As shown in Figure 4A, PTB represses exon 3 by binding to UCUU sites present in the PPT at the 3' splice site and in two overlapping copies in the DRE.[114] PTB appears to prevent spliceosome assembly by interfering with U2AF binding at the PPT,[115] although this model does not take into account the importance of the additional downstream PTB binding sites.[116] The need for multiple PTB binding sites has also been observed in other model pre-mRNAs.[117-120]

In the *c-src* pre-mRNA (Fig. 4B), inclusion of the N1 exon is repressed in all but neuronal cells through the cooperative binding of PTB to CUCUCU elements upstream and downstream of this exon.[121,122] Although the position of high-affinity PTB binding sites flanking the alternative N1 exon resembles that of the α*-TM* exon 3, repression of N1 splicing by PTB goes beyond the mere masking of splice sites. In *c-src*, PTB is an essential part of a multiprotein complex comprising hnRNP F,[68] hnRNP H,[69] KSRP[123] associating downstream of exon N1 in a cluster of regulatory elements called the downstream control sequence (DCS).[124,125] PTB does not affect

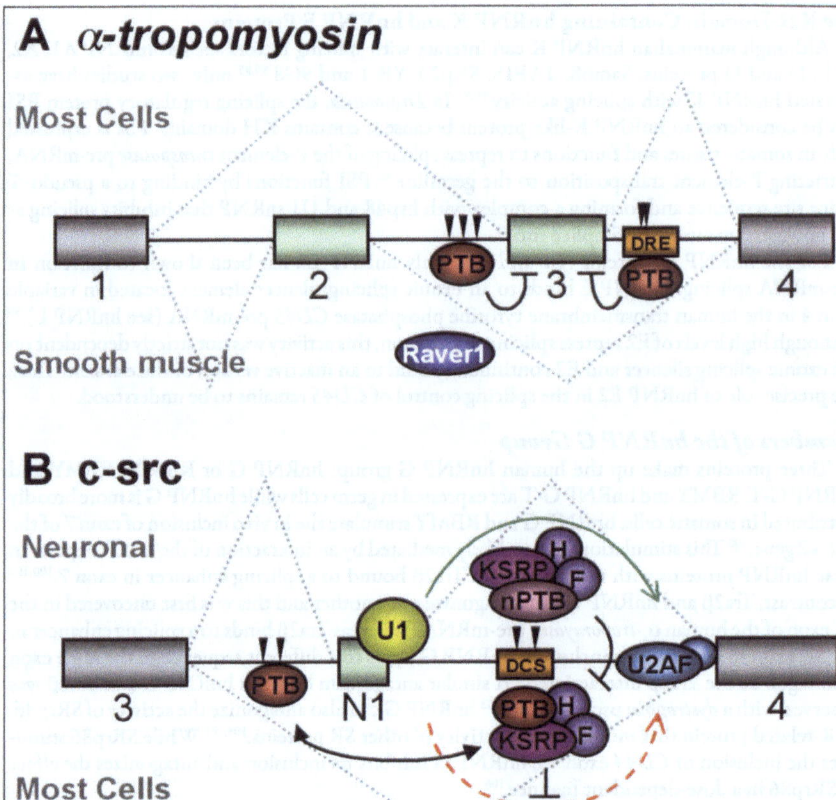

Figure 4. Repression of alternative splicing by hnRNP I/PTB. A) α-*TM* exon 3 is included in most cells, except in smooth muscle cells where it is repressed in a PTB-dependent manner. It is not yet known how tissue-specific repression is achieved. Repression is stronger when PTB associates with the Raver1 protein. B) Cooperative binding of PTB around the *c-src* alternative exon N1 represses N1 inclusion in most cells by inhibiting definition of the downstream intron. In neurons, a less repressive complex is formed when nPTB replaces PTB in the complex comprising KSRP, hnRNP F and hnRNP H. Arrows indicate PTB/nPTB binding sites.

U1 binding to the exon N1 5' splice site, but instead hinders U2AF binding at the 3' splice site of the downstream exon and interferes with the interactions that occur between U1 snRNP and U2AF, as part of the intron definition events.[121,126] In neuronal cells, a brain-specific paralog of PTB, nPTB (also known as brPTB or PTBP2)[127] can replace PTB in the complex bound to the DCS, thereby relieving the PTB-mediated repression. Consistent with this view, nPTB displays enhanced binding to CUCUCU elements but is less repressive than PTB.[127] Additional tissue-specific PTB paralogs have been identified (ROD1[128] and the rodent smPTB[129]), but their function remains unknown.

In the case of Fas receptor pre-mRNA, PTB binds to a single site in alternative exon 6. This binding does not alter the initial binding of U1 snRNP and U2AF to the flanking splicing signals but prevents the U1-mediated stimulation of U2AF65 binding that normally occurs during exon definition.[130]

To explain observations pointing to cooperation between PTB molecules bound in different introns,[122] models invoking propagative binding and looping out were proposed.[109,120] Propagative binding now appears less likely given that in both *c-src* and *Fas*, PTB does not prevent the basal interaction of splicing factors but rather compromises subsequent steps in exon or intron definition. In the looping out model, PTB molecules bound to distinct sites would interact with one another in a manner similar to what has been proposed for hnRNP A1 and F/H (Fig. 2). Alternatively and in accord with the possibility of a strictly monomeric nature of PTB,[131,132] a single PTB molecule could bind to each of these sites through its multiple RRMs.[133] Although looping out an alternative exon may not affect the initial recognition of splice sites, the conformational changes that it imposes on the pre-mRNA may explain why PTB antagonizes intron definition and stimulates splicing between the distal exons.

Notably, tissue or developmentally regulated proteins have been identified as modulators of PTB activity. For example, changes in the ratio between PTB and the antagonistic CELF proteins (CUG-BP and ETR-3-like factors) are involved in the regulation of *cTnT* and α-*actinin* splicing.[134,135] The same can be said of PTB and TIA proteins in the splicing of the *MYPT1* pre-mRNA during smooth muscle differentiation.[136] Also intriguing is the contribution of the hnRNP-like protein Raver1 which interacts with PTB,[137] and whose overexpression enhances the skipping of α-*TM* exon 3 in a PTB-dependent manner.[138] A recent study suggests that Raver1 may promote the looping of large portions of a pre-mRNA by interacting with distinct PTB molecules bound at different sites.[139] Raver2 is a novel member of this emerging family of PTB-associating proteins but it remains to be seen if Raver2 can act as a corepressor with PTB.

Interestingly, the *PTB* pre-mRNA is alternatively spliced to produce a variety of isoforms,[140,141] one of which is promoted by PTB itself to yield a mRNA that is subject to NMD degradation.[141] This appears to be a common consequence of alternative splicing and is the focus the of the chapter by Lareau et al.

hnRNP L

The first link between hnRNP L and pre-mRNA splicing control was established when hnRNP L was shown to function as a specific activator of human *endothelial nitric oxide synthase* (*eNOS*) splicing.[142] Intron 13 of human *eNOS* contains a polymorphic CA-rich region of 14 to 44 repeats located 80 nt downstream from the 5' splice site. hnRNP L binds specifically to these CA-repeats and activates the upstream 5' splice site. CA repeats are prevalent throughout the human genome and hnRNP L stimulates the splicing of many other exons that contain these repeats in close proximity to their 5' splice site.[143] CA repeats and presumably hnRNP L, can even activate cryptic intronic 5' splice sites when the repeats are positioned appropriately downstream of this site.[143] A similar function for hnRNP L was uncovered in the polymorphic mouse *Itga2* pre-mRNA.[144] In this case, the repeat is located approximately 250 nt downstream from the 5' splice site. The 21 CA repeats characteristic of haplotype 1 are associated with more efficient splicing than the 6 CA repeats of haplotype 2. It is now known that noncoding variations between individuals can have important consequences on pre-mRNA splicing and contribute to human disease (refer to chapter by Orengo and Cooper).

hnRNP L also controls alternative splicing of the human *CD45* pre-mRNA by binding to exonic elements in variable exons 4, 5 and 6 (refer to chapter by Lynch). Antigen activation promotes exon skipping by activating the exonic splicing silencers.[99,145,146] While hnRNP L, PTB and hnRNP E2 associate with the splicing silencer of exon 4 in *CD45*, only hnRNP L binding is decreased by mutations that inactivate the silencer element.[99] The role of hnRNP L was confirmed by showing that hnRNP L promotes exon skipping in a silencer-dependent manner and that its depletion increases exon inclusion. hnRNP L also binds to the silencer elements found in exons 5 and 6 and contributes to their activity.[147]

The molecular mechanisms by which hnRNP L mediates the activity of the intronic CA repeats in *eNOS* remain to be clarified. However, in the case of the exonic repressor elements in *CD45*, a recent report indicates that repression does not occur by inhibiting the binding of splicing factors to exon 4. Rather, hnRNP L favors the formation of an ATP-dependent complex containing

U1 and U2 snRNPs. This complex forms across exon 4 and it inhibits the formation of splicing complex B.[148]

hnRNP L has been shown to interact with PTB.[149] Interestingly, this interaction may be mediated by a PTB-binding motif also present in Raver1.[139] Moreover, the PTB portion involved in this interaction is part of the minimal repressor domain of PTB.[150] Thus, hnRNP L and PTB cooperate to control the splicing of some pre-mRNAs.

hnRNP M

The first hint of a role for hnRNP M in splicing came from the observation that antibodies specific for hnRNP M could inhibit splicing in vitro.[82] Curiously, splicing-deficient nuclear extracts prepared from heat-shocked cells contain no hnRNP M, a situation that can be explained by noting that heat shock promotes a strong association of hnRNP M with the insoluble nuclear matrix fraction.[82] Normally however, isoforms of hnRNP M associate with early spliceosome complexes.[151] The orthologous protein in *Chironomus tentans*, Hrp59, is recruited cotranscriptionally to nascent pre-mRNAs through specific purine-rich sequences that resemble exonic splicing enhancers.[152]

hnRNP P

At least three proteins with similar electrophoretic properties have been identified as hnRNP P.[16] However, only hnRNP P2 has been analyzed so far and it corresponds to the oncogenic TLS/FUS protein. P2 can enhance distal 5' splice site selection on the adenovirus *E1a* pre-mRNA.[153] Although it can interact with hnRNP A1, SF1 and SR proteins,[154-156] the mechanism by which hnRNP P2 modulates splice site selection is unknown.

hnRNP Q and R

hnRNP Q (also called GRY-RBP, NSAP1 or SYNCRIP) has been found in purified functional spliceosomes,[157] and has been described as an SMN-interacting protein.[158] hnRNP Q also interacts with pre-mRNAs, intron-containing splicing intermediates and mature mRNAs produced during in vitro splicing. Depleting hnRNP Q and the structurally related hnRNP R from nuclear extracts reduces splicing efficiency, while adding back purified proteins restores splicing activity.[158] The specific function of hnRNP Q in alternative splicing has not yet been investigated.

Splicing Lessons from the Study of hnRNP Proteins

Different Strategies for Controlling Splicing

Although much remains to be known about the mechanism by which hnRNP proteins function, some basic mechanistic principles are emerging. The binding of an hnRNP protein to an exon rarely improves the interaction of spliceosomal components to the adjacent splice sites. More often, hnRNP binding blocks splicing by occluding the interaction of a spliceosomal component to an overlapping or adjacent site (Fig. 5A). For example, hnRNP A1, PTB and hnRNP H interfere with the binding of SR proteins, U2AF and U1 snRNP, respectively, to the pre-mRNA.

A modified version of the occlusion mechanism involves the propagation of hnRNP protein along the pre-mRNA. This appears to occur when one or several high-affinity sites are flanked by adjacent sites of lower affinity (Fig. 5B). Cooperative interactions would then create a zone of local repression that would prevent the interaction of spliceosomal components to splice sites located far from the high affinity hnRNP protein binding site(s) though it remains unclear how far propagation can extend from the site of high-affinity.[34] However, it is noteworthy that many functionally relevant binding sites for hnRNP proteins exist at least in pairs. Cooperative binding of an hnRNP protein to multiple sites would increase the affinity and specificity of these proteins that have rather relaxed binding preferences (reviewed in ref. 159).

A different way in which hnRNP proteins can alter splicing decisions is by modulating interactions between splicing factors that occur after the initial events of spliceosome assembly. For example, in some pre-mRNAs, PTB does not interfere with the initial binding of spliceosomal

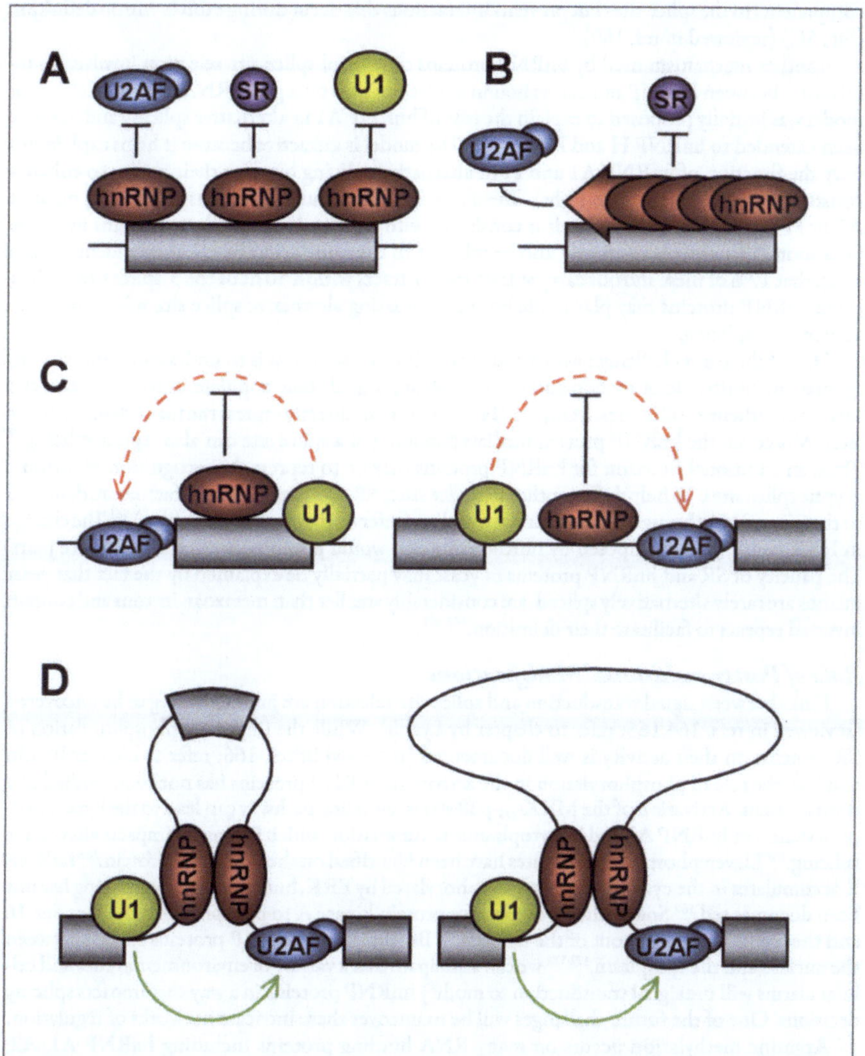

Figure 5. Strategies used by hnRNP proteins to control splice site selection. A) Binding of hnRNP proteins close to splicing signals can occlude the binding of factors to these signals. Likewise, the binding of hnRNP proteins to exonic sequences can antagonize the interactions of SR proteins with exonic splicing enhancer sequences. B) The propagation of hnRNP binding from a site of high-affinity located in an exon (in the case of hnRNP A1) may occlude the binding of splicing factors. A similar situation has been proposed to occur when PTB binds to an intron sequence. C) As documented for PTB (see text), an hnRNP can inhibit exon definition when bound to an exon (left), or can inhibit intron definition when bound to an intron (right). D) As proposed for hnRNP A1, H and PTB, interactions between bound hnRNP proteins may loop out portions of a pre-mRNA to promote exon skipping (left) or stimulate intron definition (right).

components to the splice sites but prevents interactions that occur during exon or intron definition (Fig. 5C) (reviewed in ref. 160).

Another mechanism used by hnRNP proteins to control splice site selection involves an interaction between hnRNP molecules bound at distinct sites on a pre-mRNA (Fig. 5D, left). This model was initially proposed to explain the role of hnRNP A1 in alternative splicing and has now been extended to hnRNP H and PTB.[54,74,109] The model is attractive because it helps explain not only the function of hnRNP A1 and H in alternative splicing but also their ability to enhance constitutive splicing (Fig. 5D, right). The sheer number of human introns containing putative A1 or H binding sites at both ends is consistent with a general role for these proteins in intron definition.[54] This mechanism may also be relevant to the minor class of U12-dependent introns given that 17% of these introns carry at least two G-tracts within 50 nt of the 5' splice site.[79] Thus, these hnRNP proteins may play a role both in regulating alternative splice site selection and in constitutive splicing.

One of the main challenges associated with splice site selection is to understand how the cell ignores the multitude of pseudo and cryptic splicing signals that populate introns yet correctly identifies authentic splice sites. Many hnRNP proteins can directly repress the use of flanking splice sites. Moreover, the hnRNP protein-mediated looping of a splice site can also repress splicing.[76] Thus, an additional function for hnRNP proteins may be to repress the recognition of intronic cryptic splice sites. To help define authentic splice sites, SR proteins and other factors are delivered to the pre-mRNA through transcriptional coupling (refer to chapter by Kornblihtt).[161] The change in RNA conformation imposed by hnRNP proteins would justapose authentic splice site pairs. The paucity of SR and hnRNP proteins in yeast may partially be explained by the fact that yeast introns are rarely alternatively spliced, are considerably smaller than metazoan introns and contain inverted repeats to facilitate their definition.[162,163]

Role of Posttranslational Modifications

Links between signal transduction and splice site selection are just beginning to be uncovered (reviewed in refs. 164,165; refer to chapter by Lynch). While the impact of phosphorylation of SR proteins on their activity is well documented (reviewed in ref. 166; refer to chapter by Lin and Fu), the role of phosphorylation in the activity of hnRNP proteins has not been studied to a similar extent. Activation of the MKK-$_{3/6}$-p38 stress-signaling pathway can lead to the hyperphosphorylation of hnRNP A1 and its cytoplasmic accumulation which indirectly impacts alternative splicing.[167] Eleven phosphorylation sites have been identified on the hnRNP K protein.[168] hnRNP K accumulates in the cytoplasm when phosphorylated by ERK, but an impact on splicing has not been documented.[169] Some stimuli can induce protein kinase A to phosphorylate PTB at Ser-16 and this event forces PTB out of the nucleus.[170] Because many hnRNP proteins shuttle between the nucleus and the cytoplasm,[171,172] we can anticipate that a variety of environmental cues and cellular alarms will use signal transduction to modify hnRNP proteins in a way that impacts splicing decisions. One of the future challenges will be to uncover these intricate networks of regulation.

Arginine methylation occurs on many RNA binding proteins including hnRNP A1, A2, K and U proteins.[173-177] Moreover, the transient overexpression of an isoform of the CARM1 methyltransferase can affect splice site selection of the adenovirus *E1a* pre-mRNA.[178] Because many hnRNP proteins contain arginine-rich domains that could potentially be methylated, this modification may be used to modulate the RNA binding and protein interaction properties of hnRNP proteins. Also, hnRNP A1, C, K, H and M can also be modified by SUMO,[179-180] but the exact role of this modification on the activity of these proteins is unknown.

Potential Role of hnRNP Proteins in Transcriptional Coupling

It is now becoming apparent that splicing decisions occur cotranscriptionally (reviewed in refs. 181, 182; refer to chapter by Kornblihtt). The coupling between transcription and splicing implies that many aspects of transcription could impact splice site selection. For example, complexes formed at promoters may recruit specific splicing factors that determine the splicing pattern of the pre-mRNA to be transcribed. Moreover, for splicing units containing multiple competing splice sites, the intrinsic speed of the RNA polymerase, the presence of pausing sites and the extent of chromatin packing may affect splice site selection. Many hnRNP proteins have been associated with transcription. hnRNP K can bind to the single-stranded and double-stranded pyrimidine-rich DNA motifs at various promoters, associate with the transcription factors Sp1 and TBP and modulate the transcription of several genes.[183,184] Chromatin-immunoprecipitation assays have revealed a preferential association of PTB with promoters.[185] hnRNP P2 can function as a transcriptional activator[186,187] and hnRNP A1 has been implicated in transcriptional control.[188-190] hnRNP C proteins are components of the chromatin remodeling complex of the β-globin locus control region.[191] hnRNP F has been found to interact with TBP.[192] hnRNP U (also known as SAF-A) can associate with transcription factors including the phosphorylated CTD of RNA polymerase II,[193,194] and can affect transcription.[195-197] These findings strongly indicate that an important function of hnRNP proteins is to coordinate splicing decisions in a cotranscriptional manner.

Concluding Remarks

Metazoan hnRNP proteins were operationally defined as a limited group of abundant proteins that efficiently crosslink to pre-messenger RNAs. The number of hnRNP proteins is now steadily increasing because splice variants and paralogs of the original members are being identified. However, the rules that determine membership to the hnRNP protein family are ambiguous; newly described proteins are regularly added to the group simply based on their structural similarity, whereas other related proteins are not. For example, the hnRNP protein family could be expanded to include the KH-domain containing proteins Nova-1, Nova-2, KSRP, FMR1 and related proteins, as well as the RRM-domain containing CELF, TIA and Raver proteins.[137,198-201] Likewise, a more general definition would incorporate hnRNP-like proteins of the yeast *Saccharomyces cerevisiae* such as Yra1p, Yra2p, Hrp1/Nab4, Nab2, Npl3 and Tho1.[202-206] In the not too distant future, it would be useful to lay down clear principles that determine membership to the hnRNP family and other classes of RNA binding proteins. Using structural attributes should facilitate this classification, but would completely change our current definition of what is an hnRNP protein. Nonetheless, it is clear that the term "hnRNP protein" is not a functional definition.

The numerous studies that have examined the association of metazoan hnRNP proteins with pre-messenger RNAs have uncovered diverse and specific functions for these proteins in the control of RNA splicing. Abundantly expressed hnRNP proteins often contribute to alternative pre-mRNA splicing by inhibiting splice site utilization. This mode of action may also be important to repress the use of pseudo and cryptic splice sites that are found in the introns of complex metazoan genomes. Although many hnRNP proteins can repress splice site usage, the number of examples of a positive role for hnRNP proteins in constitutive and alternative splicing is increasing. In some cases, hnRNP proteins may promote a change in pre-mRNA conformation that encourages the use of specific splice sites while simultaneously repressing others.

Although alternative splicing and different types of posttranslational modifications expand the repertoire of hnRNP protein isoforms, the functional impact of this diversity remains largely unexplored. Likewise, our understanding of the cellular strategies that use hnRNP proteins to control splice site choice in normal tissues and in pathological conditions is still extremely rudimentary. Nonetheless, we expect that applying combinations of high-throughput methods (refer to chapter by Calarco et al) to these important questions will rapidly produce an insightful description of the global contribution of hnRNP proteins to biological mechanisms. This achievement will likely be accompanied by many surprises and by new opportunities for applying this knowledge to the benefit of human health.

Acknowledgements

We thank Sherif Abou Elela for comments on the manuscript. Work on mammalian alternative splicing in the Chabot laboratory is supported by a grant from the Canadian Institute of Health Research. B. Chabot holds the Canada Research Chair in functional genomics.

References

1. Johnson JM, Castle J, Garrett-Engele P et al. Genome-wide survey of human alternative pre-mRNA splicing with exon junction microarrays. Science 2003; 302(5653):2141-2144.
2. Lipscombe D. Neuronal proteins custom designed by alternative splicing. Curr Opin Neurobiol. 2005;15(3):358-363.
3. Garcia-Blanco MA, Baraniak AP, Lasda EL. Alternative splicing in disease and therapy. Nat Biotechnol. 2004;22(5):535-546.
4. Pagani F, Baralle FE. Genomic variants in exons and introns: identifying the splicing spoilers. Nat Rev Genet. 2004;5(5):389-396.
5. Li HR, Wang-Rodriguez J, Nair TM et al. Two-dimensional transcriptome profiling: identification of messenger RNA isoform signatures in prostate cancer from archived paraffin-embedded cancer specimens. Cancer Res 2006; 66(8):4079-4088.
6. Zhang C, Li HR, Fan JB et al. Profiling alternatively spliced mRNA isoforms for prostate cancer classification. BMC Bioinformatics 2006; 7:202.
7. Will CL, Luhrmann R. Spliceosomal UsnRNP biogenesis, structure and function. Curr Opin Cell Biol. 2001;13(3):290-301.
8. Black DL. Mechanisms of alternative pre-messenger RNA splicing. Annual Rev. Biochem. 2003;72:291-336.
9. Lavigueur A, La Branche H, Kornblihtt AR et al. A splicing enhancer in the human fibronectin alternate ED1 exon interacts with SR proteins and stimulates U2 snRNP binding. Genes Dev 1993; 7(12A):2405-2417.
10. Wang Z, Hoffmann HM, Grabowski PJ. Intrinsic U2AF binding is modulated by exon enhancer signals in parallel with changes in splicing activity. RNA 1995; 1(1):21-35.
11. Graveley BR, Hertel KJ, Maniatis T. The role of U2AF35 and U2AF65 in enhancer-dependent splicing. RNA 2001; 7(6):806-818.
12. Kohtz JD, Jamison SF, Will CL et al. Protein-protein interactions and 5'-splice-site recognition in mammalian mRNA precursors. Nature 1994; 368(6467):119-124.
13. Zuo P, Maniatis T. The splicing factor U2AF35 mediates critical protein-protein interactions in constitutive and enhancer-dependent splicing. Genes Dev 1996; 10(11):1356-1368.
14. Dreyfuss G, Matunis MJ, Pinol-Roma S, Burd CG. hnRNP proteins and the biogenesis of mRNA. Annu Rev Biochem. 1993;62:289-321.
15. Kiledjian M, Burd CG, Görlach M et al. Structure and function of hnRNP proteins. In: Mattaj KNaI, ed. RNA-protein interactions: Frontiers in Molecular Biology. Oxford: Oxford University Press; 1994:127-149.
16. Pinol-Roma S, Choi YD, Matunis MJ et al. Immunopurification of heterogeneous nuclear ribonucleoprotein particles reveals an assortment of RNA-binding proteins. Genes Dev 1988; 2(2):215-227.
17. Matunis EL, Matunis MJ, Dreyfuss G. Characterization of the major hnRNP proteins from Drosophila melanogaster. J Cell Biol 1992; 116(2):257-269.
18. Matunis MJ, Matunis EL, Dreyfuss G. Isolation of hnRNP complexes from Drosophila melanogaster. J Cell Biol 1992; 116(2):245-255.
19. Raychaudhuri G, Haynes SR, Beyer AL. Heterogeneous nuclear ribonucleoprotein complexes and proteins in Drosophila melanogaster. Mol Cell Biol 1992; 12(2):847-855.
20. Choi YD, Dreyfuss G. Isolation of the heterogeneous nuclear RNA-ribonucleoprotein complex (hnRNP): a unique supramolecular assembly. Proc Natl Acad Sci USA 1984; 81(23):7471-7475.
21. Matunis EL, Matunis MJ, Dreyfuss G. Association of individual hnRNP proteins and snRNPs with nascent transcripts. J Cell Biol 1993; 121(2):219-228.
22. Wurtz T, Kiseleva E, Nacheva G et al. Identification of two RNA-binding proteins in Balbiani ring pre-messenger ribonucleoprotein granules and presence of these proteins in specific subsets of heterogeneous nuclear ribonucleoprotein particles. Mol Cell Biol 1996; 16(4):1425-1435.
23. Bennett M, Pinol-Roma S, Staknis D et al. Differential binding of heterogeneous nuclear ribonucleoproteins to mRNA precursors prior to spliceosome assembly in vitro. Mol Cell Biol 1992; 12(7):3165-3175.
24. Michaud S, Reed R. An ATP-independent complex commits pre-mRNA to the mammalian spliceosome assembly pathway. Genes Dev 1991; 5(12B):2534-2546.
25. Mayeda A, Helfman DM, Krainer AR. Modulation of exon skipping and inclusion by heterogeneous nuclear ribonucleoprotein A1 and pre-mRNA splicing factor SF2/ASF. Mol Cell Biol 1993; 13(5):2993-3001.
26. Krecic AM, Swanson MS. hnRNP complexes: composition, structure, and function. Curr Opin Cell Biol. 1999;11(3):363-371.
27. Mayeda A, Krainer AR. Regulation of alternative pre-mRNA splicing by hnRNP A1 and splicing factor SF2. Cell 1992; 68(2):365-375.

28. Caceres JF, Stamm S, Helfman DM et al. Regulation of alternative splicing in vivo by overexpression of antagonistic splicing factors. Science 1994; 265(5179):1706-1709.
29. Yang X, Bani MR, Lu SJ et al. The A1 and A1B proteins of heterogeneous nuclear ribonucleoparticles modulate 5' splice site selection in vivo. Proc Natl Acad Sci USA 1994; 91(15):6924-6928.
30. Del Gatto-Konczak F, Olive M, Gesnel MC et al. hnRNP A1 recruited to an exon in vivo can function as an exon splicing silencer. Mol Cell Biol 1999; 19(1):251-260.
31. Caputi M, Mayeda A, Krainer AR et al. hnRNP A/B proteins are required for inhibition of HIV-1 pre-mRNA splicing. EMBO J 1999; 18(14):4060-4067.
32. Bilodeau PS, Domsic JK, Mayeda A et al. RNA splicing at human immunodeficiency virus type 1 3' splice site A2 is regulated by binding of hnRNP A/B proteins to an exonic splicing silencer element. J Virol 2001; 75(18):8487-8497.
33. Tange TO, Damgaard CK, Guth S et al. The hnRNP A1 protein regulates HIV-1 tat splicing via a novel intron silencer element. EMBO J 2001; 20(20):5748-5758.
34. Zhu J, Mayeda A, Krainer AR. Exon identity established through differential antagonism between exonic splicing silencer-bound hnRNP A1 and enhancer-bound SR proteins. Mol Cell 2001; 8(6):1351-1361.
35. Matter N, Marx M, Weg-Remers S et al. Heterogeneous ribonucleoprotein A1 is part of an exon-specific splice-silencing complex controlled by oncogenic signaling pathways. J Biol Chem 2000; 275(45):35353-35360.
36. Rooke N, Markovtsov V, Cagavi E et al. Roles for SR proteins and hnRNP A1 in the regulation of c-src exon N1. Mol Cell Biol 2003; 23(6):1874-1884.
37. Hou VC, Lersch R, Gee SL et al. Decrease in hnRNP A/B expression during erythropoiesis mediates a pre-mRNA splicing switch. EMBO J 2002; 21(22):6195-6204.
38. Venables JP, Bourgeois CF, Dalgliesh C et al. Up-regulation of the ubiquitous alternative splicing factor Tra2beta causes inclusion of a germ cell-specific exon. Hum Mol Genet 2005; 14(16):2289-2303.
39. Pollard AJ, Krainer AR, Robson SC et al. Alternative splicing of the adenylyl cyclase stimulatory G-protein G alpha(s) is regulated by SF2/ASF and heterogeneous nuclear ribonucleoprotein A1 (hnRNPA1) and involves the use of an unusual TG 3'-splice Site. J Biol Chem 2002; 277(18):15241-15251.
40. Arikan MC, Memmott J, Broderick JA et al. Modulation of the membrane-binding projection domain of tau protein: splicing regulation of exon 3. Brain Res Mol Brain Res 2002; 101(1-2):109-121.
41. Expert-Bezancon A, Sureau A, Durosay P et al. hnRNP A1 and the SR proteins ASF/SF2 and SC35 have antagonistic functions in splicing of beta-tropomyosin exon 6B. J Biol Chem 2004; 279(37):38249-38259.
42. Guil S, Gattoni R, Carrascal M et al. Roles of hnRNP A1, SR proteins and p68 helicase in c-H-ras alternative splicing regulation. Mol Cell Biol 2003; 23(8):2927-2941.
43. Zhao X, Rush M, Schwartz S. Identification of an hnRNP A1-dependent splicing silencer in the human papillomavirus type 16 L1 coding region that prevents premature expression of the late L1 gene. J Virol 2004; 78(20):10888-10905.
44. Princler GL, Julias JG, Hughes SH et al. Roles of viral and cellular proteins in the expression of alternatively spliced HTLV-1 pX mRNAs. Virology 2003; 317(1):136-145.
45. Kress E, Baydoun HH, Bex F et al. Critical role of hnRNP A1 in HTLV-1 replication in human transformed T-lymphocytes. Retrovirology 2005; 2(1):8.
46. Disset A, Bourgeois CF, Benmalek N et al. An exon skipping-associated nonsense mutation in the dystrophin gene uncovers a complex interplay between multiple antagonistic splicing elements. Hum Mol Genet 2006; 15(6):999-1013.
47. Mayeda A, Munroe SH, Caceres JF et al. Function of conserved domains of hnRNP A1 and other hnRNP A/B proteins. EMBO J 1994; 13(22):5483-5495.
48. Hutchison S, LeBel C, Blanchette M et al. Distinct sets of adjacent heterogeneous nuclear ribonucleoprotein (hnRNP) A1/A2 binding sites control 5' splice site selection in the hnRNP A1 mRNA precursor. J Biol Chem 2002; 277(33):29745-29752.
49. Chen CD, Kobayashi R, Helfman DM. Binding of hnRNP H to an exonic splicing silencer is involved in the regulation of alternative splicing of the rat beta-tropomyosin gene. Genes Dev 1999; 13(5):593-606.
50. Fogel BL, McNally MT. A cellular protein, hnRNP H, binds to the negative regulator of splicing element from Rous sarcoma virus. J Biol Chem 2000; 275(41):32371-32378.
51. Jacquenet S, Mereau A, Bilodeau PS et al. A second exon splicing silencer within human immunodeficiency virus type 1 tat exon 2 represses splicing of Tat mRNA and binds protein hnRNP H. J Biol Chem 2001; 276(44):40464-40475.
52. Han K, Yeo G, An P et al. A combinatorial code for splicing silencing: UAGG and GGGG motifs. PLoS Biol 2005; 3(5):e158.

53. Pagani F, Buratti E, Stuani C et al. Missense, nonsense and neutral mutations define juxtaposed regulatory elements of splicing in cystic fibrosis transmembrane regulator exon 9. J Biol Chem 2003; 278(29):26580-26588.
54. Martinez-Contreras R, Fisette JF, Nasim FU et al. Intronic binding sites for hnRNP A/B and hnRNP F/H proteins stimulate pre-mRNA splicing. PLoS Biol 2006; 4(2):e21.
55. Paul S, Dansithong W, Kim D et al. Interaction of muskleblind, CUG-BP1 and hnRNP H proteins in DM1-associated aberrant IR splicing. EMBO J 2006; 25(18):4271-4283.
56. Wang Z, Rolish ME, Yeo G et al. Systematic identification and analysis of exonic splicing silencers. Cell 2004; 119(6):831-845.
57. Sironi M, Menozzi G, Riva L et al. Silencer elements as possible inhibitors of pseudoexon splicing. Nucleic Acids Res 2004; 32(5):1783-1791.
58. Romano M, Marcucci R, Buratti E et al. Regulation of 3' splice site selection in the 844ins68 polymorphism of the cystathionine Beta -synthase gene. J Biol Chem 2002; 277(46):43821-43829.
59. Buratti E, Baralle M, De Conti L et al. hnRNP H binding at the 5' splice site correlates with the pathological effect of two intronic mutations in the NF-1 and TSHbeta genes. Nucleic Acids Res 2004; 32(14):4224-4236.
60. Siebel CW, Admon A, Rio DC. Soma-specific expression and cloning of PSI, a negative regulator of P element pre-mRNA splicing. Genes Dev 1995; 9(3):269-283.
61. Zahler AM, Damgaard CK, Kjems J et al. SC35 and heterogeneous nuclear ribonucleoprotein A/B proteins bind to a juxtaposed exonic splicing enhancer/exonic splicing silencer element to regulate HIV-1 tat exon 2 splicing. J Biol Chem 2004; 279(11):10077-10084.
62. Hallay H, Locker N, Ayadi L et al. Biochemical and NMR study on the competition between proteins SC35, SRp40 and hnRNP A1 at the HIV-1 Tat exon 2 splicing site. J Biol Chem 2006.
63. Kashima T, Manley JL. A negative element in SMN2 exon 7 inhibits splicing in spinal muscular atrophy. Nat Genet 2003; 34(4):460-463.
64. Cartegni L, Hastings ML, Calarco JA et al. Determinants of exon 7 splicing in the spinal muscular atrophy genes, SMN1 and SMN2. Am J Hum Genet 2006; 78(1):63-77.
65. Cartegni L, Krainer AR. Disruption of an SF2/ASF-dependent exonic splicing enhancer in SMN2 causes spinal muscular atrophy in the absence of SMN1. Nat Genet 2002; 30(4):377-384.
66. Crawford JB, Patton JG. Activation of {alpha} -tropomyosin exon 2 is regulated by the SR protein 9G8 and heterogeneous nuclear ribonucleoproteins H and F. Mol Cell Biol 2006.
67. Domsic JK, Wang Y, Mayeda A et al. Human immunodeficiency virus type 1 hnRNP A/B-dependent exonic splicing silencer ESSV antagonizes binding of U2AF65 to viral polypyrimidine tracts. Mol Cell Biol 2003; 23(23):8762-8772.
68. Min H, Chan RC, Black DL. The generally expressed hnRNP F is involved in a neural-specific pre-mRNA splicing event. Genes Dev 1995; 9(21):2659-2671.
69. Chou MY, Rooke N, Turck CW et al. hnRNP H is a component of a splicing enhancer complex that activates a c-src alternative exon in neuronal cells. Mol Cell Biol 1999; 19(1):69-77.
70. Garneau D, Revil T, Fisette JF et al. Heterogeneous nuclear ribonucleoprotein F/H proteins modulate the alternative splicing of the apoptotic mediator Bcl-x. J Biol Chem 2005; 280(24):22641-22650.
71. Hastings ML, Wilson CM, Munroe SH. A purine-rich intronic element enhances alternative splicing of thyroid hormone receptor mRNA. RNA 2001; 7(6):859-874.
72. Caputi M, Zahler AM. Determination of the RNA binding specificity of the heterogeneous nuclear ribonucleoprotein (hnRNP) H/H'/F/2H9 family. J Biol Chem 2001; 276(47):43850-43859.
73. Chabot B, Blanchette M, Lapierre I et al. An intron element modulating 5' splice site selection in the hnRNP A1 pre-mRNA interacts with hnRNP A1. Mol Cell Biol 1997; 17(4):1776-1786.
74. Blanchette M, Chabot B. Modulation of exon skipping by high-affinity hnRNP A1-binding sites and by intron elements that repress splice site utilization. EMBO J 1999; 18(7):1939-1952.
75. Chabot B, LeBel C, Hutchison S et al. Heterogeneous nuclear ribonucleoprotein particle A/B proteins and the control of alternative splicing of the mammalian heterogeneous nuclear ribonucleoprotein particle A1 pre-mRNA. Prog Mol Subcell Biol 2003; 31:59-88.
76. Nasim FU, Hutchison S, Cordeau M et al. High-affinity hnRNP A1 binding sites and duplex-forming inverted repeats have similar effects on 5' splice site selection in support of a common looping out and repression mechanism. RNA 2002; 8(8):1078-1089.
77. Yeo G, Hoon S, Venkatesh B et al. Variation in sequence and organization of splicing regulatory elements in vertebrate genes. Proc Natl Acad Sci USA 2004; 101(44):15700-15705.
78. McCullough AJ, Berget SM. G triplets located throughout a class of small vertebrate introns enforce intron borders and regulate splice site selection. Mol Cell Biol 1997; 17(8):4562-4571.
79. McNally LM, Yee L, McNally MT. Heterogeneous nuclear ribonucleoprotein H is required for optimal U11 small nuclear ribonucleoprotein binding to a retroviral RNA-processing control element: implications for U12-dependent RNA splicing. J Biol Chem 2006; 281(5):2478-2488.

80. Gamberi C, Izaurralde E, Beisel C et al. Interaction between the human nuclear cap-binding protein complex and hnRNP F. Mol Cell Biol 1997; 17(5):2587-2597.
81. Mahe D, Mahl P, Gattoni R et al. Cloning of human 2H9 heterogeneous nuclear ribonucleoproteins. Relation with splicing and early heat shock-induced splicing arrest. J Biol Chem 1997; 272(3):1827-1836.
82. Gattoni R, Mahe D, Mahl P et al. The human hnRNP-M proteins: structure and relation with early heat shock-induced splicing arrest and chromosome mapping. Nucleic Acids Res 1996; 24(13):2535-2542.
83. Zu K, Sikes ML, Haynes SR et al. Altered levels of the Drosophila HRB87F/hrp36 hnRNP protein have limited effects on alternative splicing in vivo. Mol Biol Cell 1996; 7(7):1059-1073.
84. Park JW, Parisky K, Celotto AM et al. Identification of alternative splicing regulators by RNA interference in Drosophila. Proc Natl Acad Sci USA 2004.
85. Blanchette M, Green RE, Brenner SE et al. Global analysis of positive and negative pre-mRNA splicing regulators in Drosophila. Genes Dev 2005; 19(11):1306-1314.
86. Sierakowska H, Szer W, Furdon PJ et al. Antibodies to hnRNP core proteins inhibit in vitro splicing of human beta-globin pre-mRNA. Nucleic Acids Res 1986; 14(13):5241-5254.
87. Choi YD, Grabowski PJ, Sharp PA et al. Heterogeneous nuclear ribonucleoproteins: role in RNA splicing. Science 1986; 231(4745):1534-1539.
88. Staknis D, Reed R. SR proteins promote the first specific recognition of pre-mRNA and are present together with the U1 small nuclear ribonucleoprotein particle in a general splicing enhancer complex. Mol Cell Biol 1994; 14(11):7670-7682.
89. Garcia-Blanco MA, Jamison SF, Sharp PA. Identification and purification of a 62,000-dalton protein that binds specifically to the polypyrimidine tract of introns. Genes Dev 1989; 3(12A):1874-1886.
90. Swanson MS, Dreyfuss G. RNA binding specificity of hnRNP proteins: a subset bind to the 3' end of introns. EMBO J 1988; 7(11):3519-3529.
91. Roscigno RF, Weiner M, Garcia-Blanco MA. A mutational analysis of the polypyrimidine tract of introns. Effects of sequence differences in pyrimidine tracts on splicing. J Biol Chem 1993; 268(15):11222-11229.
92. Sebillon P, Beldjord C, Kaplan JC et al. A T to G mutation in the polypyrimidine tract of the second intron of the human beta-globin gene reduces in vitro splicing efficiency: evidence for an increased hnRNP C interaction. Nucleic Acids Res 1995; 23(17):3419-3425.
93. Williamson DJ, Banik-Maiti S, DeGregori J et al. hnRNP C is required for postimplantation mouse development but Is dispensable for cell viability. Mol Cell Biol 2000; 20(11):4094-4105.
94. Mikula M, Dzwonek A, Karczmarski J et al. Landscape of the hnRNP K protein-protein interactome. Proteomics 2006; 6(8):2395-2406.
95. Bomsztyk K, Denisenko O, Ostrowski J. hnRNP K: one protein multiple processes. Bioessays 2004; 26(6):629-638.
96. Expert-Bezancon A, Le Caer JP, Marie J. Heterogeneous nuclear ribonucleoprotein (hnRNP) K is a component of an intronic splicing enhancer complex that activates the splicing of the alternative exon 6A from chicken beta-tropomyosin pre-mRNA. J Biol Chem 2002; 277(19):16614-16623.
97. Ule J, Stefani G, Mele A et al. An RNA map predicting Nova-dependent splicing regulation. Nature 2006; 444(7119):580-586.
98. Siebel CW, Fresco LD, Rio DC. The mechanism of somatic inhibition of Drosophila P-element pre-mRNA splicing: multiprotein complexes at an exon pseudo-5' splice site control U1 snRNP binding. Genes Dev 1992; 6(8):1386-1401.
99. Rothrock CR, House AE, Lynch KW. HnRNP L represses exon splicing via a regulated exonic splicing silencer. EMBO J 2005; 24(15):2792-2802.
100. Hofmann Y, Wirth B. hnRNP-G promotes exon 7 inclusion of survival motor neuron (SMN) via direct interaction with Htra2-beta1. Hum Mol Genet 2002; 11(17):2037-2049.
101. Hofmann Y, Lorson CL, Stamm S et al. Htra2-beta 1 stimulates an exonic splicing enhancer and can restore full-length SMN expression to survival motor neuron 2 (SMN2). Proc Natl Acad Sci USA 2000; 97(17):9618-9623.
102. Venables JP, Elliott DJ, Makarova OV et al. RBMY, a probable human spermatogenesis factor and other hnRNP G proteins interact with Tra2beta and affect splicing. Hum Mol Genet 2000; 9(5):685-694.
103. Nasim MT, Chernova TK, Chowdhury HM et al. HnRNP G and Tra2beta: opposite effects on splicing matched by antagonism in RNA binding. Hum Mol Genet 2003; 12(11):1337-1348.
104. Barnard DC, Li J, Peng R et al. Regulation of alternative splicing by SRrp86 through coactivation and repression of specific SR proteins. RNA 2002; 8(4):526-533.
105. Barnard DC, Patton JG. Identification and characterization of a novel serine-arginine-rich splicing regulatory protein. Mol Cell Biol 2000; 20(9):3049-3057.
106. Li J, Hawkins IC, Harvey CD et al. Regulation of alternative splicing by SRrp86 and its interacting proteins. Mol Cell Biol 2003; 23(21):7437-7447.

107. Valcarcel J, Gebauer F. Post-transcriptional regulation: the dawn of PTB. Curr Biol. 1997;7(11): R705-708.
108. Spellman R, Rideau A, Matlin A, et al. Regulation of alternative splicing by PTB and associated factors. Biochem Soc Trans. 2005;33(Pt 3):457-460.
109. Wagner EJ, Garcia-Blanco MA. Polypyrimidine tract binding protein antagonizes exon definition. Mol Cell Biol 2001; 21(10):3281-3288.
110. Lou H, Helfman DM, Gagel RF et al. Polypyrimidine tract-binding protein positively regulates inclusion of an alternative 3'-terminal exon. Mol Cell Biol 1999; 19(1):78-85.
111. Smith CW, Nadal-Ginard B. Mutually exclusive splicing of alpha-tropomyosin exons enforced by an unusual lariat branch point location: implications for constitutive splicing. Cell 1989; 56(5):749-758.
112. Mullen MP, Smith CW, Patton JG et al. Alpha-tropomyosin mutually exclusive exon selection: competition between branchpoint/polypyrimidine tracts determines default exon choice. Genes Dev 1991; 5(4):642-655.
113. Gooding C, Roberts GC, Moreau G et al. Smooth muscle-specific switching of alpha-tropomyosin mutually exclusive exon selection by specific inhibition of the strong default exon. EMBO J 1994; 13(16):3861-3872.
114. Perez I, Lin CH, McAfee JG et al. Mutation of PTB binding sites causes misregulation of alternative 3' splice site selection in vivo. RNA 1997; 3(7):764-778.
115. Lin CH, Patton JG. Regulation of alternative 3' splice site selection by constitutive splicing factors. RNA 1995; 1(3):234-245.
116. Gooding C, Roberts GC, Smith CW. Role of an inhibitory pyrimidine element and polypyrimidine tract binding protein in repression of a regulated alpha-tropomyosin exon. RNA 1998; 4(1):85-100.
117. Wagner EJ, Garcia-Blanco MA. RNAi-mediated PTB depletion leads to enhanced exon definition. Mol Cell 2002; 10(4):943-949.
118. Southby J, Gooding C, Smith CW. Polypyrimidine tract binding protein functions as a repressor to regulate alternative splicing of alpha-actinin mutally exclusive exons. Mol Cell Biol 1999; 19(4):2699-2711.
119. Ashiya M, Grabowski PJ. A neuron-specific splicing switch mediated by an array of pre-mRNA repressor sites: evidence of a regulatory role for the polypyrimidine tract binding protein and a brain-specific PTB counterpart. RNA 1997; 3(9):996-1015.
120. Amir-Ahmady B, Boutz PL, Markovtsov V et al. Exon repression by polypyrimidine tract binding protein. RNA 2005; 11(5):699-716.
121. Chan RC, Black DL. The polypyrimidine tract binding protein binds upstream of neural cell-specific c-src exon N1 to repress the splicing of the intron downstream. Mol Cell Biol 1997; 17(8):4667-4676.
122. Chou MY, Underwood JG, Nikolic J et al. Multisite RNA binding and release of polypyrimidine tract binding protein during the regulation of c-src neural-specific splicing. Mol Cell 2000; 5(6):949-957.
123. Min H, Turck CW, Nikolic JM et al. A new regulatory protein, KSRP, mediates exon inclusion through an intronic splicing enhancer. Genes Dev 1997; 11(8):1023-1036.
124. Black DL. Activation of c-src neuron-specific splicing by an unusual RNA element in vivo and in vitro. Cell 1992; 69(5):795-807.
125. Modafferi EF, Black DL. A complex intronic splicing enhancer from the c-src pre-mRNA activates inclusion of a heterologous exon. Mol Cell Biol 1997; 17(11):6537-6545.
126. Sharma S, Falick AM, Black DL. Polypyrimidine tract binding protein blocks the 5' splice site-dependent assembly of U2AF and the prespliceosomal E complex. Mol Cell 2005; 19(4):485-496.
127. Markovtsov V, Nikolic JM, Goldman JA et al. Cooperative assembly of an hnRNP complex induced by a tissue-specific homolog of polypyrimidine tract binding protein. Mol Cell Biol 2000; 20(20):7463-7479.
128. Yamamoto H, Tsukahara K, Kanaoka Y et al. Isolation of a mammalian homologue of a fission yeast differentiation regulator. Mol Cell Biol 1999; 19(5):3829-3841.
129. Gooding C, Kemp P, Smith CW. A novel polypyrimidine tract-binding protein paralog expressed in smooth muscle cells. J Biol Chem 2003; 278(17):15201-15207.
130. Izquierdo JM, Majos N, Bonnal S et al. Regulation of Fas alternative splicing by antagonistic effects of TIA-1 and PTB on exon definition. Mol Cell 2005; 19(4):475-484.
131. Monie TP, Hernandez H, Robinson CV et al. The polypyrimidine tract binding protein is a monomer. RNA 2005; 11(12):1803-1808.
132. Petoukhov MV, Monie TP, Allain FH et al. Conformation of polypyrimidine tract binding protein in solution. Structure 2006; 14(6):1021-1027.
133. Oberstrass FC, Auweter SD, Erat M et al. Structure of PTB bound to RNA: specific binding and implications for splicing regulation. Science 2005; 309(5743):2054-2057.
134. Charlet BN, Logan P, Singh G et al. Dynamic antagonism between ETR-3 and PTB regulates cell type-specific alternative splicing. Mol Cell 2002; 9(3):649-658.

135. Gromak N, Matlin AJ, Cooper TA et al. Antagonistic regulation of alpha-actinin alternative splicing by CELF proteins and polypyrimidine tract binding protein. RNA 2003; 9(4):443-456.
136. Shukla S, Del Gatto-Konczak F, Breathnach R et al. Competition of PTB with TIA proteins for binding to a U-rich cis-element determines tissue-specific splicing of the myosin phosphatase targeting subunit 1. RNA 2005; 11(11):1725-1736.
137. Huttelmaier S, Illenberger S, Grosheva I et al. Raver1, a dual compartment protein, is a ligand for PTB/hnRNPI and microfilament attachment proteins. J Cell Biol 2001; 155(5):775-786.
138. Gromak N, Rideau A, Southby J et al. The PTB interacting protein raver1 regulates alpha-tropomyosin alternative splicing. EMBO J 2003; 22(23):6356-6364.
139. Rideau AP, Gooding C, Simpson PJ et al. A peptide motif in Raver1 mediates splicing repression by interaction with the PTB RRM2 domain. Nat Struct Mol Biol 2006; 13(9):839-848.
140. Wollerton MC, Gooding C, Robinson F et al. Differential alternative splicing activity of isoforms of polypyrimidine tract binding protein (PTB). RNA 2001; 7(6):819-832.
141. Wollerton MC, Gooding C, Wagner EJ et al. Autoregulation of polypyrimidine tract binding protein by alternative splicing leading to nonsense-mediated decay. Mol Cell 2004; 13(1):91-100.
142. Hui J, Stangl K, Lane WS et al. HnRNP L stimulates splicing of the eNOS gene by binding to variable-length CA repeats. Nat Struct Biol 2003; 10(1):33-37.
143. Hui J, Hung LH, Heiner M et al. Intronic CA-repeat and CA-rich elements: a new class of regulators of mammalian alternative splicing. EMBO J 2005; 24(11):1988-1998.
144. Cheli Y, Kunicki TJ. hnRNP L regulates differences in expression of mouse integrin alpha2beta1. Blood 2006; 107(11):4391-4398.
145. Rothrock C, Cannon B, Hahm B et al. A conserved signal-responsive sequence mediates activation-induced alternative splicing of CD45. Mol Cell 2003; 12(5):1317-1324.
146. Lynch KW, Weiss A. A model system for activation-induced alternative splicing of CD45 pre-mRNA in T-cells implicates protein kinase C and Ras. Mol Cell Biol 2000; 20(1):70-80.
147. Tong A, Nguyen J, Lynch KW. Differential expression of CD45 isoforms is controlled by the combined activity of basal and inducible splicing-regulatory elements in each of the variable exons. J Biol Chem 2005; 280(46):38297-38304.
148. House AE, Lynch KW. An exonic splicing silencer represses spliceosome assembly after ATP-dependent exon recognition. Nat Struct Mol Biol 2006; 13(10):937-944.
149. Hahm B, Cho OH, Kim JE et al. Polypyrimidine tract-binding protein interacts with HnRNP L. FEBS Lett 1998; 425(3):401-406.
150. Robinson F, Smith CW. A splicing repressor domain in polypyrimidine tract-binding protein. J Biol Chem 2006; 281(2):800-806.
151. Kafasla P, Patrinou-Georgoula M, Lewis JD et al. Association of the 72/74-kDa proteins, members of the heterogeneous nuclear ribonucleoprotein M group, with the pre-mRNA at early stages of spliceosome assembly. Biochem J 2002; 363(Pt 3):793-799.
152. Kiesler E, Hase ME, Brodin D et al. Hrp59, an hnRNP M protein in Chironomus and Drosophila,binds to exonic splicing enhancers and is required for expression of a subset of mRNAs. J Cell Biol 2005; 168(7):1013-1025.
153. Hallier M, Lerga A, Barnache S et al. The transcription factor Spi-1/PU.1 interacts with the potential splicing factor TLS. J Biol Chem 1998; 273(9):4838-4842.
154. Yang L, Embree LJ, Tsai S et al. Oncoprotein TLS interacts with serine-arginine proteins involved in RNA splicing. J Biol Chem 1998; 273(43):27761-27764.
155. Zhang D, Paley AJ, Childs G. The transcriptional repressor ZFM1 interacts with and modulates the ability of EWS to activate transcription. J Biol Chem 1998; 273(29):18086-18091.
156. Zinszner H, Albalat R, Ron D. A novel effector domain from the RNA-binding protein TLS or EWS is required for oncogenic transformation by CHOP. Genes Dev 1994; 8(21):2513-2526.
157. Neubauer G, King A, Rappsilber J et al. Mass spectrometry and EST-database searching allows characterization of the multi-protein spliceosome complex. Nat Genet 1998; 20(1):46-50.
158. Mourelatos Z, Abel L, Yong J et al. SMN interacts with a novel family of hnRNP and spliceosomal proteins. EMBO J 2001; 20(19):5443-5452.
159. Singh R, Valcarcel J. Building specificity with nonspecific RNA-binding proteins. Nat Struct Mol Biol. 2005;12(8):645-653.
160. Spellman R, Smith CW. Novel modes of splicing repression by PTB. Trends Biochem Sci. 2006;31(2):73-76.
161. Das R, Dufu K, Romney B et al. Functional coupling of RNAPII transcription to spliceosome assembly. Genes Dev 2006; 20(9):1100-1109.
162. Charpentier B, Rosbash M. Intramolecular structure in yeast introns aids the early steps of in vitro spliceosome assembly. RNA 1996; 2(6):509-522.

163. Newman A. Specific accessory sequences in Saccharomyces cerevisiae introns control assembly of pre-mRNAs into spliceosomes. EMBO J 1987; 6(12):3833-3839.
164. Shin C, Manley JL. Cell signalling and the control of pre-mRNA splicing. Nat Rev Mol Cell Biol. 2004;5(9):727-738.
165. Blaustein M, Pelisch F, Tanos T, et al. Concerted regulation of nuclear and cytoplasmic activities of SR proteins by AKT. Nat Struct Mol Biol. 2005;12(12):1037-1044.
166. Graveley BR. Sorting out the complexity of SR protein functions. RNA. 2000;6(9):1197-1211.
167. van der Houven van Oordt W, Diaz-Meco MT, Lozano J et al. The MKK(3/6)-p38-signaling cascade alters the subcellular distribution of hnRNP A1 and modulates alternative splicing regulation. J Cell Biol 2000; 149(2):307-316.
168. Mikula M, Karczmarski J, Dzwonek A et al. Casein kinases phosphorylate multiple residues spanning the entire hnRNP K length. Biochim Biophys Acta 2006; 1764(2):299-306.
169. Habelhah H, Shah K, Huang L et al. ERK phosphorylation drives cytoplasmic accumulation of hnRNP-K and inhibition of mRNA translation. Nat Cell Biol 2001; 3(3):325-330.
170. Xie J, Lee JA, Kress TL et al. Protein kinase A phosphorylation modulates transport of the polypyrimidine tract-binding protein. Proc Natl Acad Sci USA 2003; 100(15):8776-8781.
171. Pinol-Roma S, Dreyfuss G. Shuttling of pre-mRNA binding proteins between nucleus and cytoplasm. Nature 1992; 355(6362):730-732.
172. Michael WM, Choi M, Dreyfuss G. A nuclear export signal in hnRNP A1: a signal-mediated, temperature-dependent nuclear protein export pathway. Cell 1995; 83(3):415-422.
173. Kim S, Merrill BM, Rajpurohit R et al. Identification of N(G)-methylarginine residues in human heterogeneous RNP protein A1: Phe/Gly-Gly-Gly-Arg-Gly-Gly-Gly/Phe is a preferred recognition motif. Biochemistry 1997; 36(17):5185-5192.
174. Liu Q, Dreyfuss G. In vivo and in vitro arginine methylation of RNA-binding proteins. Mol Cell Biol 1995; 15(5):2800-2808.
175. Herrmann F, Bossert M, Schwander A et al. Arginine methylation of scaffold attachment factor A by heterogeneous nuclear ribonucleoprotein particle-associated PRMT1. J Biol Chem 2004; 279(47):48774-48779.
176. Ostareck-Lederer A, Ostareck DH, Rucknagel KP et al. Asymmetric arginine dimethylation of heterogeneous nuclear ribonucleoprotein K by protein-arginine methyltransferase 1 inhibits its interaction with c-Src. J Biol Chem 2006; 281(16):11115-11125.
177. Nichols RC, Wang XW, Tang J et al. The RGG domain in hnRNP A2 affects subcellular localization. Exp Cell Res 2000; 256(2):522-532.
178. Ohkura N, Takahashi M, Yaguchi H et al. Coactivator-associated arginine methyltransferase 1, CARM1, affects pre-mRNA splicing in an isoform-specific manner. J Biol Chem 2005; 280(32):28927-28935.
179. Vassileva MT, Matunis MJ. SUMO modification of heterogeneous nuclear ribonucleoproteins. Mol Cell Biol 2004; 24(9):3623-3632.
180. Li T, Evdokimov E, Shen RF et al. Sumoylation of heterogeneous nuclear ribonucleoproteins, zinc finger proteins and nuclear pore complex proteins: a proteomic analysis. Proc Natl Acad Sci USA 2004; 101(23):8551-8556.
181. Kornblihtt AR. Promoter usage and alternative splicing. Curr Opin Cell Biol. 2005;17(3):262-268.
182. Bentley DL. Rules of engagement: co-transcriptional recruitment of pre-mRNA processing factors. Curr Opin Cell Biol. 2005;17(3):251-256.
183. Wei CC, Zhang SL, Chen YW et al. Heterogeneous nuclear ribonucleoprotein k modulates angiotensinogen gene expression in kidney cells. J Biol Chem 2006; 281(35):25344-25355.
184. Moumen A, Masterson P, O'Connor MJ et al. hnRNP K: an HDM2 target and transcriptional coactivator of p53 in response to DNA damage. Cell 2005; 123(6):1065-1078.
185. Swinburne IA, Meyer CA, Liu XS et al. Genomic localization of RNA binding proteins reveals links between pre-mRNA processing and transcription. Genome Res 2006; 16(7):912-921.
186. Uranishi H, Tetsuka T, Yamashita M et al. Involvement of the pro-oncoprotein TLS (translocated in liposarcoma) in nuclear factor-kappa B p65-mediated transcription as a coactivator. J Biol Chem 2001; 276(16):13395-13401.
187. Law WJ, Cann KL, Hicks GG. TLS, EWS and TAF15: a model for transcriptional integration of gene expression. Brief Funct Genomic Proteomic 2006; 5(1):8-14.
188. Gao C, Guo H, Mi Z et al. Transcriptional regulatory functions of heterogeneous nuclear ribonucleoprotein-U and -A/B in endotoxin-mediated macrophage expression of osteopontin. J Immunol 2005; 175(1):523-530.
189. Xia H. Regulation of gamma-fibrinogen chain expression by heterogeneous nuclear ribonucleoprotein A1. J Biol Chem 2005; 280(13):13171-13178.

190. Das S, Ward SV, Markle D et al. DNA damage-binding proteins and heterogeneous nuclear ribonucleoprotein A1 function as constitutive KCS element components of the interferon-inducible RNA-dependent protein kinase promoter. J Biol Chem 2004; 279(8):7313-7321.
191. Mahajan MC, Narlikar GJ, Boyapaty G et al. Heterogeneous nuclear ribonucleoprotein C1/C2, MeCP1 and SWI/SNF form a chromatin remodeling complex at the beta-globin locus control region. Proc Natl Acad Sci USA 2005; 102(42):15012-15017.
192. Yoshida T, Makino Y, Tamura T. Association of the rat heterogeneous nuclear RNA-ribonucleoprotein F with TATA-binding protein. FEBS Lett 1999; 457(2):251-254.
193. Mattern KA, van Goethem RE, de Jong L et al. Major internal nuclear matrix proteins are common to different human cell types. J Cell Biochem 1997; 65(1):42-52.
194. Hager GL, Nagaich AK, Johnson TA et al. Dynamics of nuclear receptor movement and transcription. Biochim Biophys Acta 2004; 1677(1-3):46-51.
195. Kim MK, Nikodem VM. hnRNP U inhibits carboxy-terminal domain phosphorylation by TFIIH and represses RNA polymerase II elongation. Mol Cell Biol 1999; 19(10):6833-6844.
196. Kukalev A, Nord Y, Palmberg C et al. Actin and hnRNP U cooperate for productive transcription by RNA polymerase II. Nat Struct Mol Biol 2005; 12(3):238-244.
197. Spraggon L, Dudnakova T, Slight J et al. hnRNP-U directly interacts with WT1 and modulates WT1 transcriptional activation. Oncogene 2006.
198. Beck AR, Medley QG, O'Brien S et al. Structure, tissue distribution and genomic organization of the murine RRM-type RNA binding proteins TIA-1 and TIAR. Nucleic Acids Res 1996; 24(19):3829-3835.
199. Barreau C, Paillard L, Mereau A et al. Mammalian CELF/Bruno-like RNA-binding proteins: molecular characteristics and biological functions. Biochimie 2006; 88(5):515-525.
200. Han J, Cooper TA. Identification of CELF splicing activation and repression domains in vivo. Nucleic Acids Res 2005; 33(9):2769-2780.
201. Kleinhenz B, Fabienke M, Swiniarski S et al. Raver2, a new member of the hnRNP family. FEBS Lett 2005; 579(20):4254-4258.
202. Kim Guisbert K, Duncan K, Li H et al. Functional specificity of shuttling hnRNPs revealed by genome-wide analysis of their RNA binding profiles. RNA 2005; 11(4):383-393.
203. Stutz F, Bachi A, Doerks T et al. REF, an evolutionary conserved family of hnRNP-like proteins, interacts with TAP/Mex67p and participates in mRNA nuclear export. RNA 2000; 6(4):638-650.
204. Zenklusen D, Vinciguerra P, Strahm Y et al. The yeast hnRNP-Like proteins Yra1p and Yra2p participate in mRNA export through interaction with Mex67p. Mol Cell Biol 2001; 21(13):4219-4232.
205. Preker PJ, Guthrie C. Autoregulation of the mRNA export factor Yra1p requires inefficient splicing of its pre-mRNA. RNA 2006; 12(6):994-1006.
206. Jimeno S, Luna R, Garcia-Rubio M et al. Tho1, a novel hnRNP and Sub2 provide alternative pathways for mRNP biogenesis in yeast THO mutants. Mol Cell Biol 2006; 26(12):4387-4398.
207. Burd CG, Dreyfuss G. RNA binding specificity of hnRNP A1: significance of hnRNP A1 high-affinity binding sites in pre-mRNA splicing. EMBO J 1994; 13(5):1197-1204.
208. Del Gatto F, Gesnel MC, Breathnach R. The exon sequence TAGG can inhibit splicing. Nucleic Acids Res 1996; 24(11):2017-2021.
209. Gorlach M, Burd CG, Dreyfuss G. The determinants of RNA-binding specificity of the heterogeneous nuclear ribonucleoprotein C proteins. J Biol Chem 1994; 269(37):23074-23078.
210. Wilusz J, Shenk T. A uridylate tract mediates efficient heterogeneous nuclear ribonucleoprotein C protein-RNA cross-linking and functionally substitutes for the downstream element of the polyadenylation signal. Mol Cell Biol 1990; 10(12):6397-6407.
211. Soltaninassab SR, McAfee JG, Shahied-Milam L et al. Oligonucleotide binding specificities of the hnRNP C protein tetramer. Nucleic Acids Res 1998; 26(14):3410-3417.
212. Laroia G, Cuesta R, Brewer G et al. Control of mRNA decay by heat shock-ubiquitin-proteasome pathway. Science 1999; 284(5413):499-502.
213. Xu N, Chen CY, Shyu AB. Versatile role for hnRNP D isoforms in the differential regulation of cytoplasmic mRNA turnover. Mol Cell Biol 2001; 21(20):6960-6971.
214. Ishikawa F, Matunis MJ, Dreyfuss G et al. Nuclear proteins that bind the pre-mRNA 3' splice site sequence r(UUAG/G) and the human telomeric DNA sequence d(TTAGGG)n. Mol Cell Biol 1993; 13(7):4301-4310.
215. Reimann I, Huth A, Thiele H et al. Suppression of 15-lipoxygenase synthesis by hnRNP E1 is dependent on repetitive nature of LOX mRNA 3'-UTR control element DICE. J Mol Biol 2002; 315(5):965-974.

216. Thisted T, Lyakhov DL, Liebhaber SA. Optimized RNA targets of two closely related triple KH domain proteins, heterogeneous nuclear ribonucleoprotein K and alphaCP-2KL, suggest Distinct modes of RNA recognition. J Biol Chem 2001; 276(20):17484-17496.
217. Datar KV, Dreyfuss G, Swanson MS. The human hnRNP M proteins: identification of a methionine/arginine-rich repeat motif in ribonucleoproteins. Nucleic Acids Res 1993; 21(3):439-446.
218. Lerga A, Hallier M, Delva L et al. Identification of an RNA binding specificity for the potential splicing factor TLS. J Biol Chem 2001; 276(9):6807-6816.
219. Blanc V, Navaratnam N, Henderson JO et al. Identification of GRY-RBP as an apolipoprotein B RNA-binding protein that interacts with both apobec-1 and apobec-1 complementation factor to modulate C to U editing. J Biol Chem 2001; 276(13):10272-10283.
220. Rossoll W, Kroning AK, Ohndorf UM et al. Specific interaction of Smn, the spinal muscular atrophy determining gene product, with hnRNP-R and gry-rbp/hnRNP-Q: a role for Smn in RNA processing in motor axons? Hum Mol Genet 2002; 11(1):93-105.
221. Kiledjian M, Dreyfuss G. Primary structure and binding activity of the hnRNP U protein: binding RNA through RGG box. EMBO J 1992; 11(7):2655-2664.
222. Fackelmayer FO, Dahm K, Renz A et al. Nucleic-acid-binding properties of hnRNP-U/SAF-A, a nuclear-matrix protein which binds DNA and RNA in vivo and in vitro. Eur J Biochem 1994; 221(2):749-757.
223. Wu JY, Maniatis T. Specific interactions between proteins implicated in splice site selection and regulated alternative splicing. Cell 1993; 75(6):1061-1070.
224. Akindahunsi AA, Bandiera A, Manzini G. Vertebrate 2xRBD hnRNP proteins: a comparative analysis of genome, mRNA and protein sequences. Comput Biol Chem 2005; 29(1):13-23.
225. Buvoli M, Cobianchi F, Bestagno MG et al. Alternative splicing in the human gene for the core protein A1 generates another hnRNP protein. EMBO J 1990; 9(4):1229-1235.
226. Blanchette M, Chabot B. A highly stable duplex structure sequesters the 5' splice site region of hnRNP A1 alternative exon 7B. RNA 1997; 3(4):405-419.
227. Burd CG, Swanson MS, Gorlach M et al. Primary structures of the heterogeneous nuclear ribonucleoprotein A2, B1 and C2 proteins: a diversity of RNA binding proteins is generated by small peptide inserts. Proc Natl Acad Sci USA 1989; 86(24):9788-9792.
228. Siomi H, Matunis MJ, Michael WM et al. The pre-mRNA binding K protein contains a novel evolutionarily conserved motif. Nucleic Acids Res 1993; 21(5):1193-1198.
229. Biamonti G, Ruggiu M, Saccone S et al. Two homologous genes, originated by duplication, encode the human hnRNP proteins A2 and A1. Nucleic Acids Res 1994; 22(11):1996-2002.
230. Gorlach M, Wittekind M, Beckman RA et al. Interaction of the RNA-binding domain of the hnRNP C proteins with RNA. EMBO J 1992; 11(9):3289-3295.
231. Wan L, Kim JK, Pollard VW et al. Mutational definition of RNA-binding and protein-protein interaction domains of heterogeneous nuclear RNP C1. J Biol Chem 2001; 276(10):7681-7688.
232. McAfee JG, Shahied-Milam L, Soltaninassab SR et al. A major determinant of hnRNP C protein binding to RNA is a novel bZIP-like RNA binding domain. RNA 1996; 2(11):1139-1152.
233. Shahied-Milam L, Soltaninassab SR, Iyer GV et al. The heterogeneous nuclear ribonucleoprotein C protein tetramer binds U1, U2 and U6 snRNAs through its high affinity RNA binding domain (the bZIP-like motif). J Biol Chem 1998; 273(33):21359-21367.
234. Tan JH, Kajiwara Y, Shahied L et al. The bZIP-like motif of hnRNP C directs the nuclear accumulation of pre-mRNA and lethality in yeast. J Mol Biol 2001; 305(4):829-838.
23. Nakielny S, Dreyfuss G. The hnRNP C proteins contain a nuclear retention sequence that can override nuclear export signals. J Cell Biol 1996; 134(6):1365-1373.
236. Makeyev AV, Chkheidze AN, Liebhaber SA. A set of highly conserved RNA-binding proteins, alphaCP-1 and alphaCP-2, implicated in mRNA stabilization, are coexpressed from an intronless gene and its intron-containing paralog. J Biol Chem 1999; 274(35):24849-24857.
237. Kiledjian M, Wang X, Liebhaber SA. Identification of two KH domain proteins in the alpha-globin mRNP stability complex. EMBO J 1995; 14(17):4357-4364.
238. Leffers H, Dejgaard K, Celis JE. Characterisation of two major cellular poly(rC)-binding human proteins, each containing three K-homologous (KH) domains. Eur J Biochem 1995; 230(2):447-453.
239. Van Seuningen I, Ostrowski J, Bustelo XR et al. The K protein domain that recruits the interleukin 1-responsive K protein kinase lies adjacent to a cluster of c-Src and Vav SH3-binding sites. Implications that K protein acts as a docking platform. J Biol Chem 1995; 270(45):26976-26985.
240. Makeyev AV, Liebhaber SA. Identification of two novel mammalian genes establishes a subfamily of KH-domain RNA-binding proteins. Genomics 2000; 67(3):301-316.
241. Dejgaard K, Leffers H. Characterisation of the nucleic-acid-binding activity of KH domains. Different properties of different domains. Eur J Biochem 1996; 241(2):425-431.

242. Makeyev AV, Liebhaber SA. The poly(C)-binding proteins: a multiplicity of functions and a search for mechanisms. RNA 2002; 8(3):265-278.
243. Michael WM, Eder PS, Dreyfuss G. The K nuclear shuttling domain: a novel signal for nuclear import and nuclear export in the hnRNP K protein. EMBO J 1997; 16(12):3587-3598.
244. Honore B, Rasmussen HH, Vorum H et al. Heterogeneous nuclear ribonucleoproteins H, H' and F are members of a ubiquitously expressed subfamily of related but distinct proteins encoded by genes mapping to different chromosomes. J Biol Chem 1995; 270(48):28780-28789.
245. Honore B. The hnRNP 2H9 gene, which is involved in the splicing reaction, is a multiply spliced gene. Biochim Biophys Acta 2000; 1492(1):108-119.
246. Qian Z, Wilusz J. GRSF-1: a poly(A) + mRNA binding protein which interacts with a conserved G-rich element. Nucleic Acids Res 1994; 22(12):2334-2343.
247. Soulard M, Della Valle V, Siomi MC et al. hnRNP G: sequence and characterization of a glycosylated RNA-binding protein. Nucleic Acids Res 1993; 21(18):4210-4217.
248. Le Coniat M, Soulard M, Della Valle V et al. Localization of the human gene encoding heterogeneous nuclear RNA ribonucleoprotein G (hnRNP-G) to chromosome 6p12. Hum Genet 1992; 88(5):593-595.
249. Lingenfelter PA, Delbridge ML, Thomas S et al. Expression and conservation of processed copies of the RBMX gene. Mamm Genome 2001; 12(7):538-545.
250. Elliott DJ, Ma K, Kerr SM et al. An RBM homologue maps to the mouse Y chromosome and is expressed in germ cells. Hum Mol Genet 1996; 5(7):869-874.
251. Venables JP, Vernet C, Chew SL et al. T-STAR/ETOILE: a novel relative of SAM68 that interacts with an RNA-binding protein implicated in spermatogenesis. Hum Mol Genet 1999; 8(6):959-969.
252. Elliott DJ. RBMY genes and AZFb deletions. J Endocrinol Invest 2000; 23(10):652-658.
253. Delbridge ML, Ma K, Subbarao MN et al. Evolution of mammalian HNRPG and its relationship with the putative azoospermia factor RBM. Mamm Genome 1998; 9(2):168-170.
254. Ghetti A, Pinol-Roma S, Michael WM et al. hnRNP I, the polypyrimidine tract-binding protein: distinct nuclear localization and association with hnRNAs. Nucleic Acids Res 1992; 20(14):3671-3678.
255. Polydorides AD, Okano HJ, Yang YY et al. A brain-enriched polypyrimidine tract-binding protein antagonizes the ability of Nova to regulate neuron-specific alternative splicing. Proc Natl Acad Sci USA 2000; 97(12):6350-6355.
256. Hahm B, Kim YK, Kim JH et al. Heterogeneous nuclear ribonucleoprotein L interacts with the 3' border of the internal ribosomal entry site of hepatitis C virus. J Virol 1998; 72(11):8782-8788.
257. Kafasla P, Patrinou-Georgoula M, Guialis A. The 72/74-kDa polypeptides of the 70-110 S large heterogeneous nuclear ribonucleoprotein complex (LH-nRNP) represent a discrete subset of the hnRNP M protein family. Biochem J 2000; 350 Pt 2:495-503.
258. Iko Y, Kodama TS, Kasai N et al. Domain architectures and characterization of an RNA-binding protein, TLS. J Biol Chem 2004; 279(43):44834-44840.
259. Kipp M, Schwab BL, Przybylski M et al. Apoptotic cleavage of scaffold attachment factor A (SAF-A) by caspase-3 occurs at a noncanonical cleavage site. J Biol Chem 2000; 275(7):5031-5036.

Chapter 9

Functional and Mechanistic Insights From Genome-Wide Studies of Splicing Regulation in the Brain

Jernej Ule and Robert B. Darnell*

Abstract

We review here results arising from the systematic functional analysis of Nova, a neuron-specific RNA binding protein targeted in an autoimmune neurological disorder associated with cancer. We have developed a combination of biochemical, genetic and bioinformatic methods to generate a global understanding of Nova's role as a splicing regulator. Genome-wide identification and validation of Nova target RNAs has yielded unexpected insights into the protein's mechanism of action and into the functionally coherent role of Nova in the biology of the neuronal synapse. These studies provide us with a platform for understanding the role of RNA binding proteins in tissue-specific splicing regulation and in disease.

Introduction

Since all cell types in our body share the same DNA, differential function mainly depends on cell type-specific gene expression. The roles of DNA-binding proteins in tissue-specific functions are generally better characterized than those of RNA-binding proteins.[1,2] However, RNA offers greater complexity of regulation than DNA. Alternative splicing has the potential to produce large numbers of protein isoforms from a single gene, more than generally found to be produced by alternative promoter usage, as is exemplified by the case of ***neurexin*** genes.[3] In addition, regulation of RNA localization and localized translation in the cytoplasm allows for precise regulation of protein expression in space and time.[4] Given the complexities of neuronal function, it is natural to wonder whether neurons might possess specific systems for RNA regulation.

At the Intersection between Cancer Cells and Neurons

The first clue that neurons might have a unique system for regulating RNA metabolism came from studies of the intersection between cancer cells and neurons. Ron Evans and colleagues used a medullary carcinoma of the thyroid tumor cell line as a model for comparative gene expression studies. They found that a unique transcript of the ***calcitonin*** gene was expressed in the tumor cell line and in further studies, that this was an alternatively spliced isoform normally expressed in the brain. The protein product of this brain- and tumor-specific isoform was named CGRP (calcitonin gene related peptide).[5] This study suggested for the first time that the brain possesses its own system to regulate alternative splicing. In addition, it had been shown that analysis of tumor-specific gene expression might paradoxically yield insight into neuron-specific gene expression.

*Corresponding Author: Robert B. Darnell—Laboratory of Molecular Neuro-Oncology, The Rockefeller University, Box 226, 1230 York Avenue, New York, NY 10021, USA. Email: darnelr@rockefeller.edu

Alternative Splicing in the Postgenomic Era, edited by Benjamin J. Blencowe and Brenton R. Graveley. ©2007 Landes Bioscience and Springer Science+Business Media.

This second insight was extended in a more systematic fashion through studies of a group of neurological disorders termed the paraneoplastic neurological disorders (PNDs).[6,7] PND patients present specific neurodegenerative syndromes, which can vary widely between patients and include memory loss, blindness, cerebellar dysfunction and motor or sensory disorders. Each set of neurological symptoms is associated with the occurrence of characteristic tumors, present in the patient's body, which express brain-specific proteins. Because of the blood-brain barrier, brain-specific proteins are foreign to the immune system, which is thus able to mount an effective immune response to neuronal antigens expressed in peripheral tumors.[8] While the details of disease pathogenesis remain under investigation,[9] the Darnell laboratory established methods to use the high titer antibodies in PND patients to screen expression cDNA libraries and identify the genes encoding a number of target PND antigens.[8,10]

One set of PND antigens that tumor cells consistently express are neuron-specific RNA binding proteins. Two families of such proteins were discovered by using PND antisera to screen cDNA libraries: the Nova proteins,[11] targeted in patients harboring lung or gynecologic cancers and manifesting neurological symptoms of excess motor movements ("POMA") and the Hu proteins.[12] While the functions of the Hu proteins in the brain are still incompletely understood,[13] we have been able to establish that Nova is only expressed in central nervous system neurons, where it regulates alternative splicing in an interesting subset of pre-mRNAs.[14] A combination of genetic, biochemical and bioinformatic studies of Nova-regulated RNAs has now established a crude template for attempting to understand RNA binding protein function on a genome-wide scale, which we will discuss below.

The Three Main Benefits of Splicing Regulation in the Brain

Alternative splicing enables the limited number of existing genes to generate the proteomic diversity required for neuronal structure and function. For example, each of the three neurexin genes contains multiple alternative exons and one alternative promoter, allowing for the generation of over a thousand different isoforms.[3] Specific alternative ***neurexin*** isoforms at the presynaptic neuron preferentially bind to a specific postsynaptic neuroligin isoform and this specificity of interaction contributes to the specificity of synaptic connections A receptor 1 (NR1) subunit mRNA, with the proximal 3' splice site (C2' variants) predominating upon activity blockade and the distal 3' splice site (C2 variants) upon increased activity.[15] Splicing of *NR1* exon 20 could contribute to the regulation of synaptic connections.[16,17] With the use of splicing microarrays (refer to chapter by Calarco et al) it became clear that brain-specific regulation of alternative splicing is at least as precisely regulated as transcription.[18,19] Analysis of splicing changes in $Nova1^{-/-}$ and $Nova2^{-/-}$ mouse brains using a custom Affymetrix splicing microarray revealed that most exons were unchanged, while a small set showed profound changes (over 20% change in exon inclusion). Statistical analysis supported the conclusion that Nova acts on a small, nonrandom subset of genes.[18] Most of these genes are ubiquitously expressed and in nonneuronal cells mainly function in cell-cell communication. In neurons, almost all of these genes encode proteins with synapse-related functions, suggesting that by changing the splicing patterns of these genes, Nova is able to reconfigure the ubiquitous cell-cell communication functions for the purpose of synaptic communication.[18]

A second benefit of alternative splicing in the brain is the expansion of the spectrum of responses to neural activity. Splicing of numerous alternative exons has already been found to respond to neuronal activity. One such example is the alternative utilization of 3' splice sites of exon 20 (previously referred to as exon 22) of the *NMDA receptor 1* (*NR1*) subunit, with the proximal 3' splice site (C2' variants) predominating upon activity blockade and the distal 3' splice site (C2 variants) upon increased activity.[15] Splicing of *NR1* exon 20 could contribute to the regulation of synaptic plasticity since it affects trafficking of the NMDA receptor from the ER to the synapse.[15] Interestingly, exon 20 is preceded by an alternatively spliced exon 19 (C1 domain), which also regulates NMDA receptor localization when transfected into mouse fibroblasts.[20] Two other exons shown to be regulated in response to activity are the stress axis-regulated exon (STREX) in *BK* (previously referred to as *Slo*) potassium channel mRNA, which changes the firing properties of the channel[21,22] and an exon in *Apolipoprotein E receptor 2* mRNA *(Apoer2)*, which is required for Reelin-dependent enhancement of Long Term Potentiation (LTP).[23]

A third benefit of alternative splicing in the brain is the contribution to functional diversification among neuronal sub-populations.[24] One such example is differential splicing of calcium channel *N-type* $Ca_V2.2$ (*Cacna1b*) in nociceptive sensory neurons. This type of sensory neuron expresses an isoform $Ca_V2.2e(37a)$, containing a unique sequence in its C-terminus due to inclusion of exon 37a. $Ca_V2.2e(37a)$ channels remain open for longer and form a higher density of functional channels than the isoform $Ca_V2.2e(37b)$, which is predominantly expressed in other neuronal types.[24,25] The best studied example of splicing-dependent generation of neuronal diversity is alternative splicing of *Dscam* in *Drosophila*, which can potentially generate over 38,000 different isoforms (refer to chapter by Park and Graveley).[26,27] Different neurons express unique repertoires of *Dscam* isoforms[28] and this diversity is required to ensure fidelity and precision of neuronal connectivity.[29]

Understanding the Nature of Protein-RNA Interactions

An essential foundation in approaching RNA regulation is a detailed understanding of the nature of regulatory protein-RNA interactions (Fig. 1). As a first step, we approached this problem by undertaking in vitro RNA selection experiments, using protocols established by Jack Szostak and colleagues.[30,31] An idealized Nova target RNA was identified using long random RNA libraries and full-length recombinant Nova protein. This led to identification of a stem-loop RNA harboring three repeats of a core motif, $(UCAU)_3$.[32] Mutagenesis studies identified the CA dinucleotide as a critical invariant component of binding, with some flexibility allowing the flanking U nucleotide to be replaced by a C nucleotide, generating the consensus Nova binding motif, YCAY. These studies were complemented by the results from a collaboration with the crystallography laboratory of Stephen Burley, whose lab delimited a core protease-resistant region surrounding the KH domains that successfully formed crystals suitable for structure determination. Repeating these studies in the presence of RNAs demonstrated that the Nova2 KH3 domain is protease resistant only when Nova is bound to RNA, suggesting a role for this domain in RNA binding.[33] RNA selection experiments were used again to optimize an RNA target for crystallography with the Nova KH3 domain,[34] which led to a high resolution X-ray structure of the Nova KH3 domain bound to a short hairpin containing the UCAC motif.[35] This structure demonstrated that the Nova KH domain folds to position side-chain amino acids to precisely contribute hydrogen bond donor/acceptors in a manner that mimics a second nucleic acid strand.

Identification of a YCAY cluster as the in vitro Nova binding site encouraged us to search for brain transcripts containing such clusters. Our first very crude approach used Microsoft Word as a search tool to examine by hand intronic and exonic sequences in a database of 350 neuronal transcripts, which had been established at CSHL by Stamm and Helfman.[36] We identified one YCAY cluster in this database within an intronic sequence upstream of an alternatively spliced exon (E3A) of the inhibitory gene *glycine receptor 2* (*GlyR2*). Generation of a minigene encoding this element and the surrounding exons demonstrated that in transfected tissue culture cells, Nova was able to mediate an increase in inclusion of E3A, which required the intact YCAY cluster.[32] *GlyR2* splicing was later assessed in $Nova1^{-/-}$ mice and a consistent 2-fold decrease in E3A utilization was observed.[37]

To assay the specificity of Nova action, splicing of a small, randomly chosen pool of alternatively spliced transcripts was analyzed in the $Nova1^{-/-}$ mouse brain. We found that splicing of the tested exons were unaffected, with the exception of a second fortuitously identified Nova-regulated exon, the alternatively spliced 2L exon of the $GABA_A$ transcript.[37] Identification of the 2L exon was thus independent of a search for YCAY clusters and this prompted a biochemical characterization of the element in the $GABA_A$ transcript able to mediate the action of Nova. These studies identified a 24 nucleotide element within the intron downstream of the 2 L exon able to confer Nova-dependent splicing on a heterologous transcript and sequencing of this element revealed that it contains four YCAY elements.[38] Biochemical studies of this and one additional Nova regulated exon (an

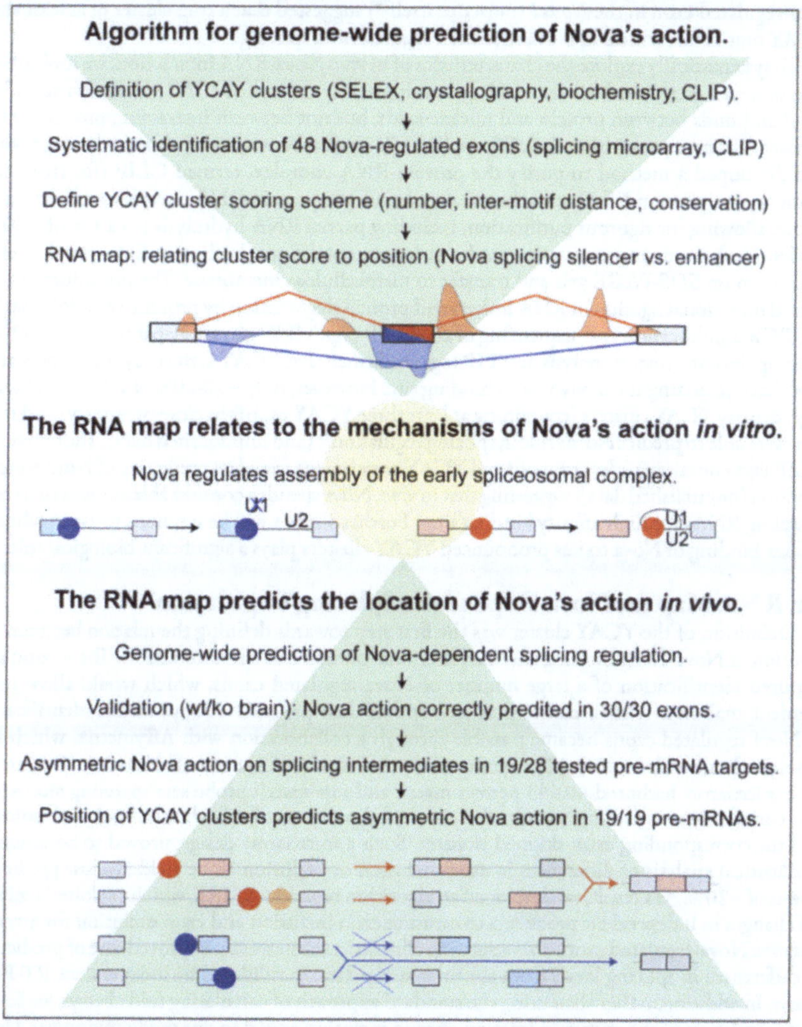

Figure 1. General scheme to define and validate the RNA map and relate it to the mechanisms of Nova-dependent splicing regulation. Prior work established YCAY clusters as legitimate target sequences for analysis, based on RNA selection, biochemistry/X-Ray crystallography[32] and new methods for identification and genetic validation of Nova splicing targets.[18,37,41] These targets were analyzed by a computational procedure to define the Nova RNA map of splicing regulation (upper triangle). The RNA map was related to the mechanism of Nova action through biochemical studies in a reconstituted splicing system in vitro (circle). Two examples illustrate Nova inhibition of exon inclusion by binding NESS2 element and blocking U1 snRNP assembly on the pre-mRNA and Nova upregulation of exon inclusion by binding NISE2/3 element to enhance spliceosome assembly.[14] The RNA map was able to predict the action of Nova on new splicing targets (bottom triangle). Furthermore, quantification of splicing intermediates in *Nova* double knockout brain detected 19 cases of asymmetric Nova action on removal of the two introns flanking the alternative exons. In all of these cases, the Nova binding site predicted by the RNA map locates to the Nova-regulated intron.[14]

auto-regulated exon in the *Nova1* transcript itself[39]) suggested that a core cluster of at least three YCAY motifs was critical in Nova-mediated regulation of splicing.

To systematically explore the characteristics of in vivo Nova-RNA interaction, we took advantage of a well established biochemical approach, namely the use of UV-B irradiation to induce covalent bonds between protein and nucleic acids, but not between interacting proteins, when contact distances are within ~1 Å. We applied UV-irradiation to acutely dissected mouse brains and developed a method to purify the protein-RNA complex, termed CLIP (for *c*ross-*l*ink-*i*mmuno*p*recipitation.[40,41] Due to the covalent bond, the protein-RNA complexes are extremely stable, allowing for rigorous purification, including partial RNA hydrolysis (to a size of ~40-70 nucleotides), immunoprecipitation under stringent conditions, boiling in SDS-sample buffer, separation on SDS-PAGE gels and transfer to nitrocellulose membrane. The procedure also involved directional ligation of RNA linkers and protein degradation by proteinase K, followed by RT-PCR amplification and sequencing of the CLIP tags. CLIP gives a snapshot of in vivo Nova binding; bioinformatics analysis of CLIP tags confirmed that YCAY is the only over-represented motif characterizing the in vivo Nova binding site. However, only ~10-20% of CLIP tags harbor high density YCAY clusters (containing at least three YCAY motifs in close proximity) and these tags were able to predict Nova's role in splicing regulation[41] (and unpublished data). The remaining CLIP tags containing a less pronounced YCAY enrichment were less predictive of Nova splicing function (unpublished data), suggesting that in vivo, Nova spends a considerable amount of its time sampling RNAs for high affinity binding sites. Further studies will be required to assess whether weaker binding of Nova to less pronounced YCAY clusters plays a significant biological role.

An RNA Map for Nova-Dependent Splicing Regulation

Definition of the YCAY cluster was the first step towards defining the relation between the position of Nova-RNA binding and the functional outcome of this interaction. The second step required identification of a large number of Nova-regulated exons, which would allow independent analysis of YCAY cluster positions within their pre-mRNAs. Systematic identification of Nova-regulated exons became possible through a collaboration with Affymetrix, which had developed a new alternative splicing microarray that was itself in need of validation. This prototype microarray harbored 40,443 perfect match and mis-match probe sets spanning alternative and constitutive exons; importantly, these included probe sets for both exon-included isoforms and the corresponding exon-skipped isoform. Such a microarray design proved to be essential, as statistical analysis of differences in exon inclusion or exclusion alone yielded a low predictive power of ~20%. As a result, we developed an algorithm termed ASPIRE, which exploited reciprocal changes in independent probe sets to measure exon inclusion and exon exclusion for any one putative Nova-regulated exon. This approach afforded a substantially improved rate of prediction of differences in splicing levels from the microarray data, as validated by independent RT-PCR assays. In addition, rather than using the standard approach of calculating fold change, we found that calculation of the change in fraction of exon inclusion added to our predictive power. Using ASPIRE, we were able to validate 49/49 of our top predicted Nova regulated exons identified by comparing *Nova2*$^{-/-}$ and wildtype postnatal day 7 neocortex RNA.[18]

Identification of 49 additional Nova-regulated exons allowed us to determine genome-wide rules that correlate the position of YCAY clusters with Nova action, which we termed an RNA map (Fig. 1).[14] We developed an algorithm to score transcripts as potential Nova targets on the basis of YCAY cluster density and location within their RNA map and used it to search a library of alternative exons to predict 51 Nova-regulated exons based on high YCAY scores. When tested in a *Nova1*$^{-/-}$/*Nova2*$^{-/-}$ brain, 30 of these exons showed splicing defects. In support of the RNA map validity, the position of YCAY clusters correlated well with the action of Nova on alternative exon inclusion or exclusion.[14] Clusters upstream (NISS1, NISS2) or within alternate exons (NESS1, NESS2) predicted Nova-dependent inhibition of exon inclusion, whereas clusters downstream of alternative exons (NISE2, NISE3) predicted Nova-dependent enhancement of exon inclusion (Figs. 1, 2). Thus, a well-defined RNA binding map was generated, in which both the density and position of Nova binding elements predicted the functional outcome of Nova binding.

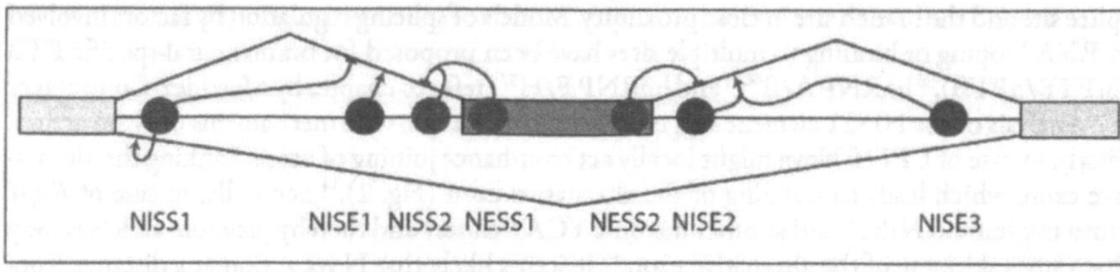

Figure 2. Mechanistic implications of the RNA map. The schematic RNA map indicates the proposed mechanisms of Nova action when binding in the vicinity or at a distance from the alternative exon. When Nova binds in vicinity of alternative exon, local enhancement of the proximal splice site leads to exon inclusion (via NISE1 or NISE2) and local inhibition leads to exon skipping (via NISS2, NESS1 or NESS2). However, Nova action via distant sites might involve various mechanisms, as is suggested by the cases of NISE3 and NISS1 elements. Nova action via NISE3 elements may require cooperatively with NISE2 to promote formation of an RNA loop that brings the 5` splice site and the branch site in close proximity. Analysis of the NISS1 elements suggests at least two possible mechanisms: Nova might locally act to enhance joining of exons flanking the alternative exon, or promote formation of an RNA loop that leads to skipping of the alternative exon.[14] Further biochemical analysis will be required to understand the potential mechanisms of Nova action via distant binding sites.

Relating the RNA Map to Mechanisms of Splicing Regulation

The correlation between the position of YCAY clusters and Nova's action to silence or enhance alternative exon inclusion was shown to relate to Nova's mechanism of action (Figs. 1, 2). Analysis of Nova's action on a NESS2 splicing silencer demonstrated that Nova interferes specifically with U1 snRNP, consistent with the location of the YCAY cluster in the vicinity of the 5' splice site, while U2 snRNP is able to assemble on these blocked transcripts.[14] A model whereby Nova binding to YCAY clusters blocks access to the constitutive splicing machinery is reminiscent of the mechanisms by which Sxl,[42] hnRNPA1[43,44] and PTB[45,46] displace U2AF[65] or U2 snRNP, a mechanism Nova might also use when binding in the vicinity of the 3' splice site (NISS2 and NESS1 elements; Figs. 1, 2).

Nova-dependent splicing enhancer elements are most often located in introns, either downstream of the 5' splice site (NISE2; Fig. 2) or upstream of the branch site (NISE1, NISE3; Fig. 2). In vitro splicing assays demonstrate that Nova acts on an NISE element to promote spliceosome assembly.[14] While splicing enhancers have most often been associated with exonic sequences,[47-50] in some instances proteins such as TIA1,[45] KSRP, hnRNP F and H,[51,52] Fox1 and Fox2[53] promote spliceosome assembly by binding to intronic sequences downstream of the exon. Interestingly, Nova-regulated exons flanked by NISE2 elements are generally shorter than 50 nt,[14] consistent with previous suggestions[54] that intronic enhancer elements are particularly important for regulating short exons, which themselves have a lower chance of harboring exonic splicing enhancers.

To further explore the relation of the RNA map to the location of Nova action in vivo, we quantified splicing intermediates in *Nova1*$^{-/-}$/*Nova2*$^{-/-}$ brain. These experiments demonstrated that Nova acts on the intron harboring or proximate to YCAY binding elements, while generally having little or no effect on splicing of the other intron flanking the alternative exon (Fig. 1). These results support the in vitro finding that Nova acts locally on spliceosome assembly, enhancing splicing of the 3' and 5' intron via NISE1 and NISE2/NISE3, respectively and silencing splicing of the 3' and 5' intron via NISS2/NESS1 and NESS2, respectively (Fig. 2).[14]

Thus when Nova binds in the immediate vicinity of an alternative exon, the mechanistic model is straightforward: local enhancement of the proximal splice site leads to exon inclusion and local inhibition leads to exon skipping (Fig. 2). However, when Nova binds at a distant site, its action may be more complex, as suggested by the cases of NISE3 and NISS1 elements (Fig. 2). Of the 13 Nova-target pre-mRNAs with NISE3 elements, 10 also contained a NISE2 element,[14] suggesting that Nova could bind two sites and multimerize[33,55] to form an RNA loop, thereby bringing the 5' splice site and the branch site in close proximity. Models of splicing regulation by factors involved in RNA looping or binding to multiple sites have been proposed for brain/neural-specific PTB (brPTB/nPTB),[56] hnRNP A/B[44,57] and hnRNP F/H[57] (refer to chapter by Martinez-Contreras et al). Analysis of the NISS1 element suggests two additional possible mechanisms of Nova action. Firstly, in case of *CP110* Nova might locally act to enhance joining of exons flanking the alternative exon, which leads to skipping of the alternative exon (Fig. 2).[14] Secondly, in case of *Rap1* Nova might bind NISS1 and another intronic YCAY cluster and thereby promote an RNA loop that causes skipping of the alternative exon.[14] It seems likely that Nova action at a distance from the alternative exon probably involves various mechanisms and that their full understanding will require further biochemical analysis.

Pre-mRNAs containing exons that are silenced in the presence of Nova most often display some asymmetry of splicing intermediates even in the absence of Nova. In each such instance, YCAY cluster location allows Nova to inhibit a predetermined rate-limiting step for exon inclusion. These data support previous biochemical studies indicating that pre-mRNAs often follow a preferential splicing order[58-62] and suggest that such a preferential splicing order may allow proteins such as Nova to change a splicing outcome by regulating the excision of one of the two introns that flank an alternative exon. As a general approach, studies of Nova suggest that a detailed understanding of the nature of the protein-RNA interaction, together with identification of a large set of RNA targets validated in a genetic system, can be combined with bioinformatics and biochemistry to identify the sites and mechanisms of action of alternative splicing factors.

Is Combinatorial Splicing Regulation Cooperative or Additive?

Regulation of neuronal splicing often involves a combination of several RNA-binding proteins, including ubiquitous factors, acting in concert to determine the inclusion level of an alternative exon.[63-65] There are several well studied examples of combinatorial control of neuronal splicing, including the identification of up to 6 factors that promote neuronal usage of the exon N1 in *c-src* proto-oncogene mRNA[46,66] and several factors involved in the regulation of *gamma-aminobutyric acid A receptor, gamma 2* (*GABAR-γ2*) or *NMDAR1* exon 19.[18,38,52,67,68] In vivo studies have shown that the brain-specific RNA-binding protein Nova2 is required for the inclusion of *NR1* exon 19, as this exon is rarely included in the *NR1* mRNA detected in *Nova2*$^{-/-}$ forebrain.[18] Evidence from minigene assays in cell cultures suggests that ubiquitous heterogeneous nuclear ribonucleoproteins A1 and H (hnRNP A1 and hnRNP H) and neuroblastoma apoptosis-related RNA-binding protein (NAPOR, also called CUGBP2 or ETR-3[68]) also regulate *NR1* exon 19 inclusion.[52] Loss of forebrain-specific splicing of *NR1* exon 19 in *Nova2*$^{-/-}$ brain thus suggests that the combinatorial control of *NR1* exon 19 acts in a cooperative, rather than additive manner, allowing the activity of a single factor to cause a major splicing change.[2] Another example of combinatorial control is the ability of brPTB to antagonize the action of Nova on *GlyRα2* exon 3A by binding adjacent to the Nova binding site.[69] Even though the large splicing changes in *Nova1*$^{-/-}$/*Nova2*$^{-/-}$ neocortex tissue suggest that Nova might be the primary or sole regulatory factor for many of its target exons, the case of *NR1* exon 19 serves as an example for other exons which Nova might regulate in a cooperative manner together with other splicing factors.

Modular Structure of Coregulated Transcripts

Cellular protein networks can be thought of as composed of functional units, or modules.[70,71] The modules encapsulate specific functions and contain the control circuits, composed both of negative and positive feedback, whereby the output is monitored and regulated. Indications of modular regulation of alternative splicing were initially offered by studies of sex determination in *Drosophila melanogaster*.[72-75] The early studies identifying *GlyR2* and *GABA*$_A$ transcripts as Nova targets suggested the possibility that Nova acts on a biologically restricted set of RNAs and that those RNAs might relate to the pathogenesis of the inhibitory motor dysfunction evident in patients with the paraneoplastic POMA syndrome.[6] Unbiased identification and validation of a large number of Nova-regulated alternative exons via CLIP,[41] splicing microarray[18] and bioinformatics[14] confirmed this prediction, demonstrating that alternative splicing in mammals can be regulated in a functionally coherent, modular fashion. Analysis of splicing microarray data showed that all of the proteins encoded by Nova target RNAs with known functions in the brain act either in the synapse (34 out of 40) and/or in axon guidance (8 out of 40). Moreover, of the 35 proteins with known interaction partners, 74% (26 out of 35) interact with at least one other protein from the same group, suggesting that they form synaptic functional modules.[18]

To explore the biological relevance of Nova target transcripts, we collaborated with L. Jan's group who discovered a new type of synaptic plasticity: long-term potentiation of slow inhibitory postsynaptic currents (sIPSC) mediated by $GABA_B$ receptors and G protein-activated inward rectifier potassium channel 2 (GIRK2) channels,[76] two proteins encoded by RNAs that are Nova CLIP targets.[41] Moreover, LTP of sIPSC can be potentiated by activation of NMDA-Rs and CaMKII[76] and the RNAs encoding these proteins are Nova splicing targets.[18] Electrophysiological analysis of CA1 hippocampal neurons showed that LTP of sIPSC was specifically lost in *Nova-2*$^{-/-}$ neurons, while basal synaptic transmission for excitation, inhibition and LTP of excitatory postsynaptic currents was normal.[76] Taken together, the study of *Nova2*$^{-/-}$ brain shows that a splicing factor can regulate a specific aspect of tissue-specific function via a set of functionally coherent RNAs.[2]

Evolutionary Considerations

In the RNA world-view, RNA molecules were the first informatic and enzymatic dual-function molecules to arise in evolution.[77] How then did proteins evolve to harness the power of RNA? This question is directly approached when considering how RNA-binding proteins evolved to regulate alternative splicing. Our studies with Nova point out some interesting issues in considering this problem. The Nova binding site is of rather low complexity; a cluster of three YCAY motifs within 7 nucleotides of each other should occur on average approximately once every 5,000 nucleotides. Given that the affinity of Nova for RNA increases gradually with an increased number of YCAY motifs and that varying the position of YCAY cluster within the RNA map can change Nova's activity, Nova splicing silencer and enhancers could evolve by step-wise mutations. This step-wise process would enable exons to gradually sample the consequences of evolving a Nova binding site and achieve an optimum outcome by changing the position of the site and the density of YCAY motifs.

Preliminary analysis of the evolutionary conservation of exons that were identified as Nova targets in the mouse brain has shown that approximately one half has gained Nova-binding sites during recent vertebrate evolution from fish to mammals.[80] Analysis of the splicing of Nova-regulated exons in different species suggests that the introduction of YCAY clusters into the pre-mRNA might often be the primary mutation that enabled brain-specific splicing of these exons.[80] Alternative splicing of *Dab1* exons 7b and 7c (termed 555* in a previous study[78]) provides an example of the correlation between evolution of YCAY clusters and the corresponding exons (Fig. 3). The peptide sequences encoded by the two exons are highly related, suggesting a duplication event during evolution;[78] interestingly, YCAY clusters were duplicated together with the two exons (Fig. 3). Exons 7b and 7c are included in mRNA of proliferating neuronal precursors, but are absent from mRNA of differentiated neurons, which correlates with the induction of Nova expression (Buckanovich, Park and Darnell, unpublished observation). We observe that the YCAY motifs within NISS2 clusters are more conserved than the sequence of exons 7b and 7c themselves (Fig. 3), indicating the high selective pressure for preserving ability of Nova to silence inclusion of these exons in brain, which might be a prerequisite for the function of Dab1 in neuronal outgrowth.[79]

Flexibility in position and sequence of functional YCAY clusters may have enabled evolution to refine the biologic coherence of transcripts harboring Nova binding sites. On the other hand, while individual YCAY clusters are flexible to evolve, the Nova RNA binding domain is under a stricter evolutionary constraint. This constraint is essential, as mutations altering the recognition motif of Nova would simultaneously destroy the regulation of an array of crucial alternate exon information. In fact, the Nova KH domains are 98% identical between Nova1 and Nova2 of all vertebrate species with sequenced genomes[11] (and data not shown), suggesting that the ability to

```
mouse   GTCATCATTAAACTCATAAATCCATGTCATTTGTGTGTCATTCACTCCCTCCTCATTCATCATTCCCTCG
human   GTCATCATTAAACTCATAAATCCATGTCATTTGTGTGTCATTCACTTCATCCTCATTCATCATCCCTTC-
opossum CTCATCATC---CTCATAAATTCATGTCATG-GTATGTC--TCATTTCATGTCCATTCATCCTGCCTTC-
chick   AGATTTAGTACGGATAAATGGTGTATTTCCCCAATCTTGCAGCATTTAAAATCGTTTTGACTGTGTGATT
         TCATCAT    CTCATAAATYCATGTCAT  GT TGTC  TCAYT C T  YCATTCATC T CC TC

mouse   CACTTGCAGAATAGCCAGCCGCTGGAGGATTTCGACTCGCGCTTTGCCGCAGCCACGCCGGTAAGAATGC
human   CACTTTCAGAATGGCTATTCGTTTGAGGATTTTGAAGAACGGTTTGCTGCAGCCACCCCGGTAAGAATT-
opossum CTGTTCCAGAATGGTCACTCCTTTGATGACTTTGAGGAGCGCTTTGCAGCAGCCATACCGGTAAGGCTGG
chick   TCGTATCTGTACTCACTGCAAGATGGAAAGGGTTTTCGTTTCAATAGTGGAGTCAATAATTACGGAGGTT
        C  TT CAGAAT G  A  C  T GA GA TT GA    CG TTTGC GCAGCCA  CCGGTAAG  T

mouse   GAGGGCCTTGTCTGTTTAACTGTTTCATGTTGTCGGACCATCACTCATTCTGTGCTCATCCATCATGCCT
human   -ACAGCCCTCTCTGGTTAACTGTTTCATGTCATTTTCTCATCATTCATTTCATGCTCATCAGTCATGCCT
opossum GAAGTCCCTCTTCAATTCATTTATTGCAATTGTCTTCATGTCATTCATTTCATTTTCATACATCATGCCT
chick   ATCATTTTCATGCTCAGAAATTCATGTCATTTTCATAAAATCATTCATTTCATT-TCATAAATCATGCTT
                     T       A T   T    T T      TCAYTCATT   T  TCAT   TCATGC T

mouse   TGTGGTTGCAGAACAGGAACCTGTCAATGGACTTTGATGAGCTTCTCGAGGCAACCA---AGGTGAGTCT
human   TGTGGTTACAGAACAGAAACCTGCCCACAGACTTTGATGAGATTTTTGAGGCAACGA---AGGTAAGGCC
opossum TGTGGTTTCAGAACCGAAACTCAATGGTGGACTTTGACGAGTGTTACAGGGTCGAAAATCAGGTAAGACT
chick   TACAGCTGGAGAATGGAAACTTATTGCTGGACATTGATGAAAATCTTGGTTCAGTCACTCAGGTAAGGTT
        T   G T  AGAA  G AAC          GAC TTGA GA   T             A   AGGT AG
```

Figure 3. Evolutionary alignment of YCAY clusters in *Dab1* pre-mRNA. Alignment of orthologous sequences containing exons 7b and 7c (termed 555* in a previous study[78]) and surrounding intronic sequence containing NISS2 elements in *Dab1* pre-mRNA. YCAY motifs are gray, conserved motifs are bold gray, and exons 7b and 7c are shaded. In human, mouse and opossum, these two exons are ~50 bp long, separated by an intron of ~90 bp. The chick genome lacks the first of the two exons, as well the corresponding the NISS2 YCAY cluster, suggesting that this YCAY cluster has co-evolved with the exon 7b in mammals.

bind a YCAY motif became a powerful and unalterable facet of Nova-RNA regulation early in evolution. This turns RNA regulation on its head in a way that is in harmony with the idea of the RNA world: RNA remains the powerful emerging evolutionary force, while the protein regulators take on roles as relatively inert drones to mediate the regulation that RNA demands.

An interesting question for the future will be to explore the extent to which Nova might also contribute to two other benefits of alternative splicing in neurons discussed at the beginning of this review, i.e., the response to neuronal activity and generation of neuronal complexity. Analysis of *Nova1*$^{-/+}$ and *Nova2*$^{-/+}$ brains showed that Nova-regulated exons respond to the presence of Nova in a dose-dependent manner and that different exons have a different threshold for Nova action,[37] (Ruggiu, Wang and Darnell, unpublished observations). Thus, titration of Nova levels within different neurons may generate variety in synaptic function between different types of neurons. Correlation of sensitivity to Nova-dependent regulation with YCAY cluster scores may provide a means of evaluating this notion. Furthermore, there are many thousands of synapses within a single neuron and the question arises as to whether Nova regulation might contribute to the activity-dependent plasticity of specific synapses. The exciting finding that Nova regulates LTP of sIPSC,[76] together with the finding that Nova is present at neuronal synapses (Darnell, Triller et al unpublished observations), suggests that there may remain yet undiscovered dimensions to the ways in which Nova-dependent regulation of alternative splicing in combination with other steps in RNA regulation may regulate neuronal complexity.

Acknowledgements

The authors wish to acknowledge support from the NIH (R01 NS34389 and NS40955) and the Howard Hughes Medical Institute. RBD is an Investigator of the Howard Hughes Medical Institute.

References

1. Hong EJ, West AE, Greenberg ME. Transcriptional control of cognitive development. Curr Opin Neurobiol 2005; 15(1):21-28.
2. Ule J, Darnell RB. RNA binding proteins and the regulation of neuronal synaptic plasticity. Curr Opin Neurobiol 2006; 16(1):102-110.
3. Ullrich B, Ushkaryov YA, Sudhof TC. Cartography of neurexins: more than 1000 isoforms generated by alternative splicing and expressed in distinct subsets of neurons. Neuron 1995; 14(3):497-507.
4. Darnell RB. RNA logic in time and space. Cell 2002; 110(5):545-550.
5. Rosenfeld MG, Mermod JJ, Amara SG et al. Production of a novel neuropeptide encoded by the calcitonin gene via tissue-specific RNA processing. Nature 1983; 304(5922):129-135.
6. Darnell RB, Posner JB. Paraneoplastic syndromes involving the nervous system. N Engl J Med 2003; 349(16):1543-1554.
7. Darnell RB, Posner JB. Paraneoplastic syndromes affecting the nervous system. Semin Oncol 2006; 33(3):270-298.
8. Darnell RB. Onconeural antigens and the paraneoplastic neurologic disorders: at the intersection of cancer, immunity and the brain. Proc Natl Acad Sci USA 1996; 93(10):4529-4536.
9. Albert ML, Darnell RB. Paraneoplastic neurological degenerations: keys to tumour immunity. Nat Rev Cancer 2004; 4(1):36-44.
10. Newman LS, McKeever MO, Okano HJ et al. Beta-NAP, a cerebellar degeneration antigen, is a neuron-specific vesicle coat protein. Cell 1995; 82(5):773-783.
11. Buckanovich RJ, Posner JB, Darnell RB. Nova, the paraneoplastic Ri antigen, is homologous to an RNA-binding protein and is specifically expressed in the developing motor system. Neuron 1993; 11(4):657-672.
12. Szabo A, Dalmau J, Manley G et al. HuD, a paraneoplastic encephalomyelitis antigen, contains RNA-binding domains and is homologous to Elav and Sex-lethal. Cell 1991; 67(2):325-333.
13. Musunuru K, Darnell RB. Paraneoplastic neurologic disease antigens: RNA-binding proteins and signaling proteins in neuronal degeneration. Annu Rev Neurosci 2001; 24:239-262.
14. Ule J, Stefani G, Mele A et al. An RNA map predicting Nova-dependent splicing regulation. Nature 2006; 444(7119):580-586.
15. Mu Y, Otsuka T, Horton AC et al. Activity-dependent mRNA splicing controls ER export and synaptic delivery of NMDA receptors. Neuron 2003; 40(3):581-594.
16. Chih B, Gollan L, Scheiffele P. Alternative splicing controls selective trans-synaptic interactions of the neuroligin-neurexin complex. Neuron 2006; 51(2):171-178.

17. Boucard AA, Chubykin AA, Comoletti D et al. A splice code for trans-synaptic cell adhesion mediated by binding of neuroligin 1 to alpha- and beta-neurexins. Neuron 2005; 48(2):229-236.
18. Ule J, Ule A, Spencer J et al. Nova regulates brain-specific splicing to shape the synapse. Nat Genet 2005; 37(8):844-852.
19. Pan Q, Shai O, Misquitta C et al. Revealing global regulatory features of mammalian alternative splicing using a quantitative microarray platform. Mol Cell 2004; 16(6):929-941.
20. Ehlers MD, Tingley WG, Huganir RL. Regulated subcellular distribution of the NR1 subunit of the NMDA receptor. Science 1995; 269(5231):1734-1737.
21. Xie J, Black DL. A CaMK IV responsive RNA element mediates depolarization-induced alternative splicing of ion channels. Nature 2001; 410(6831):936-939.
22. Xie J, McCobb DP. Control of alternative splicing of potassium channels by stress hormones. Science 1998; 280(5362):443-446.
23. Beffert U, Weeber EJ, Durudas A et al. Modulation of synaptic plasticity and memory by Reelin involves differential splicing of the lipoprotein receptor Apoer 2. Neuron 2005; 47(4):567-579.
24. Lipscombe D. Neuronal proteins custom designed by alternative splicing. Curr Opin Neurobiol 2005; 15(3):358-363.
25. Castiglioni AJ, Raingo J, Lipscombe D. Alternative splicing in the C-terminus of CaV2.2 controls expression and gating of N-type calcium channels. J Physiol 2006.
26. Celotto AM, Graveley BR. Alternative splicing of the Drosophila Dscam pre-mRNA is both temporally and spatially regulated. Genetics 2001; 159(2):599-608.
27. Schmucker D, Clemens JC, Shu H et al. Drosophila Dscam is an axon guidance receptor exhibiting extraordinary molecular diversity. Cell 2000; 101(6):671-684.
28. Zhan XL, Clemens JC, Neves G et al. Analysis of Dscam diversity in regulating axon guidance in Drosophila mushroom bodies. Neuron 2004; 43(5):673-686.
29. Chen BE, Kondo M, Garnier A et al. The molecular diversity of Dscam is functionally required for neuronal wiring specificity in Drosophila. Cell 2006; 125(3):607-620.
30. Szostak JW, Ellington AD. In vitro selection of functional RNA sequences: Cold Spring Harbor Laboratory Press, 1993.
31. Green R, Ellington AD, Bartel DP et al. In vitro genetic analysis: selection and amplification of rare functional nucleic acids. Methods Compan Methods Enzymol 1991; 2:75-86.
32. Buckanovich RJ, Darnell RB. The neuronal RNA binding protein Nova-1 recognizes specific RNA targets in vitro and in vivo. Mol Cell Biol 1997; 17(6):3194-3201.
33. Lewis HA, Chen H, Edo C et al. Crystal structures of Nova-1 and Nova-2 K-homology RNA-binding domains. Structure Fold Des 1999; 7(2):191-203.
34. Jensen KB, Musunuru K, Lewis HA et al. The tetranucleotide UCAY directs the specific recognition of RNA by the Nova K-homology 3 domain. Proc Natl Acad Sci USA 2000; 97(11):5740-5745.
35. Lewis HA, Musunuru K, Jensen KB et al. Sequence-specific RNA binding by a Nova KH domain: implications for paraneoplastic disease and the fragile X syndrome. Cell 2000; 100(3):323-332.
36. Stamm S, Zhang MQ, Marr TG et al. A sequence compliation and comparison of exons that are alternatively spliced in neurons. Nucl Acids Res 1994; 9:1515-1526.
37. Jensen KB, Dredge BK, Stefani G et al. Nova-1 regulates neuron-specific alternative splicing and is essential for neuronal viability. Neuron 2000; 25(2):359-371.
38. Dredge BK, Darnell RB. Nova regulates GABA(A) receptor gamma2 alternative splicing via a distal downstream UCAU-rich intronic splicing enhancer. Mol Cell Biol 2003; 23(13):4687-4700.
39. Dredge BK, Stefani G, Engelhard CC et al. Nova autoregulation reveals dual functions in neuronal splicing. EMBO J 2005; 24(8):1608-1620.
40. Ule J, Jensen K, Mele A et al. CLIP: A method for identifying protein-RNA interaction sites in living cells. Methods 2005; 37(4):376-386.
41. Ule J, Jensen KB, Ruggiu M et al. CLIP identifies Nova-regulated RNA networks in the brain. Science 2003; 302(5648):1212-1215.
42. Valcarcel J, Singh R, Zamore PD et al. The protein Sex-lethal antagonizes the splicing factor U2AF to regulate alternative splicing of transformer pre-mRNA. Nature 1993; 362(6416):171-175.
43. Del Gatto-Konczak F, Olive M, Gesnel MC et al. hnRNP A1 recruited to an exon in vivo can function as an exon splicing silencer. Mol Cell Biol 1999; 19(1):251-260.
44. Zhu J, Mayeda A, Krainer AR. Exon identity established through differential antagonism between exonic splicing silencer-bound hnRNP A1 and enhancer-bound SR proteins. Mol Cell 2001; 8(6):1351-1361.
45. Izquierdo JM, Majos N, Bonnal S et al. Regulation of Fas alternative splicing by antagonistic effects of TIA-1 and PTB on exon definition. Mol Cell 2005; 19(4):475-484.
46. Sharma S, Falick AM, Black DL. Polypyrimidine Tract Binding Protein Blocks the 5' Splice Site-Dependent Assembly of U2AF and the Prespliceosomal E Complex. Mol Cell 2005; 19(4):485-496.

47. Fairbrother WG, Yeh RF, Sharp PA et al. Predictive identification of exonic splicing enhancers in human genes. Science 2002; 297(5583):1007-1013.
48. Liu HX, Zhang M, Krainer AR. Identification of functional exonic splicing enhancer motifs recognized by individual SR proteins. Genes Dev 1998; 12(13):1998-2012.
49. Goren A, Ram O, Amit M et al. Comparative analysis identifies exonic splicing regulatory sequences—The complex definition of enhancers and silencers. Mol Cell 2006; 22(6):769-781.
50. Coulter LR, Landree MA, Cooper TA. Identification of a new class of exonic splicing enhancers by in vivo selection. Mol Cell Biol 1997; 17(4):2143-2150.
51. Chou MY, Rooke N, Turck CW et al. hnRNP H is a component of a splicing enhancer complex that activates a c-src alternative exon in neuronal cells. Mol Cell Biol 1999; 19(1):69-77.
52. Han K, Yeo G, An P et al. A combinatorial code for splicing silencing: UAGG and GGGG motifs. PLoS Biol 2005; 3(5):e.158
53. Underwood JG, Boutz PL, Dougherty JD et al. Homologues of the Caenorhabditis elegans Fox-1 protein are neuronal splicing regulators in mammals. Mol Cell Biol 2005; 25(22):10005-10016.
54. Carlo T, Sterner DA, Berget SM. An intron splicing enhancer containing a G-rich repeat facilitates inclusion of a vertebrate micro-exon. RNA 1996; 2(4):342-353.
55. Ramos A, Hollingworth D, Major SA et al. Role of dimerization in KH/RNA complexes: the example of Nova KH3. Biochemistry 2002; 41(13):4193-4201.
56. Chou MY, Underwood JG, Nikolic J et al. Multisite RNA binding and release of polypyrimidine tract binding protein during the regulation of c-src neural-specific splicing. Mol Cell 2000; 5(6):949-957.
57. Martinez-Contreras R, Fisette JF, Nasim FU et al. Intronic Binding Sites for hnRNP A/B and hnRNP F/H Proteins Stimulate pre-mRNA Splicing. PLoS Biol 2006; 4(2):e21.
58. Gattoni R, Keohavong P, Stevenin J. Splicing of the E2A pre-messenger RNA of adenovirus serotype 2. Multiple pathways in spite of excision of the entire large intron. J Mol Biol 1986; 187(3):379-397.
59. Gee SL, Aoyagi K, Lersch R et al. Alternative splicing of protein 4.1R exon 16: ordered excision of flanking introns ensures proper splice site choice. Blood 2000; 95(2):692-699.
60. Kessler O, Jiang Y, Chasin LA. Order of intron removal during splicing of endogenous adenine phosphoribosyltransferase and dihydrofolate reductase pre-mRNA. Mol Cell Biol 1993; 13(10):6211-6222.
61. Lang KM, Spritz RA. In vitro splicing pathways of pre-mRNAs containing multiple intervening sequences? Mol Cell Biol 1987; 7(10):3428-3437.
62. Tsai MJ, Ting AC, Nordstrom JL et al. Processing of high molecular weight ovalbumin and ovomucoid precursor RNAs to messenger RNA. Cell 1980; 22(1 Pt 1):219-230.
63. Caceres JF, Stamm S, Helfman DM et al. Regulation of alternative splicing in vivo by overexpression of antagonistic splicing factors. Science 1994; 265(5179):1706-1709.
64. Black DL. Mechanisms of alternative pre-messenger RNA splicing. Annu Rev Biochem 2003; 72:291-336.
65. Singh R, Valcarcel J. Building specificity with nonspecific RNA-binding proteins. Nat Struct Mol Biol 2005; 12(8):645-653.
66. Rooke N, Markovtsov V, Cagavi E et al. Roles for SR proteins and hnRNP A1 in the regulation of c-src exon N1. Mol Cell Biol 2003; 23(6):1874-1884.
67. Zhang L, Liu W, Grabowski PJ. Coordinate repression of a trio of neuron-specific splicing events by the splicing regulator PTB. RNA 1999; 5(1):117-130.
68. Zhang W, Liu H, Han K et al. Region-specific alternative splicing in the nervous system: implications for regulation by the RNA-binding protein NAPOR. RNA 2002; 8(5):671-685.
69. Polydorides AD, Okano HJ, Yang YY et al. A brain-enriched polypyrimidine tract-binding protein antagonizes the ability of Nova to regulate neuron-specific alternative splicing. Proc Natl Acad Sci USA 2000; 97(12):6350-6355.
70. Hartwell LH, Hopfield JJ, Leibler S et al. From molecular to modular cell biology. Nature 1999; 402(6761 Suppl):C47-52.
71. Teichmann SA, Babu MM. Gene regulatory network growth by duplication. Nat Genet 2004; 36(5):492-496.
72. Nagoshi RN, McKeown M, Burtis KC et al. The control of alternative splicing at genes regulating sexual differentiation in D. melanogaster. Cell 1988; 53(2):229-236.
73. Hedley ML, Maniatis T. Sex-specific splicing and polyadenylation of dsx pre-mRNA requires a sequence that binds specifically to tra-2 protein in vitro. Cell 1991; 65(4):579-586.
74. Demir E, Dickson BJ. fruitless splicing specifies male courtship behavior in Drosophila. Cell 2005; 121(5):785-794.
75. Manoli DS, Foss M, Villella A et al. Male-specific fruitless specifies the neural substrates of Drosophila courtship behaviour. Nature 2005; 436(7049):395-400.
76. Huang CS, Shi SH, Ule J et al. Common molecular pathways mediate long-term potentiation of synaptic excitation and slow synaptic inhibition. Cell 2005; 123(1):105-118.

77. Joyce GF. The antiquity of RNA-based evolution. Nature 2002; 418(6894):214-221.
78. Bar I, Tissir F, Lambert de Rouvroit C et al. The gene encoding disabled-1 (DAB1), the intracellular adaptor of the Reelin pathway, reveals unusual complexity in human and mouse. J Biol Chem 2003; 278(8):5802-5812.
79. Katyal S, Godbout R. Alternative splicing modulates Disabled-1 (Dab1) function in the developing chick retina. EMBO J 2004; 23(8):1878-1888.
80. Jelen N, Ule J, Zivin M and Darnell RB. Evolution of Nova-dependent splicing regulation in the brain. PLoS Genet 2007; 3(10):e173 [Epub ahead of print].

Chapter 10

Regulation of Alternative Splicing by Signal Transduction Pathways

Kristen W. Lynch*

Abstract

Alternative splicing is now recognized as a ubiquitous mechanism for controlling gene expression in a tissue-specific manner. A growing body of work from the past few years has begun to also highlight the existence of networks of signal-responsive alternative splicing in a variety of cell types. While the mechanisms by which signal transduction pathways influence the splicing machinery are relatively poorly understood, a few themes have begun to emerge for how extracellular stimuli can be communicated to specific RNA-binding proteins that control splice site selection by the spliceosome. This chapter describes our current understanding of signal-induced alternative splicing with an emphasis on these emerging themes and the likely directions for future research.

Introduction

To maintain viability, most, if not all, cells within an organism must be capable of responding to a changing environment. For example, neuronal and muscle cells must respond to activation to promote behaviors and movement; cells in the liver, kidney and intestines must regulate their metabolic pathways in response to changing nutrient and hormonal environments and lymphoid cells must respond to any immune challenge to prevent or control infection. Such flexibility requires that individual cells have the ability to change function rapidly and precisely in response to a given stimulus. In general, cellular responsiveness is accomplished through the activity of signal transduction cascades that transmit signals from the cell surface to the relevant cellular machinery, often involving alterations in the protein composition of the cells.

Changes in protein expression occur through many different mechanisms and much work has focused on signal-induced regulation of transcription and translation. However, in the past few years there has been a growing recognition of the importance of signal-induced changes in alternative splicing as a mechanism for mediating biologically relevant cellular responses. Thus, the interface of the splicing and signaling fields is an emerging area of study. This chapter focuses on this interface, with a particular emphasis on the mechanisms by which signal transduction pathways affect the activity of splicing regulatory proteins. While much remains to be discovered about this process, the recent elucidation of a few pathways and growing information on several others, has begun to provide clear paradigms for how signaling pathways can impinge upon the splicing machinery and the biologic implications of such regulation. These mechanistic paradigms, described below, are grouped into categories for clarity, but it should be noted that in many cases one signal-induced perturbation will trigger another (e.g., phosphorylation and localization), so the groupings below are somewhat arbitrary and should not be viewed as "either-or", but rather as allied mechanisms that can act together to ensure a particular functional effect on alternative splicing.

*Kristen W. Lynch—Department of Biochemistry, UT Southwestern Medical Center, Dallas, Texas 75390-9038, USA. Email: kristen.lynch@utsouthwestern.edu

Alternative Splicing in the Postgenomic Era, edited by Benjamin J. Blencowe and Brenton R. Graveley.

Molecular "Hubs" Help Link the Extracellular World to the Splicing Machinery

To a first approximation, signaling cascades are thought of primarily as cytoplasmic pathways that respond directly to changing environments and stimuli at the cell surface. In contrast to this cytoplasmic activity, pre-mRNA splicing occurs in the nucleus, well separated from cell surface receptors. However, many members of the SR and hnRNP splicing regulatory protein families (refer to chapter by Lin and Fu, and Martinez-Contreras et al) shuttle between the nucleus and cytoplasm, as do other splicing factors, proteins that modify the splicing machinery and signaling proteins.[1-3] Therefore, it is not surprising that a wide variety of interactions have been described between components of the RNA processing machinery and traditional signaling molecules, many of which are discussed below. In particular, a few proteins have gained attention as potential docking "hubs" that integrate and transmit molecular information between a variety of signaling pathways and the RNA processing machinery. These molecular hubs include hnRNP K and Sam68.

HnRNP K contains three KH-type RNA binding domains and typically binds to C-rich sequences in RNA (for a review see ref. 4). Interspersed between the KH domains are proline-rich regions and sites of tyrosine phosphorylation which, respectively, bind SH3 and SH2 domains-protein-protein interaction motifs that are ubiquitous amongst signaling molecules. Not surprisingly, hnRNP K has been shown to bind to a wide variety of signaling proteins including Vav, Src-family kinases and various PKC isoforms. In addition, the Src-kinases, PKCs, Erk1/2 and JNK can all directly phosphorylate specific residues within hnRNP K and thereby regulate the various protein-protein and protein-RNA interactions involving this hnRNP protein.[4] The majority of studies related to the function of hnRNP K focus on its role in controlling mRNA stability and translation in the cytoplasm. However, this protein is present in nuclear extract preparations and has been found to interact with other hnRNPs and SR proteins that are involved in the regulation of splicing.[5,6] HnRNP K has also been shown to bind splicing enhancers and silencers within regulated pre-mRNAs.[7] While a conclusive mechanistic link has not yet been made between signaling pathways and regulated splicing via hnRNP K, it is likely that such a link exists. Indeed, phorbol esters, cytokines and hormones can all induce phosphorylation of hnRNP K and alter its ability to interact with RNAs[8,9] and at least one of the RNAs to which hnRNP K binds is *CD45*, which undergoes stimulus-regulated alternative splicing (A.A. Melton, J. Jackson and K.W.L. in preparation; see below).

Sam68 is not a classical hnRNP protein, but rather a member of the STAR (Signal-Transduction and RNA) family of proteins which contain a single RNA binding KH domain as well as multiple potential binding sites for SH2, SH3 and WW domains.[10] Sam68 was first identified as a protein that is tyrosine phosphorylated by Src during mitosis and as a protein that promotes cell cycle progression.[11-13] Further studies have demonstrated that Sam68 is also a target of serine/threonine phosphorylation, methylation and acetylation.[10] Within the cytoplasm, Sam68 is thought to function as an adaptor protein that nucleates signaling complexes proximal to several cell surface receptors. Sam68 is tyrosine phosphorylated in an inducible manner upon activation of the T-cell receptor, insulin receptor, or by stimulation of cells with leptin, leading to increased association with molecules such as PI3K, JAKs, Ras-GAP, Grb-2 and PLC-1 and the activation of downstream effector pathways.[10,14]

Sam68 also interacts with a variety of splicing factors, including several hnRNPs and other STAR proteins, as well as with proteins involved in transcription.[10,14,15] Within the nucleus, the adaptor function of Sam68 likely contributes to its ability to promote signal-induced splicing, as indicated by recent studies linking Sam68 to the signal-responsive inclusion of the *CD44* variable exon 5.[15-17] *CD44* encodes a cell surface glycoprotein that is involved in cell migration, invasion and proliferation.[18] The extracellular domain of CD44 is encoded, in part, by ten variable exons that are inducibly included upon antigen stimulation of T-cells through a pathway involving the activation of the MAP kinase pathway (Ras-Raf-MEK-Erk).[19,20] Work from several groups studying the induced inclusion of *CD44* variable exon 5 (*CD44v5*) has led to the formulation of a model in which, upon stimulation, Sam68 binds to an ESE within *CD44v5* together with the SR-related

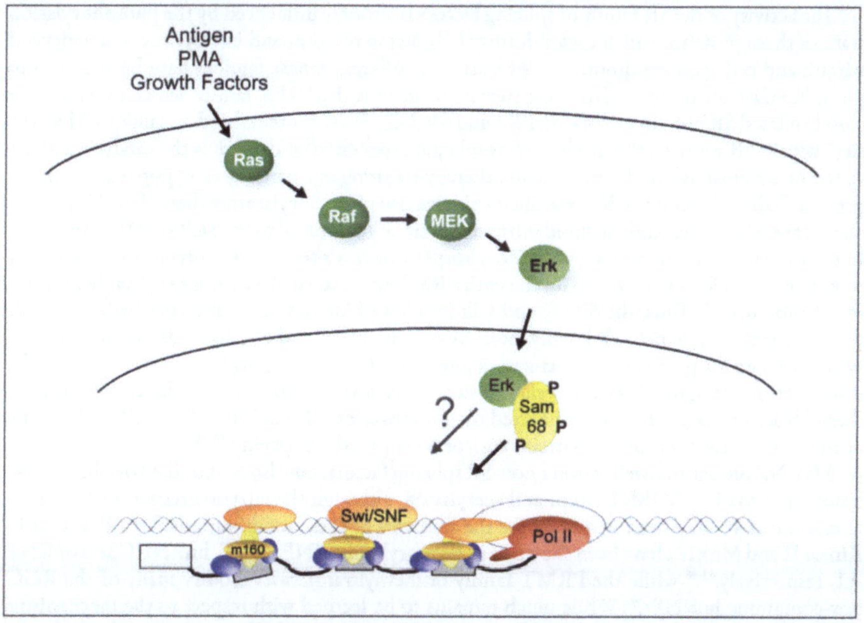

Figure 1. Model of the activity of the adaptor proteins Sam68 in the signal-induced regulation of *CD44*. Activation of the Ras-Raf-MEK-Erk pathway by various stimuli leads to phosphorylation of Sam68 by Erk. This phosphorylation of Sam68 and/or other Erk-dependent modifications (indicated by "?") leads to Sam68 binding to *CD44* variable exons (grey boxes) along with SRm160 ("m160") which promotes assembly of spliceosomal components (blue ovals) on these exons. Sam68 also interacts with the Swi/SNF chromosome remodeling complex which causes slowing of RNA polymerase II (Pol II) elongation, thus further promoting use of weak exons.

protein SRm160 (see Fig. 1).[16,17] Sam68 then also interacts with the Brm subunit of the Swi/SNF chromatin remodeling complex, thus stalling RNA polymerase II and promoting the inclusion of weak exons (see ref. 15 and discussion of transcription-coupled splicing below, and chapter by Kornblihtt). How activation of the MAP kinase pathway induces this Sam68-dependent regulation is not yet fully understood. Erk is known to phosphorylate Sam68, however mutation of the putative phosphorylation sites on Sam68 only marginally decreases its ability to enhance *CD44v5* inclusion,[16] suggesting that there must be other molecular links between Ras activation and the Sam68/SRm160/Brm complex. Finally, the Sam68-related proteins SLM-1 and SLM-2 have been shown to have activities similar to that of Sam68, both in terms of protein-protein interactions and CD44 splicing. This suggests that many or all members of the STAR family may serve as molecular links to alter splicing in response to extracellular stimuli.[10,21]

Posttranslational Modifications of Splicing Machinery

Signaling molecules can directly interact with and influence many other components of the splicing machinery. As discussed in the chapters by Lin and Fu, and Martinez-Contreras et al, the regulation of splicing is often achieved by the action of SR and hnRNP proteins. These proteins bind to sequences within and flanking alternative exons (i.e., ESEs, ESSs, ISEs, ISSs) and promote or inhibit spliceosome assembly at the nearby splice sites (refer to chapter by Chasin). It follows then that changing the activity of these splicing factors by posttranslational modifications is likely a major mechanism for altering splicing pathways.

The activity of the SR family of splicing factors is strongly influenced by the phosphorylation state of these proteins, which cycles during the splicing reaction and in response to a variety of stimuli and cell cycle conditions.[22-28] At least four different kinase families have been shown to phosphorylate SR proteins. The most specific of these is the SRPK family, which includes two closely related SR protein kinases, SRPK1 and SRPK2. These proteins bind to a unique "docking site" within SR proteins that both confers substrate specificity and restricts the catalytic activity to the N-terminal half of the RS domain, thereby resulting in a partially or hypophosphorylated protein.[29] The Clk family of dual-specificity kinases also phosphorylate members of the SR protein family, but with significantly reduced substrate specificity compared to the SRPKs.[29,30] Importantly, in contrast to the limited range of SRPK phosphorylation sites on SR proteins, the Clk family of kinases are able to phosphorylate the entire RS domain to yield a hyperphosphorylated form of SR proteins.[29,31] Thus, the SRPK and Clk families of kinases have differential effects on SR protein function (see below). Finally, both Topoisomerase I and Akt have also been shown to phosphorylate SR proteins. These enzymes phosphorylate overlapping sites that are likely to be distinct from the optimal phosphorylation sites of the SRPKs and Clks.[32-35] The activity of all of these SR kinases is presumably countered by phosphatases, with at least PP1 and PP2A having been shown to function on SR proteins and/or be required for splicing.[36-38]

HnRNP proteins, as well as other non-SR splicing factors, can also be modified by phosphorylation, methylation, SUMOylation and acetylation, although the enzymes responsible for such alterations have only been described for the first two of these modifications.[39-42] PKA, Casein Kinase II and Mnk1/2 have been shown to phosphorylate hnRNP I/PTB, hnRNP C and hnRNP A1, respectively,[43-46] while the PRMT family of methyltransferases modify many of the RGG box-containing hnRNPs.[42] While much remains to be learned with respect to the mechanisms by which these posttranslational modifications of SR, hnRNP and other splicing proteins change in response to extracellular stimuli and influence specific alternative splicing patterns, many groups have now correlated changes in the phosphorylation of SR and hnRNP proteins with the signal-induced regulation of several alternative splicing events.

A well described system in which phosphorylation of an SR protein mediates signal-induced changes in splicing is the insulin-induced inclusion of the variable βII exon within the *PKCβ* gene.[47] This induced change in the alternative splicing of *PKCβ* results in the expression of a PKCβ isozyme that is necessary for glucose uptake and is thus a critical aspect of the cellular response to insulin.[47,48] Inclusion of the βII exon is dependent on the activity of SRp40, an SR family protein, which binds to an intronic sequence downstream of the regulated exon.[34,49] Upon insulin treatment the PI-3 kinase (PI3K) pathway activates Akt which in turn phosphorylates SRp40 on a specific serine residue (Ser86) (see Fig. 2). Blocking of PI3K, Akt, SRp40, Ser86, or the binding site for SRp40 within the *PKCβ* gene all abolish the ability of insulin treatment to induce PKCβII expression.[34,49,50] However, it has yet to be determined how phosphorylation of SRp40 leads to increased exon inclusion; namely, whether phosphorylation of Ser86 increases the association of SRp40 with the *PKCβ* pre-mRNA, or rather increases the ability of SRp40 to activate exon inclusion via interactions with other splicing factors once it is bound to the pre-mRNA.

A second system in which phosphorylation of SR proteins is linked to changes in splicing is in the growth factor-induced alternative splicing of the *fibronectin* EDA exon.[51] In this case, phosphorylation of the SR proteins SF2/ASF and 9G8, again through the PI3K/Akt pathway, induces inclusion of EDA (see Fig. 2).[35] Importantly, phosphorylation of SF2/ASF or 9G8 by the Clk or SRPK family members have opposite effects on EDA splicing, thus providing evidence that the various SR kinases target different residues within SR proteins to achieve distinct functional consequences.[35] A second interesting aspect of the effect of Akt on SR proteins noted in this study is that this phosphorylation not only alters the activity of SR proteins in mediating splicing, but also influences the activity of SR proteins in translation.[35] This dual effect of Akt-dependent phosphorylation implies that alternative splicing is only one of several possible steps in RNA metabolism that can be affected by signal-induced phosphorylation of SR proteins.

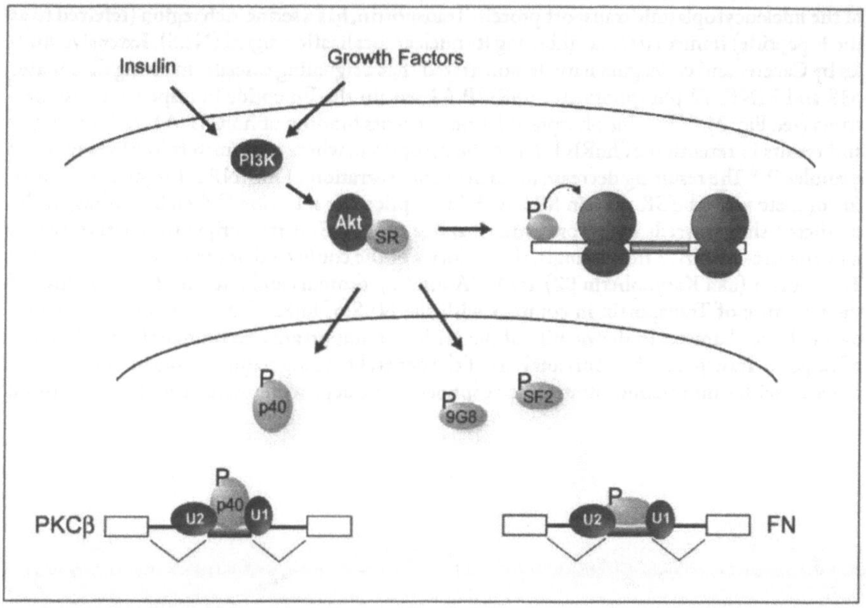

Figure 2. Model for signal-induced regulation of alternative splicing via direct phosphorylation of SR proteins. Activation of PI3K-Akt pathway by insulin or growth factors results respectively in phosphorylation of at least SRp40 or SF2/ASF and 9G8. Phosphorylation of these SR proteins is necessary for their ability to promote inclusion of weak exons in the *PKCβ* and *fibronectin* (FN) genes respectively. In addition, growth factor-dependent phosphorylation of SF2/ASF and 9G8 has been link to an increase in SR-stimulated translation of cytoplasmic mRNA. Grey shapes represent ribosomes, U1/U2 represent spliceosomal components.

Signal-Induced Changes in Localization of Splicing Factors

While posttranslational modifications may directly alter the activity of a splicing factor, the phosphorylation state of SR and hnRNP proteins can also influence their subcellular localization (see also chapters by Lin and Fu, and Martinez-Contreras et al). Since changes in the availability of splicing regulatory proteins can dramatically influence splicing patterns of specific genes, changing the localization of an SR or hnRNP protein is another potential means for achieving signal-induced alterations in splicing. Many, if not all, splicing factors localize at least to some extent in sub-nuclear foci known as "speckles". These speckles are thought to function as storage sites for proteins not actively engaged with pre-mRNA,[52,53] although speckles may be important to facilitate efficient splicing in cells.[54-56] Under normal growth conditions, splicing factors traffic between the speckles and nascent transcripts in a phosphorylation-dependent manner.[27,53] Furthermore, recent studies have shown that particular SR proteins are only recruited from the speckles to a nascent transcript when they are specifically engaged in the splicing of that transcript.[57] Interestingly, differential phosphorylation of SR proteins by overexpression of some kinases has been shown to influence their localization to speckles.[29,58,59] However, a change in speckle association is unlikely to explain all of the signal-induced changes in SR protein function since phosphorylation of SF2/ASF by Akt does not cause an apparent alteration in speckle pattern, yet can promote growth-factor induced inclusion of the *fibronectin* EDA exon as described above.[35]

A second mechanism by which phosphorylation can alter cellular localization is by altering the affinity of a protein for a nuclear transport factor. HnRNP A1, a well characterized cargo of the nucleocytoplasmic transport protein Transportin, has a serine-rich region (referred to as the F-peptide) immediately neighboring its nuclear localization signal (NLS). Extensive studies by Caceres and colleagues have demonstrated that a signaling cascade involving the kinases p38 and MNK1/2 phosphorylates hnRNP A1 within the F-peptide in response to osmotic stress (see Fig. 3).[45,60,61] This phosphorylation prevents binding of hnRNP A1 to Transportin and results in retention of hnRNP A1 in the cytoplasm, where it ultimately localizes to stress granules.[45,60] The resulting decrease in nuclear concentration of hnRNP A1 reduces its ability to compete with the SR protein SF2/ASF in 5' splice site selection,[62] thereby leading to the predicted shift towards use of proximal 5' splice sites in *E1A* transcripts from a transfected reporter pre-mRNA.[61] Interestingly, the majority of the confirmed or predicted substrates for Transportin (aka Karyopherin β2) are RNA binding proteins and a recent determination of the structure of Transportin in complex with the NLS of hnRNP A1 demonstrates that an overall basic character in the vicinity of the NLS is an important determinant for the binding of cargo to Transportin.[63,64] This analysis of the general binding requirements of Transportin, together with studies demonstrating a phosphorylation-dependent increase in the cytoplasmic

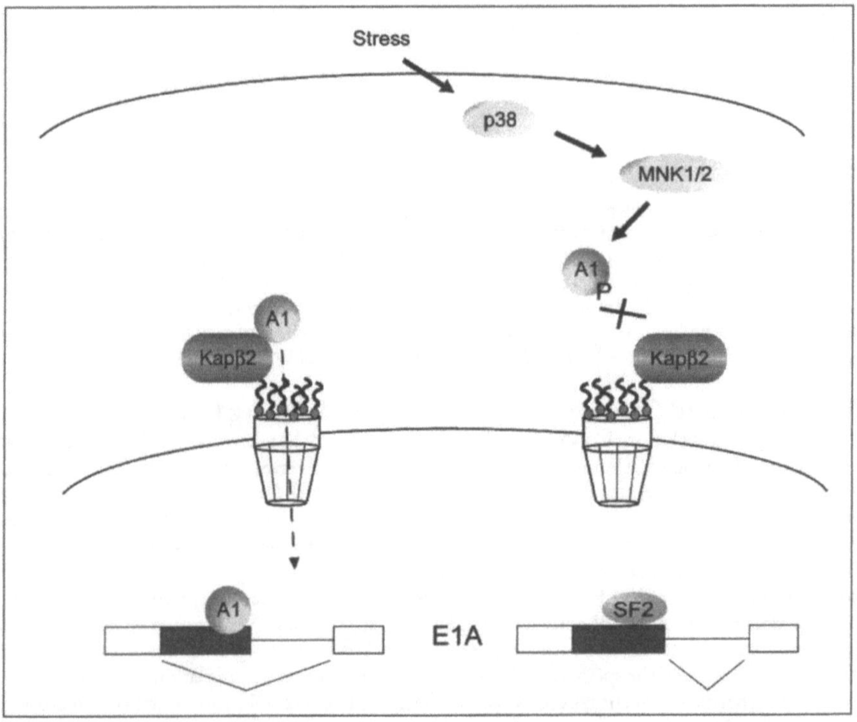

Figure 3. Model for signal-induced regulation of nuclear import. Under normal conditions Transportin (or Karyopherinβ2, Kapβ2) mediates nuclear import of hnRNP A1, which promotes use of distal 5' splice sites in the *E1A* pre-mRNA. Upon activation of the p38 stress response pathway, MNK1/2 phosphorylates hnRNP A1, thereby inhibiting binding of A1 to Transportin and preventing nuclear import of A1. Inhibition of nuclear import of A1 results in a reduction in the nuclear concentration of this proteins and allows for competing proteins, such as SF2/ASF, to preferentially bind target genes. In the case of the *E1A* pre-mRNA this results in increased utilization of proximal 5' splice sites.

localization of a few other RNA binding proteins,[46] suggests that regulation of nucleocytoplasmic transport may be a common mechanism for changing the nuclear concentration of RNA binding proteins, thus resulting in altered splicing patterns in response to extracellular cues (Fig. 3).

Other Mechanisms: Altered Protein-Protein Interactions and Protein Expression

As discussed above, both the change in nucleocytoplasmic localization of splicing proteins and the dispersion of speckles are due to a widespread disruption of protein-protein interactions via increased phosphorylation. However, posttranslational modifications can also cause specific changes in protein-protein interactions, such as those described above for the adaptor proteins hnRNP K and Sam68. While there is little direct confirmation of specific signal-induced alterations of protein-protein interactions leading to changes in splicing regulation, some recent data suggest evidence for such mechanisms. For instance, in the Sam68-dependent regulation of *CD44*, described above, signal-induced modifications to Sam68, SRm160 and Brm might influence the ability of these proteins to complex with one another. Analysis of the alternative splicing of another gene regulated in response to T-cell activation, namely *CD45*, also suggests a role of signal-regulated protein-protein interactions.

The *CD45* gene encodes a transmembrane protein tyrosine phosphatase that is involved in the regulation of signal transduction pathways in lymphocytes. In T-cells, three variable exons within *CD45* are skipped upon antigen stimulation. Recent work has shown that this signal-induced exon repression is due to the recruitment of the splicing factor PSF to an ESS within the *CD45* variable exons (Melton A, Jackson J and Lynch KW., in preparation). Interestingly, there is no difference in the nuclear concentration of PSF between resting and activated cells, nor any detectable change in the posttranslational modification of this protein. However, PSF only binds to the *CD45* ESS in response to cellular activation. PSF is known to interact with a wide spectrum of splicing factors, transcription factors and nuclear matrix proteins.[65-68] Moreover, PSF interacts with activated PKC isoforms within the cell nucleus,[69] and specific epitopes within PSF have been shown to be masked upon changing cellular conditions.[70] Therefore, a reasonable hypothesis for the signal-induced repression of *CD45* exons by PSF is that upon activation of T-cells binding partners of PSF are modified so as to either recruit PSF to the *CD45* pre-mRNA or, alternatively, to release PSF from an otherwise sequestered conformation.

Arguably, the simplest mechanism through which alternative splicing could be regulated in response to environmental cues would be through the increased or decreased expression of critical regulatory factors. Signaling pathways are known to stimulate many ubiquitous transcription factors such as NFκB, NFAT and nuclear receptors, as well as factors involved in mRNA stability/translation and proteosome-mediated degradation.[71-73] These various mechanisms typically induce broad changes in proteome expression. Not surprisingly, many SR proteins and other splicing factors have been found to be differentially expressed in a signal-dependent manner in a variety of cell types.[74-76] However, it remains to be determined whether such changes in the overall expression of splicing factors truly lead to altered splicing patterns, or whether splicing proteins are already in such excess that increased expression does not significantly alter these patterns. One example in which changes in protein expression have been shown to directly influence splicing patterns occurs during the development of erythrocytes. At a specific stage during erythropoesis there is a marked decrease in the expression of the hnRNP A/B proteins. This decrease in hnRNP A/B expression correlates temporally with the increased inclusion of exon 16 in the gene encoding the cytoskeletal protein 4.1R.[77] Since biochemical studies have shown that hnRNP A1 binds to an ESS within the *4.1R* exon 16 and causes exon skipping, the decreased expression of the hnRNP A/B proteins almost certainly is the cause of *4.1R* exon 16 inclusion in mature erythroblasts.[77]

Regulation Via Cross-Talk with Signal-Responsive Changes in Transcription

All of the paradigms for altering splicing regulation described above involve the direct manipulation of the activity or accessibility of a splicing factor. However, in the cell pre-mRNA splicing does not occur in isolation, but rather it is linked temporally, spatially and mechanistically with other mRNA production events. In particular, many recent studies have demonstrated extensive cross-talk between the transcription and splicing machineries (refer to chapter by Kornblihtt). Given the substantial effects of signaling pathways on transcription, it would not be surprising if at least some of the signal-induced regulation of transcription factors have secondary effects on alternative splicing (Fig. 4).

The primary ways in which transcription has been shown to effect splicing are summarized by two models: the "kinetic model" and the "recruitment model" (refer to chapter by Kornblihtt). The mechanism that has gained the most experimental support thus far is the kinetic model, also known colloquially as the "first come, first serve" model.[78,79] This model is based on the premise that, given the length of a typical mammalian intron, the time lag between the transcription of one exon and the transcription of the next exon is often sufficiently long that the first exon can be bound by the spliceosome before the next exon is present. This time lag potentially allows a "weak exon" (i.e., an exon with suboptimal splice sites, the absence of splicing enhancer elements and/or the presence of splicing silencer elements) to be recognized by the spliceosome without having to compete with a subsequent "strong" exon. It follows then that a reduced rate of transcriptional

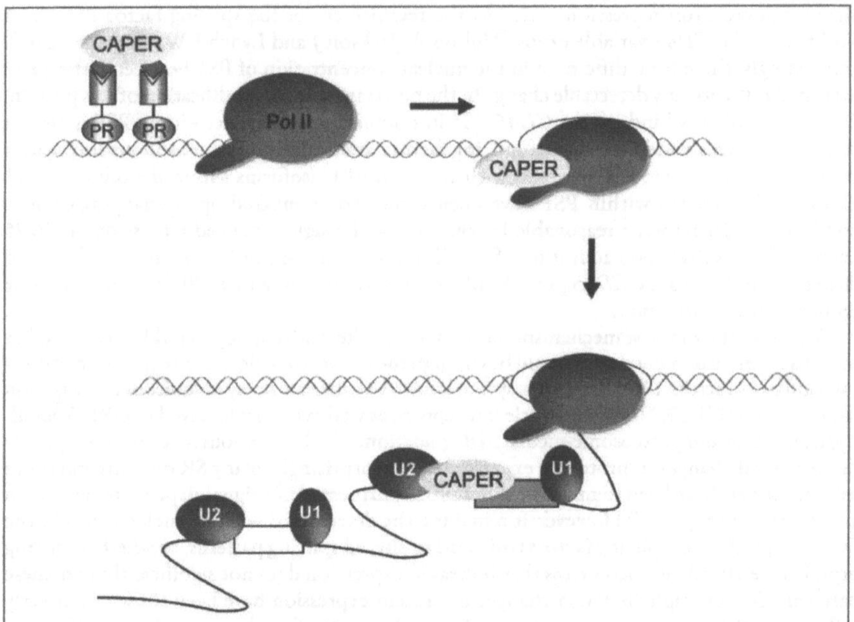

Figure 4. Model for hormone-induced alternative splicing via transcription. Binding of progesterone (green diamond) to the progesterone receptor (PR) recruits the CAPER proteins. CAPER then activates transcription by RNA polymerase II (Pol II) and promotes use of weak exons (grey box) in pre-mRNAs transcribed from the PR-dependent promoter, presumably by binding to 3' splice sites and recruiting spliceosomal components (U1/U2).

elongation favors the recognition and inclusion of weak exons, whereas an increased rate of transcription favors exon skipping.

In its simplest form, the recruitment model proposes that binding of splicing factors to the RNA polymerase II complex increases their local concentration proximal to the nascent transcript, thereby enhancing otherwise weak interactions between the splicing factors and the pre-mRNA.[80,81] However, a further complexity of the recruitment model is that transcription activators or co-activators bound at the promoters may differentially influence recruitment of splicing factors. This promoter specificity was initially suggested by studies of the *fibronectin* gene, in which the SR proteins 9G8 and SF2/ASF enhance the inclusion of the variable EDI exon by binding to an ESE, but this only occurs when the EDI containing gene is transcribed from its endogenous promoter.[82]

Sam68-dependent regulation of *CD44* splicing is one example of signal-induced regulation that relates to the kinetic model of cotranscriptional alternative splicing (see Fig. 1). In addition, the signal-induced transcription factor NFκB has been shown to increase transcription elongation,[83] suggesting that genes transcribed in an NFκB-dependent manner may also undergo signal-regulated alternative splicing through changes in transcriptional kinetics.

Signaling pathways also appear to alter splicing patterns via the recruitment model, as revealed in studies of nuclear hormone-dependent alternative splicing.[84] Work by the Berget and O'Malley groups has demonstrated that both progesterone and estrodiol can cause changes in alternative splicing profiles, but only when pre-mRNA transcription is driven by promoters that are dependent on the corresponding nuclear hormone receptors for activity.[85,86] At least in the case of progesterone-responsive alternative splicing, it was further shown that, in the presence of progesterone, the U2AF[65]-like co-activators CAPERα and β are recruited to the progesterone receptor where they induce both transcription and alternative splicing.[87] Therefore, as predicted by the recruitment model for transcription-coupled splicing, signal-induced changes in promoter occupancy can directly recruit splicing regulatory proteins that influence splicing of the transcribed pre-mRNA.

Coordinated Regulation

The primary goal of signal transduction pathways within a cell is to evoke a specialized response to any given environmental condition. Often an optimal response requires the coordinated activity of a broad spectrum of genes and proteins. For instance, neuronal depolarization induces ion trafficking across the cell membrane as well as protein and vesicle transport, whereas a T-cell must migrate, proliferate and secrete various proteins in response to antigen stimulation as part of an effective immune response. In order to achieve such a robust and comprehensive response, signaling pathways frequently activate a program of related events rather than just the expression of one individual gene or protein. For instance, the activation of NFκB by antigen stimulation of a T-cell leads to the induction of transcription of multiple genes involved in promoting cell division and inhibiting apoptosis.[71] Similarly, regulation of multiple alternative splicing events by a given extracellular stimulus could amplify potential physiological consequences. Not surprisingly, therefore, even the few examples of signal-induced alternative splicing that have been characterized demonstrate coordinated regulation. Analysis of the *CD45* gene identified a motif within the signal-responsive ESS that is present in other exons which are differentially spliced in response to T-cell activation.[88] Similarly, two regulatory sequences (intronic and exonic) have been identified as a hallmark of exons that are alternatively spliced in response to neuronal depolarization.[89,90] The identification of these signal-responsive regulatory motifs has allowed for the bioinformatic identification of novel examples of signal-induced alternative splicing and strongly suggests that genes which contain such sequences are regulated in a coordinated manner through common mechanisms.[88-91] The recent development of microarrays designed to monitor the levels of alternatively spliced isoforms has further allowed for the systematic identification of genes that undergo alternative splicing in a signal-dependent manner.[102] Subsequent studies of these signal-regulated genes are likely to reveal additional signal-responsive splicing regulatory sequences and allow for the grouping of genes into families of mechanistically-coordinated alternative splicing events.

Achieving Specificity in Signal-Responsiveness of Alternative Splicing

Despite the importance of coordinating regulation of splicing, one obvious question raised by our understanding of the mechanisms underlying signal-induced alternative splicing discussed above is regarding how specificity is achieved. That is, if Akt can phosphorylate several SR proteins and these proteins are ubiquitous splicing factors, why are the effects of these posttranslational modifications restricted from the splicing of other genes that are not regulated upon activation of Akt? Even in the case of coordinated regulation of a family of genes it is clear that some level of specificity still is at play in determining which splicing events are regulated by a particular cellular stimulus.

While we don't yet have a sufficiently clear understanding of regulated splicing in general and signal-responsive splicing in particular, to completely understand the question of specificity, current data does suggest that specificity may be conferred at the level of signaling pathways, RNA binding and/or differential sensitivity to the activity of individual splicing factors. Within the signaling field, specificity is largely understood to be conferred by location or co-association of proteins.[92] In other words, while a protein such as Sam68 may be capable of interacting with a wide range of proteins, under any given cellular condition Sam68 may only co-associate with a subset of potential partners and thus will only be able to transmit signals to certain downstream effectors. At the level of RNA binding, specificity of RNA-protein interactions is also often conferred by co-association of proteins within enhancer or silencer complexes.[93,94] Therefore, the signal-induced regulation of a particular gene may require the combinatorial effect of multiple transduction pathways, each altering the activity of one component of a larger complex. In such a scenario, a stimulus that only triggered one signaling pathway would not affect a more complex target gene. Alternatively, loss of one protein from a particular regulatory complex may be compensated for by other binding partners. A potentially related aspect of specificity is the recent discovery that a decrease in the expression of even core spliceosomal proteins has differential effects on the splicing of specific transcripts.[95] While some of this differential activity may be due to compensating protein-protein interactions, this phenomenon is primarily understood to be due to differences in the rate-limiting step of splicing for different transcripts. In other words, decreased activity of a splicing factor involved in 3' splice site selection will have the greatest effect on substrates which have weak or variable 3' splice sites.[96] Together, the specificity inherent in signaling and splicing mechanisms likely work in concert to achieve the necessary balance between strength and precision of signal-induced changes in alternative splicing.

Feed-Back and Feed-Forward

Interestingly, many of the genes that have been shown to undergo changes in splicing pattern in response to extracellular stimuli are themselves receptors or other signaling molecules. These include, among others, *CD45*, *CD44*, *NMDAR1* and *PKCIIβ*.[47,90,97] Importantly, the differential proteins expressed by all of the above-mentioned genes have been shown to have distinct signaling properties, often affecting the very signaling pathway that leads to their differential splicing pattern.[48,98-100] This strongly suggests that there is a possibility of feedback or feed-forward in which the initial activation of a signaling pathway is either promoted/maintained or turned off via signal-induced alternative splicing.

One example of such feedback is in *CD44* alternative splicing.[99] As mentioned above, activation of the Ras signaling pathway enhances inclusion of ten variable exons that encode part of the extracellular domain of CD44. Specifically, inclusion of variable exon 6 (*CD44v6*) promotes CD44 involvement in a coreceptor complex with hepatocyte growth factor and the tyrosine kinase Met that in turn promotes Ras signaling.[101] Activation of quiescent cells with growth factors leads to an initial burst of MAP kinase activation followed several hours later by a second, prolonged wave of MAP kinase activity. The inclusion of *CD44v6* that occurs in response to the initial burst of Ras activation is necessary for the subsequent wave of Ras signaling, as demonstrated by the loss of the second pulse of Ras activity when *CD44v6*-containing transcripts are specifically repressed.[99] Therefore, at least with regards to Ras signaling, alternative splicing is an important feedback mechanism to generate the sustained activation phenotype necessary to drive cells forward to proliferation.

Summary

While signal-induced alternative splicing is no doubt a prevalent phenomenon,[90,91,102] we are only just beginning to scratch the surface in terms of identifying such regulated events and in understanding the mechanisms by which they occur. The examples provided in this chapter are in no way meant to be an exhaustive list, but rather are presented as examples of how extracellular stimuli might influence the splicing machinery so as to alter splicing patterns. At present there is still only cursory data in support of many of the proposed mechanisms; however, what is clear is that there are likely numerous pathways by which changes in growth conditions or in the environment can be communicated to the splicing machinery and many critical physiological processes are influenced by signal-induced changes that impact on splicing. Clearly the questions surrounding signal-induced alternative splicing represent an important frontier that warrants major investigation.

References

1. Pinol-Roma S, Dreyfuss G. Shuttling of pre-mRNA binding proteins between nucleus and cytoplasm. Nature 1992; 355(6362):730-732.
2. Caceres JF, Screaton GR, Krainer AR. A specific subset of SR proteins shuttles continuously between the nucleus and the cytoplasm. Genes Dev 1998; 12(1):55-66.
3. Khokhlatchev AV, Canagarajah B, Wilsbacher J et al. Phosphorylation of the MAP kinase ERK2 promotes its homodimerization and nuclear translocation. Cell 1998; 93(4):605-615.
4. Bomsztyk K, Denisenko O, Ostrowski J. hnRNP K: one protein multiple processes. Bioessays 2004; 26(6):629-638.
5. Kim JH, Hahm B, Kim YK et al. Protein-protein interaction among hnRNPs shuttling between nucleus and cytoplasm. J Mol Biol 2000; 298(3):395-405.
6. Shnyreva M, Schullery DS, Suzuki H et al. Interaction of two multifunctional proteins. Heterogeneous nuclear ribonucleoprotein K and Y-box-binding protein. J Biol Chem 2000; 275(20):15498-15503.
7. Expert-Bezancon A, Le Caer JP, Marie J. Heterogeneous nuclear ribonucleoprotein (hnRNP) K is a component of an intronic splicing enhancer complex that activates the splicing of the alternative exon 6A from chicken beta-tropomyosin pre-mRNA. J Biol Chem 2002; 277(19):16614-16623.
8. Ostrowski J, Kawata Y, Schullery DS et al. Insulin alters heterogeneous nuclear ribonucleoprotein K protein binding to DNA and RNA. Proc Natl Acad Sci USA 2001; 98(16):9044-9049.
9. Ostrowski J, Schullery DS, Denisenko ON et al. Role of tyrosine phosphorylation in the regulation of the interaction of heterogenous nuclear ribonucleoprotein K protein with its protein and RNA partners. J Biol Chem 2000; 275(5):3619-3628.
10. Lukong KE, Richard S. Sam68, the KH domain-containing superSTAR. Biochim Biophys Acta 2003; 1653(2):73-86.
11. Fumagalli S, Totty NF, Hsuan JJ et al. A target for Src in mitosis. Nature 1994; 368(6474):871-874.
12. Taylor SJ, Shalloway D. An RNA-binding protein associated with Src through its SH2 and SH3 domains in mitosis. Nature 1994; 368(6474):867-871.
13. Taylor SJ, Resnick RJ, Shalloway D. Sam68 exerts separable effects on cell cycle progression and apoptosis. BMC Cell Biol 2004; 5:5.
14. Najib S, Martin-Romero C, Gonzalez-Yanes C et al. Role of Sam68 as an adaptor protein in signal transduction. Cell Mol Life Sci 2005; 62(1):36-43.
15. Batsche E, Yaniv M, Muchardt C. The human SWI/SNF subunit Brm is a regulator of alternative splicing. Nat Struct Mol Biol 2006; 13(1):22-29.
16. Matter N, Herrlich P, Konig H. Signal-dependent regulation of splicing via phosphorylation of Sam68. Nature 2002; 420(6916):691-695.
17. Cheng C, Sharp PA. Regulation of CD44 Alternative Splicing by SRm160 and Its Potential Role in Tumor Cell Invasion. Mol Cell Biol 2006; 26(1):362-370.
18. Ponta H, Sherman L, Herrlich PA. CD44: from adhesion molecules to signalling regulators. Nat Rev Mol Cell Biol 2003; 4(1):33-45.
19. Konig H, Ponta H, Herrlich P. Coupling of signal transduction to alternative pre-mRNA splicing by a composite splice regulator. EMBO J 1998; 17:2904-2913.

20. Weg-Remers S, Ponta H, Herrlich P et al. Regulation of alternative pre-mRNA splicing by the ERK MAP-kinase pathway. EMBO J 2001; 20(15):4194-4203.
21. Stoss O, Olbrich M, Hartmann AM et al. The STAR/GSG family protein rSLM-2 regulates the selection of alternative splice sites. J Biol Chem 2001; 276(12):8665-8673.
22. Xiao SH, Manley JL. Phosphorylation of the ASF/SF2 RS domain affects both protein-protein and protein-RNA interactions and is necessary for splicing. Genes Dev 1997; 11(3):334-344.
23. Xiao SH, Manley JL. Phosphorylation-dephosphorylation differentially affects activities of splicing factor ASF/SF2. EMBO J 1998; 17(21):6359-6367.
24. Sanford JR, Ellis JD, Cazalla D et al. Reversible phosphorylation differentially affects nuclear and cytoplasmic functions of splicing factor 2/alternative splicing factor. Proc Natl Acad Sci USA 2005; 102(42):15042-15047.
25. Graveley B. Sorting out the complexity of SR protein functions. RNA 2000; 6:1197-1211.
26. Cao W, Jamison SF, Garcia-Blanco MA. Both phosphorylation and dephosphorylation of ASF/SF2 are required for pre-mRNA splicing in vitro. RNA 1997; 3(12):1456-1467.
27. Misteli T, Caceres JF, Clement JQ et al. Serine phosphorylation of SR proteins is required for their recruitment to sites of transcription in vivo. J Cell Biol 1998; 143(2):297-307.
28. Huang Y, Yario TA, Steitz JA. A molecular link between SR protein dephosphorylation and mRNA export. Proc Natl Acad Sci USA 2004; 101(26):9666-9670.
29. Ngo JC, Chakrabarti S, Ding JH et al. Interplay between SRPK and Clk/Sty kinases in phosphorylation of the splicing factor ASF/SF2 is regulated by a docking motif in ASF/SF2. Mol Cell 2005; 20(1):77-89.
30. Colwill K, Feng LL, Yeakley JM et al. SRPK1 and Clk/Sty protein kinases show distinct substrate specificities for serine/arginine-rich splicing factors. J Biol Chem 1996; 271(40):24569-24575.
31. Colwill K, Pawson T, Andrews B et al. The Clk/Sty protein kinase phosphorylates SR splicing factors and regulates their intracellular distribution. EMBO J 1995; 15:265-275.
32. Labourier E, Rossi F, Gallouzi IE et al. Interaction between the N-terminal domain of human DNA topoisomerase I and the arginine-serine domain of its substrate determines phosphorylation of SF2/ASF splicing factor. Nucleic Acids Res 1998; 26(12):2955-2962.
33. Rossi F, Labourier E, Forne T et al. Specific phosphorylation of SR proteins by mammalian DNA topoisomerase I. Nature 1996; 381(6577):80-82.
34. Patel NA, Kaneko S, Apostolatos HS et al. Molecular and genetic studies imply Akt-mediated signaling promotes protein kinase CbetaII alternative splicing via phosphorylation of serine/arginine-rich splicing factor SRp40. J Biol Chem 2005; 280(14):14302-14309.
35. Blaustein M, Pelisch F, Tanos T et al. Concerted regulation of nuclear and cytoplasmic activities of SR proteins by AKT. Nat Struct Mol Biol 2005; 12(12):1037-1044.
36. Misteli T, Spector DL. Serine/threonine phosphatase 1 modulates the subnuclear distribution of pre-mRNA splicing factors. Mol Biol Cell 1996; 7(10):1559-1572.
37. Mermoud JE, Cohen P, Lamond AI. Ser/Thr-specific protein phosphatases are required for both catalytic steps of pre-mRNA splicing. Nucleic Acids Res 1992; 20(20):5263-5269.
38. Shi Y, Reddy B, Manley JL. PP1/PP2A phosphatases are required for the second step of Pre-mRNA splicing and target specific snRNP proteins. Mol Cell 2006; 23(6):819-829.
39. Krecic AM, Swanson MS. hnRNP complexes: composition, structure and function. Curr Opin Cell Biol Jun 1999; 11(3):363-371.
40. Vassileva MT, Matunis MJ. SUMO modification of heterogeneous nuclear ribonucleoproteins. Mol Cell Biol 2004; 24(9):3623-3632.
41. Kim SC, Sprung R, Chen Y et al. Substrate and functional diversity of lysine acetylation revealed by a proteomics survey. Mol Cell 2006; 23(4):607-618.
42. Bedford MT, Richard S. Arginine methylation an emerging regulator of protein function. Mol Cell 2005; 18(3):263-272.
43. Magistrelli G, Jeannin P, Herbault N et al. A soluble form of CTLA-4 generated by alternative splicing is expressed by nonstimulated human T-cells. Eur J Immunol 1999; 29(11):3596-3602.
44. Mayrand SH, Dwen P, Pederson T. Serine/threonine phosphorylation regulates binding of C hnRNP proteins to pre-mRNA. Proc Natl Acad Sci USA 1993; 90(16):7764-7768.
45. Guil S, Long JC, Caceres JF. hnRNP A1 relocalization to the stress granules reflects a role in the stress response. Mol Cell Biol 2006; 26(15):5744-5758.
46. Xie J, Lee JA, Kress TL et al. Protein kinase A phosphorylation modulates transport of the polypyrimidine tract-binding protein. Proc Natl Acad Sci USA 2003; 100(15):8776-8781.
47. Chalfant CE, Mischak H, Watson JE et al. Regulation of alternative splicing of protein kinase C beta by insulin. J Biol Chem 1995; 270(22):13326-13332.
48. Cooper DR, Watson JE, Patel N et al. Ectopic expression of protein kinase CbetaII, -delta and -epsilon, but not -betaI or -zeta, provide for insulin stimulation of glucose uptake in NIH-3T3 cells. Arch Biochem Biophys 1999; 372(1):69-79.

49. Patel NA, Chalfant CE, Watson JE et al. Insulin regulates alternative splicing of protein kinase C beta II through a phosphatidylinositol 3-kinase-dependent pathway involving the nuclear serine/arginine-rich splicing factor, SRp40, in skeletal muscle cells. J Biol Chem 2001; 276(25):22648-22654.
50. Patel NA, Apostolatos HS, Mebert K et al. Insulin regulates protein kinase CbetaII alternative splicing in multiple target tissues: development of a hormonally responsive heterologous minigene. Mol Endocrinol 2004; 18(4):899-911.
51. Blaustein M, Pelisch F, Coso OA et al. Mammary epithelial-mesenchymal interaction regulates fibronectin alternative splicing via phosphatidylinositol 3-kinase. J Biol Chem 2004; 279(20):21029-21037.
52. Lamond AI, Spector DL. Nuclear speckles: a model for nuclear organelles. Nat Rev Mol Cell Biol 2003; 4(8):605-612.
53. Misteli T, Caceres JF, Spector DL. The dynamics of a pre-mRNA splicing factor in living cells. Nature 1997; 387(6632):523-527.
54. Moen Jr PT, Johnson CV, Byron M et al. Repositioning of muscle-specific genes relative to the periphery of SC-35 domains during skeletal myogenesis. Mol Biol Cell 2004; 15(1):197-206.
55. Shopland LS, Johnson CV, Byron M et al. Clustering of multiple specific genes and gene-rich R-bands around SC-35 domains: evidence for local euchromatic neighborhoods. J Cell Biol 2003; 162(6):981-990.
56. Shopland LS, Johnson CV, Lawrence JB. Evidence that all SC-35 domains contain mRNAs and that transcripts can be structurally constrained within these domains. J Struct Biol 2002; 140(1-3):131-139.
57. Mabon SA, Misteli T. Differential recruitment of pre-mRNA splicing factors to alternatively spliced transcripts in vivo. PLoS Biol 2005; 3(11):374e.
58. Ding JH, Zhong XY, Hagopian JC et al. Regulated cellular partitioning of SR protein-specific kinases in mammalian cells. Mol Biol Cell 2006; 17(2):876-885.
59. Gui JF, Lane WS, Fu X-D. A serine kinase regulates intracellular localization of splicing factors in the cell cycle. Nature 1994; 369:678-682.
60. Allemand E, Guil S, Myers M et al. Regulation of heterogenous nuclear ribonucleoprotein A1 transport by phosphorylation in cells stressed by osmotic shock. Proc Natl Acad Sci USA 2005; 102(10):3605-3610.
61. van der Houven van Oordt W, Diaz-Meco MT, Lozano J et al. The MKK(3/6)-p38-signaling cascade alters the subcellular distribution of hnRNP A1 and modulates alternative splicing regulation. J Cell Biol 2000; 149(2):307-316.
62. Eperon IC, Makarova OV, Mayeda A et al. Selection of alternative 5' splice sites: role of U1 snRNP and models for the antagonistic effects of SF2/ASF and hnRNP A1. Mol Cell Biol 2000; 20(22):8303-8318.
63. Lee BJ, Cansizoglu AE, Suel KE et al. Rules for nuclear localization sequence recognition by karyopherin beta 2. Cell 2006; 126(3):543-558.
64. Chook YM, Blobel G. Karyopherins and nuclear import. Curr Opin Struct Biol 2001; 11(6):703-715.
65. Meissner M, Dechat T, Gerner C et al. Differential nuclear localization and nuclear matrix association of the splicing factors PSF and PTB. J Cell Biochem 2000; 76(4):559-566.
66. Rosonina E, Ip JY, Calarco JA et al. Role for PSF in mediating transcriptional activator-dependent stimulation of pre-mRNA processing in vivo. Mol Cell Biol 2005; 25(15):6734-6746.
67. Emili A, Shales M, McCracken S et al. Splicing and transcription-associated proteins PSF and p54nrb/nonO bind to the RNA polymerase II CTD. RNA 2002; 8(9):1102-1111.
68. Shav-Tal Y, Zipori D. PSF and p54(nrb)/NonO—multi-functional nuclear proteins. FEBS Lett 2002; 531(2):109-114.
69. Rosenberger U, Lehmann I, Weise C et al. Identification of PSF as a protein kinase Calpha-binding protein in the cell nucleus. J Cell Biochem 2002; 86(2):394-402.
70. Shav-Tal Y, Cohen M, Lapter S et al. Nuclear relocalization of the pre-mRNA splicing factor PSF during apoptosis involves hyperphosphorylation, masking of antigenic epitopes and changes in protein interactions. Mol Biol Cell 2001; 12(8):2328-2340.
71. Gerondakis S, Grumont R, Rourke I et al. The regulation and roles of Rel/NF-kappa B transcription factors during lymphocyte activation. Curr Opin Immunol 1998; 10(3):353-359.
72. Guhaniyogi J, Brewer G. Regulation of mRNA stability in mammalian cells. Gene 2001; 265(1-2):11-23.
73. Metivier R, Reid G, Gannon F. Transcription in four dimensions: nuclear receptor-directed initiation of gene expression. EMBO Rep 2006; 7(2):161-167.
74. Schimmer BP, Cordova M, Cheng H et al. Global profiles of gene expression induced by adrenocorticotropin in Y1 mouse adrenal cells. Endocrinology 2006; 147(5):2357-2367.
75. Uematsu F, Takahashi M, Yoshida M et al. Distinct patterns of gene expression in hepatocellular carcinomas and adjacent noncancerous, cirrhotic liver tissues in rats fed a choline-deficient, L-amino acid-defined diet. Cancer Sci 2005; 96(7):414-424.
76. Screaton GR, Caceres JF, Mayeda A et al. Identification and characterization of three members of the human SR family of pre-mRNA splicing factors. EMBO J 1995; 14:4336-4349.

77. Hou VC, Lersch R, Gee SL et al. Decrease in hnRNP A/B expression during erythropoiesis mediates a pre-mRNA splicing switch. EMBO J 2002; 21(22):6195-6204.
78. Kornblihtt AR. Chromatin, transcript elongation and alternative splicing. Nat Struct Mol Biol 2006; 13(1):5-7.
79. Kornblihtt AR, de la Mata M, Fededa JP et al. Multiple links between transcription and splicing. RNA 2004; 10(10):1489-1498.
80. Bentley DL. Rules of engagement: cotranscriptional recruitment of pre-mRNA processing factors. Curr Opin Cell Biol 2005; 17(3):251-256.
81. Maniatis T, Reed R. An extensive network of coupling among gene expression machines. Nature 2002; 416(6880):499-506.
82. Cramer P, Caceres JF, Cazalla D et al. Coupling of transcription with alternative splicing: RNA pol II promoters modulate SF2/ASF and 9G8 effects on an exonic splicing enhancer. Mol Cell 1999; 4(2):251-258.
83. West MJ, Lowe AD, Karn J. Activation of human immunodeficiency virus transcription in T-cells revisited: NF-kappaB p65 stimulates transcriptional elongation. J Virol 2001; 75(18):8524-8537.
84. Lonard DM, O'Malley BW. Expanding functional diversity of the coactivators. Trends Biochem Sci 2005; 30(3):126-132.
85. Auboeuf D, Honig A, Berget SM et al. Coordinate regulation of transcription and splicing by steroid receptor coregulators. Science 2002; 298(5592):416-419.
86. Auboeuf D, Dowhan DH, Kang YK et al. Differential recruitment of nuclear receptor coactivators may determine alternative RNA splice site choice in target genes. Proc Natl Acad Sci USA 2004; 101(8):2270-2274.
87. Dowhan DH, Hong EP, Auboeuf D et al. Steroid hormone receptor coactivation and alternative RNA splicing by U2AF65-related proteins CAPERalpha and CAPERbeta. Mol Cell 2005; 17(3):429-439.
88. Rothrock C, Cannon B, Hahm B et al. A conserved signal-responsive sequence mediates activation-induced alternative splicing of CD45. Mol Cell 2003; 12(5):1317-1324.
89. Xie J, Black DL. A CaMK IV responsive RNA element mediates depolarization-induced alternative splicing of ion channels. Nature 2001; 410(6831):936-939.
90. Lee JA, Xing Y, Nguyen D et al. Depolarization and CaM Kinase IV modulate NMDA receptor splicing through two essential RNA elements. PLoS Biol 2007; 5(2):e40.
91. An P, Grabowski PJ. Exon silencing by UAGG motifs in response to neuronal excitation. PLoS Biol 2007; 5(2):e36.
92. Ptashne M, Gann A. Signal transduction. Imposing specificity on kinases. Science 2003; 299(5609):1025-1027.
93. Lynch KW, Maniatis T. Assembly of specific SR protein complexes on distinct regulatory elements of the Drosophila doublesex splicing enhancer. Genes Dev 1996; 10:2089-2101.
94. Markovtsov V, Nikolic JM, Goldman JA et al. Cooperative assembly of an hnRNP complex induced by a tissue-specific homolog of polypyrimidine tract binding protein. Mol Cell Biol 2000; 20(20):7463-7479.
95. Park JW, Parisky K, Celotto AM et al. Identification of alternative splicing regulators by RNA interference in Drosophila. Proc Natl Acad Sci USA 2004; 101(45):15974-15979.
96. Konarska MM, Query CC. Insights into the mechanisms of splicing: more lessons from the ribosome. Genes Dev 2005; 19(19):2255-2260.
97. Lynch KW. Consequences of regulated pre-mRNA splicing in the immune system. Nat Rev Immunol 2004; 4(12):931-940.
98. Hermiston ML, Xu Z, Majeti R et al. Reciprocal regulation of lymphocyte activation by tyrosine kinases and phosphatases. J Clin Invest 2002; 109(1):9-14.
99. Cheng C, Yaffe MB, Sharp PA. A positive feedback loop couples Ras activation and CD44 alternative splicing. Genes Dev 2006; 20(13):1715-1720.
100. Carroll RC, Zukin RS. NMDA-receptor trafficking and targeting: implications for synaptic transmission and plasticity. Trends Neurosci 2002; 25(11):571-577.
101. Orian-Rousseau V, Chen L, Sleeman JP et al. CD44 is required for two consecutive steps in HGF/c-Met signaling. Genes Dev 2002; 16(23):3074-3086.
102. Ip JY, Tong A, Pan Q et al. Global analysis of alternative splicing during T-cell activation. RNA 2007; 13:563-572.

CHAPTER 11

Coupling Transcription and Alternative Splicing

Alberto R. Kornblihtt*

Abstract

Alternative splicing regulation not only depends on the interaction of splicing factors with splicing enhancers and silencers in the pre-mRNA, but also on the coupling between transcription and splicing. This coupling is possible because splicing is often cotranscriptional and promoter identity and occupation may affect alternative splicing. We discuss here the different mechanisms by which transcription regulates alternative splicing. These include the recruitment of splicing factors to the transcribing polymerase and "kinetic coupling", which involves changes in the rate of transcriptional elongation that in turn affect the timing in which splice sites are presented to the splicing machinery. The recruitment mechanism may depend on the particular features of the carboxyl terminal domain of RNA polymerase II, whereas kinetic coupling seems to be linked to how changes in chromatin structure and other factors affect transcription elongation.

Introduction

For decades RNA polymerase II (RNAPII) transcription and pre-mRNA processing have been thought to be independent steps in the pathway of eukaryotic gene expression until a series of biochemical, cytological and functional experiments demonstrated that capping, splicing and cleavage/polyadenylation are coupled to transcription.[1-7] This requires that splicing occurs cotranscriptionally. However, the existence of cotranscriptionality per se does not necessarily imply a mechanistic coupling. Indeed, splicing often occurs cotranscriptionally. Electron microscopy visualization of *Drosophila* embryo nascent transcripts (Fig. 1A) has clearly demonstrated that splicing occurs cotranscriptionally with a reasonable frequency and that splice site selection precedes polyadenylation.[8] Cotranscriptional splicing was also demonstrated in the *dystrophin* gene.[9] Since transcription of this 2.4 Mb-gene, the largest in the human genome, would take approximately 16 hours to be completed, cotranscriptional splicing of its pre-mRNA appears as a very intuitive concept. In fact, it would be very difficult to conceive that the splicing of the dozens of *dystrophin* introns would "wait" until the entire *dystrophin* pre-mRNA is synthesized. Nevertheless most biology (and even molecular biology) textbooks continue to show figures in which a fully transcribed primary transcript, with all its introns, appears as the substrate for splicing (Fig. 1B, left). A more realistic view would depict splicing as taking place while the pre-mRNA is still attached to the template DNA by RNAPII (Fig. 1B, right).

It is worth noting that cotranscriptional splicing is not obligatory. The time it takes RNAPII to synthesize each intron defines the minimal time in which splicing factors can be recruited to

*Alberto R. Kornblihtt—Laboratorio de Fisiología y Biología Molecular, Departamento de Fisiología, Biología Molecular y Celular, IFIBYNE-CONICET, Facultad de Ciencias Exactas y Naturales, Universidad de Buenos Aires. Ciudad Universitaria, Pabellón 2, (C1428EHA) Buenos Aires, Argentina. Email: ark@fbmc.fcen.uba.ar

Alternative Splicing in the Postgenomic Era, edited by Benjamin J. Blencowe and Brenton R. Graveley.

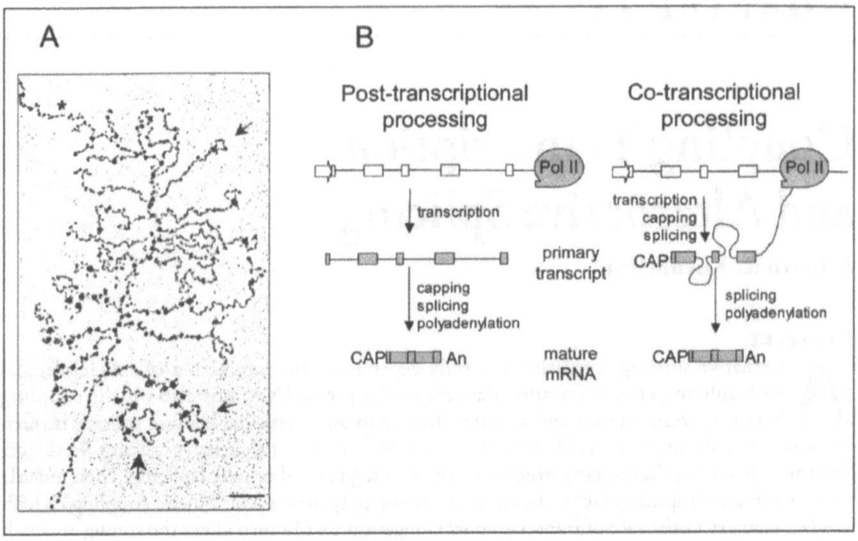

Figure 1. Splicing occurs cotranscriptionally. A) Electron micrograph of an actively transcribing *Drosophila* embryo gene shows that pre-mRNA splicing occurs on the nascent transcript. Asterisk: promoter region. Arrows: intron loops (lariats). Taken with permission from: Beyer AL, Osheim YN. Genes Dev 1988; 2(6):754-765.[8] B) Left: classical textbook picture where all pre-mRNA processing reactions are depicted as posttranscriptional. Right: pre-mRNA processing is cotranscriptional. In the depicted pre-mRNA molecule splicing of intron 1 has already occurred, introns 2 and 3 are being processed and exon 4 has not been transcribed yet.

and spliceosomes assembled upon an intron, whereas the time that it takes RNAPII to reach the end of the transcription unit and release the nascent pre-mRNA defines the maximal time in which splicing could occur cotranscriptionally.[6] For long genes, for example, some introns could be cotranscriptionally spliced whereas others could be processed well after transcription has been completed. This is indeed what happens in the Balbiani ring 1 (BR1) gene where intron 3, located 3 kb from the 5' end of the 40 kb-pre-mRNA, is mostly excised cotranscriptionally, but intron 4, located 0.6 kb from the poly A site, is excised cotranscriptionally in only 10% of the molecules.[10] For most genes, we do not know when each intron is spliced. We do not even know if a particular intron always follows the same pattern of processing. It is worth noting that if splicing were strictly cotranscriptional and that the splicing of one intron must be completed before the following intron is transcribed, alternative splicing would not exist.

Only recently has the cotranscriptional assembly of splicing factors been examined directly by chromatin immunoprecipitation (ChIP). In the budding yeast *Saccharomyces cerevisiae* it was observed that the small nuclear ribonucleoprotein particles (snRNPs) accumulate at positions along intron-containing genes that coincide with the appearance of their target splicing sequences in nascent pre-mRNA.[11-13] For instance, U1 snRNP becomes associated with the pre-mRNA shortly after the 5' splice site is transcribed, while U2 snRNP becomes associated later, after the 3' splice site has been synthesized. Studies in mammalian cells have confirmed these conclusions and have extended them by showing that in addition to general splicing factors, regulatory factors like hnRNPA1 accumulate cotranscriptionally on intron-containing genes but not on intronless ones.[14] The use of efficient in vitro transcription/splicing assays has corroborated the results obtained in living cells: nascent pre-mRNA synthesized by RNAPII is stabilized and efficiently spliced[15] apparently because it is immediately and quantitatively directed into the spliceosome assembly pathway. In contrast, nascent pre-mRNA synthesized by T7 phage

RNA polymerase is quantitatively assembled into the nonspecific hnRNP complexes which are inhibitory to spliceosome assembly, indicating that RNAPII mediates the functional coupling of transcription and splicing by directing the nascent pre-mRNA into spliceosome assembly.[16] Moreover, only genes transcribed by RNAPII encode pre-mRNAs and are efficiently recognized by the spliceosome—pre-mRNAs transcribed by RNAPI, RNAPIII or T7 RNA polymerase are poorly spliced or not at all.[17-20]

This chapter will focus on the evidence, gathered mainly during the last decade, supporting the existence of mechanisms that couple RNAPII transcription to alternative splicing and this evidence will be discussed in light of knowledge of the functional links between transcription and splicing in general. Both transcription and splicing are extremely complex processes because they involve thousands of protein factors, RNA molecules and DNA sequences. This complexity hinders any attempt at generalization and simplification. The reader should bear in mind that certain molecular interactions or kinetic constraints might be relevant for a particular gene or set of genes but not for others.

Promoters Affect Alternative Splicing

The idea that promoter regulation affects only the quantity and not the quality of the gene transcript has dominated our conception of gene expression in the past. However, the finding that promoter identity and occupation by transcription factors modulates alternative splicing[4,21,22] not only strengthened the concept of a physical and functional coupling between transcription and splicing but directed our attention towards how this coupling might affect protein expression patterns. The original observation of the promoter effect involved transient transfection of mammalian cells with reporter minigenes for the alternatively spliced extra domain I (EDI) cassette exon of ***fibronectin*** (*FN*) under the control of different RNAPII promoters. EDI is 270 bp long and contains an exonic splicing enhancer (ESE) that is recognized by the SR proteins SF2/ASF and 9G8. When transcription of the minigene is driven by the α-***globin*** promoter EDI inclusion levels in the mature mRNA are about 10 times lower than when transcription is driven by the *FN* or ***cytomegalovirus*** (*CMV*) promoters (Fig. 2). These effects are not the trivial consequence of different mRNA levels produced by each promoter (promoter strength) but depend on some qualitative properties conferred by promoters to the transcription/RNA processing machinery. This observation is consistent with microarray studies indicating that although, like global transcription profiles, global alternative splicing profiles reflect tissue identity, transcription (evaluated as promoter usage and strength) and alternative splicing act largely independently on different sets of genes to define tissue-specific expression profiles.[23]

Promoter identity has been shown independently to affect the alternative splicing of pre-mRNAs transcribed from several other genes. Reporter minigenes containing alternative exons and flanking intron regions from the ***CD44*** and the ***calcitonin gene-related product*** genes, were put under the control of steroid-sensitive promoters or promoters that do not respond to steroid hormones. Steroid hormones affected splice site selection only of pre-mRNAs produced by the steroid-sensitive promoters. As in the case of the *FN* EDI exon, promoter-dependent hormonal effects on splicing were not a consequence of an increase in transcription rate or of a saturation of the splicing machinery.[24] Promoter-dependent alternative splicing patterns have been also found in the ***cystic fibrosis transmembrane regulator***[25] and in the ***fibroblast growth factor receptor 2*** genes.[26]

The finding that promoter structure is important for alternative splicing predicts that factors that regulate alternative splicing could be acting in part through promoters and that cell-specific alternative splicing may not only result from the differential abundance of various splicing factors, but also from a more complex process involving cell-specific promoter occupation. However, promoters are not swapped in nature and since most genes have a single promoter, the only conceivable way by which promoter architecture could control alternative splicing in vivo, would be the differential occupation of promoters by transcription or splicing factors of different natures and/or mechanistic properties. Accordingly, it has been found that transcriptional activators and

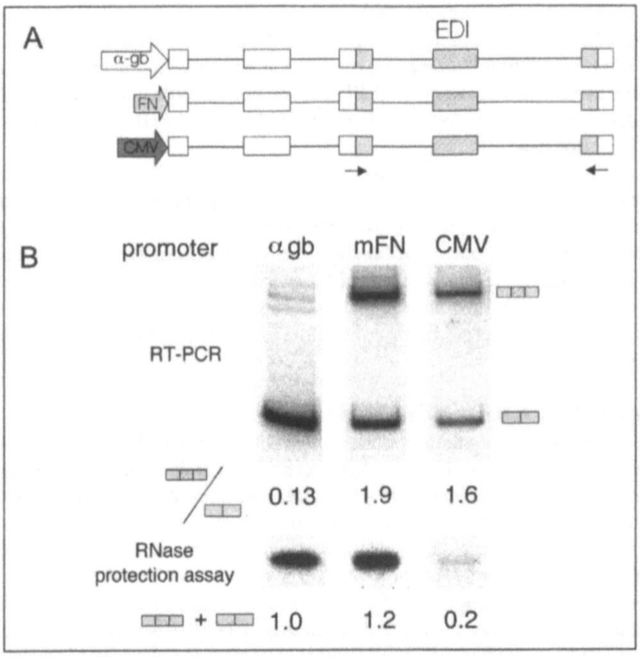

Figure 2. Promoters affect alternative splicing. A) α-*globin* (white)/*fibronectin* (gray) hybrid minigenes under the control of three different promoters, used in transient transfections of mammalian cells in culture to assess inclusion levels of the alternatively spliced EDI cassette exon (dashed) Arrows mark the positions of the primers used for RT-PCR. B. RT-PCR (top) show that inclusion levels with the *FN* and *CMV* promoters are more than 10-fold higher compared with inclusion levels with the α-*globin* promoter. RNase protection assays (bottom) show that expression levels are higher with the α-*globin* and *FN* promoters compared to the *CMV* promoter. Based on Cramer et al.[21,22]

coactivators with different actions on RNAPII initiation and elongation affect alternative splicing differentially[27,28] (see below). Alternative promoter usage is frequent in mammalian genes and could impact the usage of downstream alternative splice sites. However, the the link between alternative promoter usage and splice site selection might not be the result of transcription/splicing coupling but the consequence of deep changes in pre-mRNA secondary structure due to different first exon sequences in the pre-mRNA.

Two non-exclusive mechanisms have been proposed to explain promoter effects on alternative splicing. On the one hand, the promoter might recruit splicing factors or bifunctional factors acting on both transcription and splicing to the transcribing gene. On the other hand, the promoter might alter the rate of RNAPII elongation, affecting in turn the timing of cotranscriptional splicing. I will discuss these two modes in depth, but first will describe the features and roles of what is likely to be a central participant in these process, the carboxy terminal domain (CTD) of the large subunit of RNAPII.

RNAPII CTD and Coupling

The CTD plays a role in the nuclear distribution of components of the transcription and splicing machineries. In fact, transcriptional activation of RNAPII genes increases the association of splicing factors at sites of transcription, but this relocalization does not occur if RNAPII lacks the

CTD.[29] This is consistent with findings that overexpression of CTD-containing large subunits of RNAPII in mammalian cells induces selective nuclear reorganization of splicing factors.[30] Also consistent with the above, stimulation of transcription by strong activators is associated with increased splicing efficiency and this property of activators depends on the CTD.[51,72]

The CTD is composed of 52 tandem heptapeptide repeats in mammals (26 in yeast), with the consensus sequence YSPTSPS. The serines at positions 2 and 5 of this repeat are subject to regulatory phosphorylation. Phosphorylation of Ser5 by TFIIH is linked with transcriptional initiation, whereas phosphorylation of Ser2 by P-TEFb is associated to transcriptional elongation.[31,32] Coupling of transcription and pre-mRNA processing may in part be due to the ability of RNAPII to bind and "piggyback" some of the processing factors in a complex referred to as an "mRNA factory".[2] This concept arose from the observation by McCracken et al[33] that deletion of the CTD causes defects in capping, cleavage/polyadenylation and splicing. These authors showed that deletion of the CTD inhibits splicing of the β-*globin* gene, which is consistent with the findings that isolated CTD fragments[34] as well as purified phosphorylated RNAPII[35] are able to activate splicing in vitro. Nevertheless, isolated CTD fragments cannot duplicate the effect of the RNAPII holoenzyme unless the pre-mRNA is recognized via exon definition, i.e., it contains at least one complete internal exon with 3' and 5' splice sites. In other words, the CTD does not appear to activate splicing of pairs of splice sites across an intron. These findings support a direct role for the CTD in exon recognition and have led to speculation that the CTD functions to bring consecutive exons in proximity, thereby facilitating spliceosome assembly. Consistent with this model, Dye and Proudfoot[36] showed that exons flanking an intron that had been engineered to be cotranscriptionally cleaved by inserting a ribozyme in the middle are accurately and efficiently spliced together. These data suggest that a continuous intron transcript is not required for pre-mRNA splicing in vivo and provide evidence for a molecular tether connecting emergent splice sites in the pre-mRNA to an elongating transcription complex.

Dynamic changes in the CTD structure and phosphorylation may play significant roles in RNA processing. For instance, the peptidyl-prolyl isomerase Pin 1 stimulates CTD phosphorylation by cdc2/cyclin B and inhibits RNAPII-dependent splicing in vitro.[37] Inhibition of P-TEFb-mediated CTD phosphorylation prevents cotranscriptional splicing and 3'-end formation in *Xenopus* oocytes. In contrast, processing of injected pre-mRNA is unaffected by P-TEFb kinase inhibition, which strongly indicates that RNAPII does not participate directly in posttranscriptional processing, but phosphorylation of its CTD is required for efficient cotranscriptional processing.[38] New insights into the mechanism by which the CTD functions in splicing come from in vitro experiments with a fusion protein consisting of the CTD fused to the C-terminus of the splicing factor SF2/ASF (ASF-CTD). Compared to SF2/ASF alone, ASF-CTD increased the reaction rate during the early stages of splicing and this required both RNA-binding activity and phosphorylation of CTD in the fusion protein.[39]

It is worth mentioning that the roles of CTD in splicing may be gene-specific. For example, the absence of the CTD can affect alternative splicing of the *FN* EDI cassette exon[40] (see below) and mRNA editing of the *ADAR2* gene[41] without inhibiting general splicing.

Factor Recruitment

One model that could explain the promoter effect on alternative splicing is that the promoter itself recruits splicing factors to the site of transcription, possibly through transcription factors that bind to the promoter or to transcriptional enhancers. Some proteins, such as the transcriptional activator of the human papilloma virus,[42] and the thermogenic coactivator PGC-1, naturally function in both transcription and splicing. Interestingly, PGC-1 affects alternative splicing, but only when it is recruited to complexes associated with gene promoters[43] (Fig. 3). Another example of a bifunctional factor is the transcription factor Spi-1, required for myeloid and B lymphoid differentiation. Spi-1 is able to regulate alternative splicing of a pre-mRNA for a gene whose transcription it regulates. Guillouf et al[44] demonstrated that, similar to PGC-1, Spi-1 must bind and transactivate its cognate promoter to favor the use of a proximal 5' alternative site.

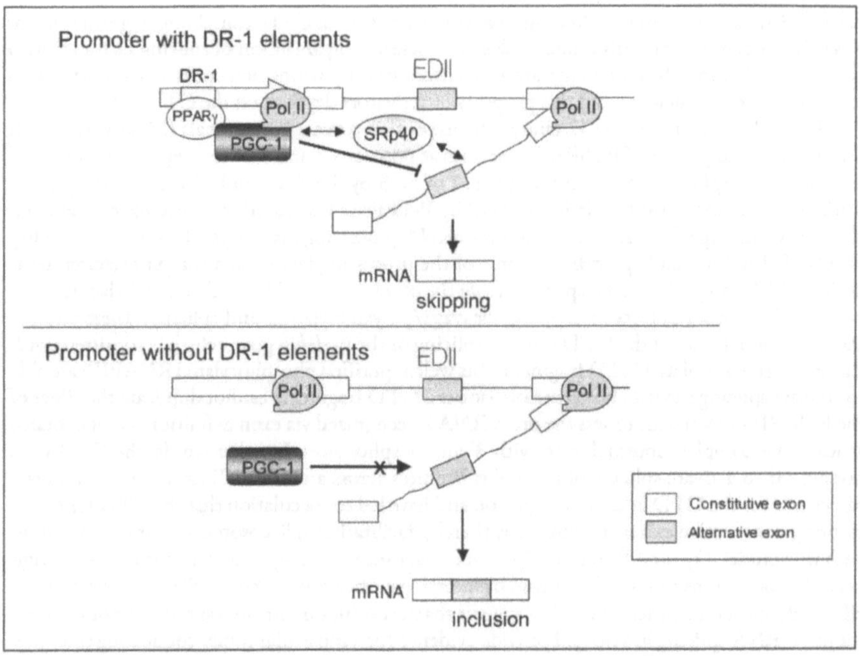

Figure 3. Example of how promoters may affect alternative splicing through recruitment of factors with dual function in transcription and splicing. A promoter with a DR-1 element binds the transcription factor PPARγ, which in turn recruits the transcriptional coactivator PGC-1. PGC-1 interacts with RNAPII and other proteins of the pre-initiation complex as well as with the splicing factor SRp40, which controls inclusion of the *fibronectin* EDII alternative exon. PGC-1 inhibits inclusion of EDII into the mature mRNA, only when targeted to a promoter. Based on Monsalve et al.[43]

Other mammalian proteins that appear to act as bifunctional factors include the product of the *WT-1* gene, which is essential for normal kidney development,[45] SAF-B, which mediates chromatin attachment to the nuclear matrix,[46] CA150, a human nuclear factor with characteristic WW and FF domains implicated in transcriptional elongation[47,48] and a group of proteins known as SCAFs (SR-like CTD associated factors) which interact with the CTD and, similarly to SR proteins, contain an RS domain and an RNA binding domain.[49] However there is no formal evidence that SCAFs function in splicing.

Transcriptional coregulators have also been implicated in the control of alternative splicing. Several coregulators of steroid hormone nuclear receptors have shown to have differential effects on alternative splicing in a promoter-dependent manner.[50] Some coregulators, such as CoAA (coactivator activator), act by recruiting coactivators. CoAA interacts with the transcriptional coregulator TRBP, which is in turn recruited to promoters through interactions with activated nuclear receptors. CoAA regulates alternative splicing in a promoter-dependent manner. It similarly enhances transcription of steroid-sensitive or -insensitive promoters, but only affects alternative splicing of transcripts synthesized from the progesterone-activated MMTV promoter.[28] In addition, transcriptional activators seem to not only modulate alternative but also constitutive splicing in a RNAPII CTD-dependent manner.[51]

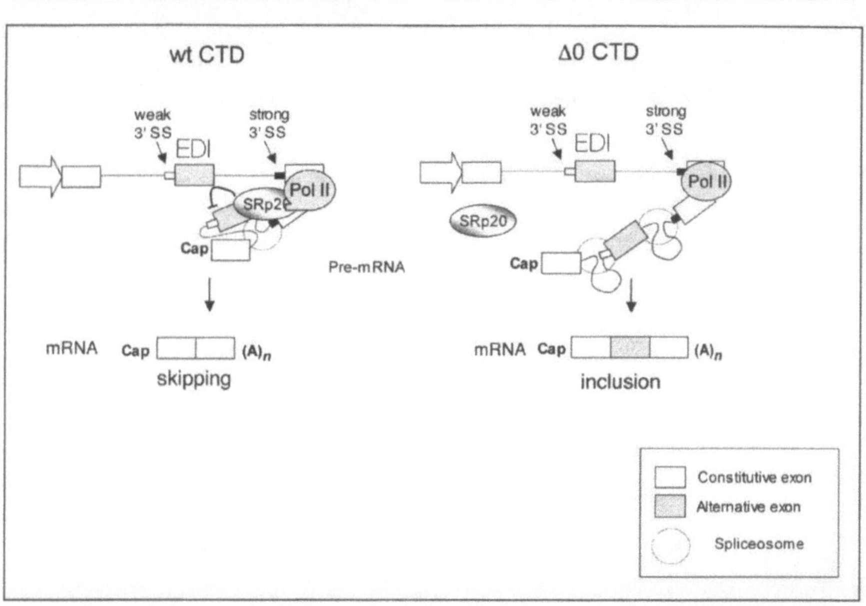

Figure 4. The carboxy terminal domain (CTD) of RNA polymerase II mediates the inhibitory effect of the SR protein SRp20 on the inclusion of the alternatively spliced *fibronectin* EDI exon. Transcription by a WT RNAPII (left) allows recruitment of SRp20 to the transcription machinery which stimulates EDI skipping. Transcription by a mutated RNAPII lacking the CTD (ΔCTD, right) causes higher EDI inclusion because SRp20 is not recruited. Based on de la Mata et al.[40]

Recruitment of SRp20, the CTD and Alternative Splicing

Transcription by an RNAPII mutant lacking the CTD provokes a dramatic enhancement in the inclusion levels of the *FN* EDI alternative cassette exon without affecting the efficiency of general splicing. Interestingly, the CTD influences alternative splicing in a way that is independent of capping and 3' end processing. Experiments using RNAPII CTD variants with different numbers of repeats revealed that the length of the CTD correlates inversely with EDI inclusion levels, with 19 heptads being the minimum number of repeats necessary to sustain normal EDI splicing. This finding is in agreement with reports showing that 22 tandem repeats are sufficient to support wild-type levels of splicing of pre-mRNAs containing constitutively spliced introns or enhancer-dependent introns.[52] Using siRNA knockdown strategies we found that whereas activation of EDI inclusion by the SR protein SF2/ASF is not affected by the absence of the CTD, inhibition of EDI inclusion by another SR protein, SRp20, is completely abolished when transcription is carried out by a ΔCTD RNAPII, suggesting that SRp20 requires the CTD to be recruited to the transcription/splicing machinery[40] (Fig. 4). We were not able to demonstrate direct physical interactions between SRp20 and any portion of the RNAPII large subunit. However, we believe that such an interaction, perhaps weak or indirect, must exist because SRp20 has been found in a transcription complex known as "mediator" together with the large subunit of RNAPII[53] and because immunocytochemical studies have shown that SRp20 preferentially associates with sites of RNAPII transcription[54] and is efficiently recruited to the *tau* gene when one of its alternative exons is included, but not when it is excluded.[55]

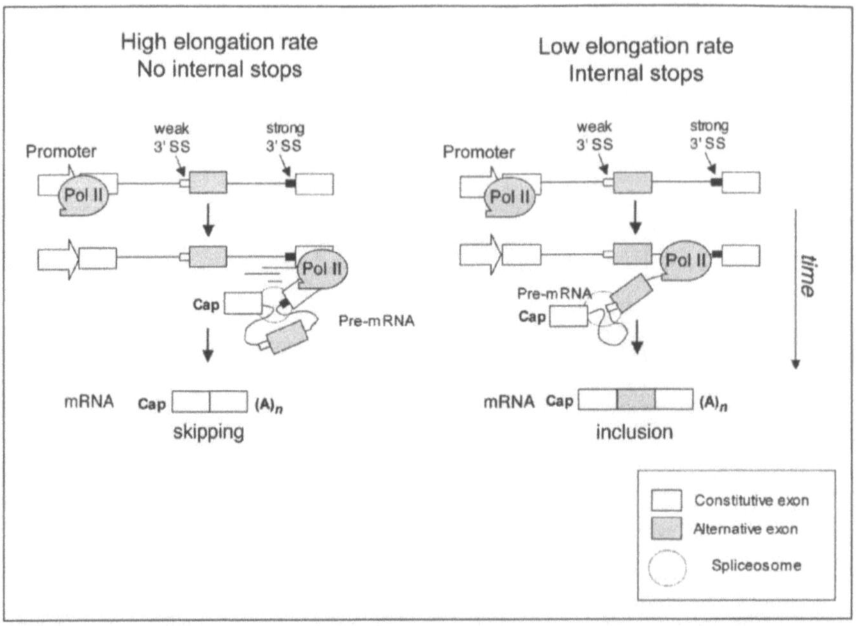

Figure 5. Kinetic coupling model for the regulation of alternative splicing by RNAPII elongation. The 3' splice site (SS) by the alternative cassette exon (white) is weaker than the 3' SS of the downstream intron (black). Low transcriptional elongation rates (right) favor exon inclusion, whereas high elongation rates (left) favor skipping.

Transcription Elongation and Alternative Splicing

Promoters can also control alternative splicing by regulating the rates of RNAPII transcription elongation. Low elongation rates or transcriptional pausing would favor the inclusion of alternative exons governed by an exon skipping mechanism, whereas rapid elongation rates or the absence of transcriptional pausing would favor exclusion of these exons. The mechanism by which elongation rates affect EDI splicing is a consequence of EDI pre-mRNA sequence. EDI exon skipping occurs because the 3' splice site of the upstream intron is weaker than the 3' splice site of the downstream intron. If the polymerase pauses between these two splice sites, the upstream intron will be spliced. Once the transcription complex resumes elongation, the downstream intron will be removed and the exon will be included. In contrast, a rapidly elongating transcription complex will transcribe both introns before the 5' splice site of the upstream intron can be used. As a result, the 5' splice site will be preferentially spliced to the strong downstream 3' splice site, rather than the weak upstream 3' splice site, resulting in exon skipping (Fig. 5). When a weak 3' splice site is followed by a strong one, as is the case in many alternative splicing events, the transcription elongation rate can affect the relative amounts of the mRNA isoforms. However, when two consecutive strong 3' splice sites occur, as in constitutive splicing, transcription elongation rates are less relevant.

A kinetic role for transcription in alternative splicing was originally suggested by Eperon et al[56] who found that the rate of RNA synthesis can affect the secondary structure of a nascent transcript surrounding a 5' splice site, which affects splicing. A similar mechanism involving a kinetic link was suggested from experiments in which transcription pause sites were found to affect alternative splicing by delaying the transcription of an essential splicing inhibitory element (DRE) required for regulation of tropomyosin exon 3.[57]

Several additional experiments indirectly support a role for transcriptipon elongation in alternative splicing:

- Transcription factors that primarily stimulate transcriptional initiation, such as Sp1 and CTF/NF1, have little effect on alternative splicing, whereas factors that stimulate elongation, such as VP16, promote skipping of the EDI exon.[27,58]
- Phosphorylation of the RNAPII CTD at Ser2 by the elongation factor P-TEFb converts the polymerase from a nonprocessive to a processive form. Inhibitors of this kinase such as DRB (dichlororibofuranosylbenzimidazole) inhibit RNAPII elongation. Cells transfected with EDI splicing reporters and treated with DRB displayed a 3-fold increase in EDI inclusion into mature mRNA compared to untreated cells.[27]
- Changes in chromatin structure also affect splicing. Treatment of cells with trichostatin A, a potent inhibitor of histone deacetylation, favors EDI skipping.[27] This finding supports the hypothesis that acetylation of the core histones would facilitate the passage of the transcribing polymerase, which is in turn consistent with the model of chromatin opening being mediated by a RNAPII transcription elongation complex piggybacking a histone acetyltransferase activity tracking along the DNA.[59] Moreover, replication of transfected minigene reporters, which compacts the chromatin structure and slows the passage of the polymerase, causes a 10 to 30-fold increase in EDI exon inclusion levels in the transcript.[58] Interestingly, it has recently been shown that DNA methylation at internal regions of a gene provokes a closed chromatin structure and reduces the efficiency of transcription elongation.[60] This suggests that alternative splicing could be indirectly modulated by the DNA methylation status not only at the promoter but also internally.
- Transcriptional regulatory elements that activate transcription elongation, such as the SV40 enhancer, promote skipping of the EDI exon.[61]
- Chromatin immunoprecipitation experiments have shown that stalled transcription elongation complexes exist more frequently upstream of the alternative EDI exon on minigenes with promoters that favor EDI inclusion (i.e., the *FN* promoter) than on minigenes with promoters that favor EDI skipping (i.e., the *α-globin* promoter).[61]
- Mutation analysis shows that the better the EDI alternative exon is recognized by the splicing machinery, the less its degree of inclusion is affected by factors that modulate transcriptional elongation.[62]
- Although dealing with general and not alternative splicing, two recent reports provide strong evidence for a kinetic link between transcription, splicing factor recruitment and splicing catalysis. Using chromatin-RNA immunoprecipitation (ChRIP), Listerman et al[14] showed that while *fos* pre-mRNA can be spliced in vivo both co- and posttranscriptionally, the topoisomerase inhibitor camptothecin, which stalls RNAPII elongation, increased co-transcriptional splicing factor accumulation and splicing in parallel. The second report by the Rosbash lab[63] elegantly shows that cleavage of an intron by a hammerhead ribozyme competes with the splicing of that intron. If splicing of this pre-mRNA is prevented by mutating the 5' splice site, the ribozyme is able to cleave the intron, while for a wild-type pre-mRNA cotranscriptional splicing occurs prior to ribozyme cleavage. These results strongly suggest that introns are recognized cotranscriptionally. Furthermore, the *DST1* gene, which encodes the transcription elongation factor TFIIS, was identified in a screen for genes required to prevent cleavage of the intronic ribozyme in a normal splicing reporter. This again provides a link between transcription elongation and pre-mRNA splicing.

Slow Polymerases and Alternative Splicing

A more direct demonstration that transcription elongation affects alternative splicing in human cells was provided by the use of a mutant form of RNAPII (called C4) that possesses a reduced elongation rate.[64] The slow polymerase stimulates the inclusion of the *fibronectin* EDI exon by 4-fold, confirming the inverse correlation between elongation rate and inclusion of this alternative exon. The C4 mutation also affected the splicing of the adenovirus *E1a* pre-mRNA, by favoring the use of the most upstream of the three alternative 5' splice sites that compete for a common 3' splice site. Most importantly and of physiological relevance, *Drosophila* flies carrying the C4 mutation show changes in the alternative splicing profile of transcripts encoded by the large *ultrabithorax* (*Ubx*) endogenous gene.[64] The observed changes are consistent with a kinetic mechanism which allows more time for early splicing events. Most interestingly, *C4* heterozygous flies display a phenotype, known as the "Ubx effect", where the halteres present a morphology that resembles the one of the *Ultrabithorax* mutation.[72]

Similar effects of RNAPII elongation rates on splicing were found in yeast. Alternative splicing is a very rare event in yeast. Mutating the branchpoint upstream of the constitutive internal exon of the *DYN2* gene creates an artificial alternatively spliced cassette exon. Skipping of this exon is prevented when expressed in a yeast mutant carrying a slow RNAPII or in the presence elongation inhibitors.[65] This supports the hypothesis that what is important to the balance between exon skipping and exon inclusion are relative rates of spliceosome formation and RNAPII processivity.

Chromatin, Elongation and Alternative Splicing

Batsché et al[66] revealed a new role in alternative splicing for the chromatin remodeling factor SWI/SNF, whose mechanism of action involves the regulation of RNAPII elongation. SWI/SNF is known to interact with RNAPII, splicing factors and spliceosome-associated proteins. Overexpression of Brahma (Brm), the key subunit of SWI/SNF, favors inclusion of a block of consecutive alternative exons in the *CD44* gene, which is a target for SWI/SNF transcriptional activation. As expected for a splicing regulator, Brm interacts with complexes containing U1 and U5 snRNAs, which are present in spliceosomes, but not with U3 snRNA, which is involved in ribosomal RNA processing. Brm also interacts with Sam68, a nuclear RNA-binding protein that in turn binds splicing regulatory elements present in the *CD44* alternative exons and stimulates their inclusion upon activation of the ERK MAP kinases. How does Brm use these multiple interactions to control alternative splicing? *CD44* contains a cluster of ten consecutive alternative exons (v1 to v10) located between constitutive exons 5 and 16. ChIP experiments have shown that Brm is not only present at the gene promoter but appears to be distributed along the whole transcriptional unit with levels that decrease gradually towards the 3' end. Although also concentrated at the promoter region, RNAPII displays a different distribution inside the gene, with a clear accumulation within the variable region peaking on exon v4. This peak disappears when Brm is knocked down by RNAi, but is higher when cells are treated with phorbol esthers that activate ERKs.

These findings strongly suggest that activation of Sam68 by ERK triggers the formation of macromolecular complexes containing Sam68, RNAPII and Brm at the central block of variable exons. This results in the stalling of RNAPII and the inclusion of the variable exons into mature mRNA, in agreement with the kinetic coupling model (Fig. 5). Interestingly, there is a dramatic change in the phosphorylation status of RNAPII at the pause site.[66] Successive ChIPs (ChIP-reChIPs) using first anti-Brm and then anti-phospho-CTD antibodies specific for either phospho-Ser5 or phospho-Ser2 revealed that within the *CD44* constant region Brm associates with phospho-Ser2 CTD RNAPII. However, Brm associates with phospho-Ser5 CTD RNAPII species at the *CD44* alternative exons. The return of the RNAPII phosphorylation status to that typical of promoters at specific sites within genes could generate internal "road blocks" to elongation (Fig. 6). In any case it is now clear that internal road blocks exist in vivo, can be regulated by external signals and are very important for alternative splicing.

Coordination Between and Polarity in Multiple Alternative Splicing Events

Soon after the discovery of splicing it became evident that many genes contained more than one region that is alternatively spliced, a feature that significantly expands the protein-encoding potential of a genome. The *fibronectin* gene is a paradigmatic example,[67] as it contains three regions of alternative splicing that display cell type- and developmental stage-specific regulation. This organization can give rise to up to 20 mRNA isoforms in humans, 12 in rodents and 8 in

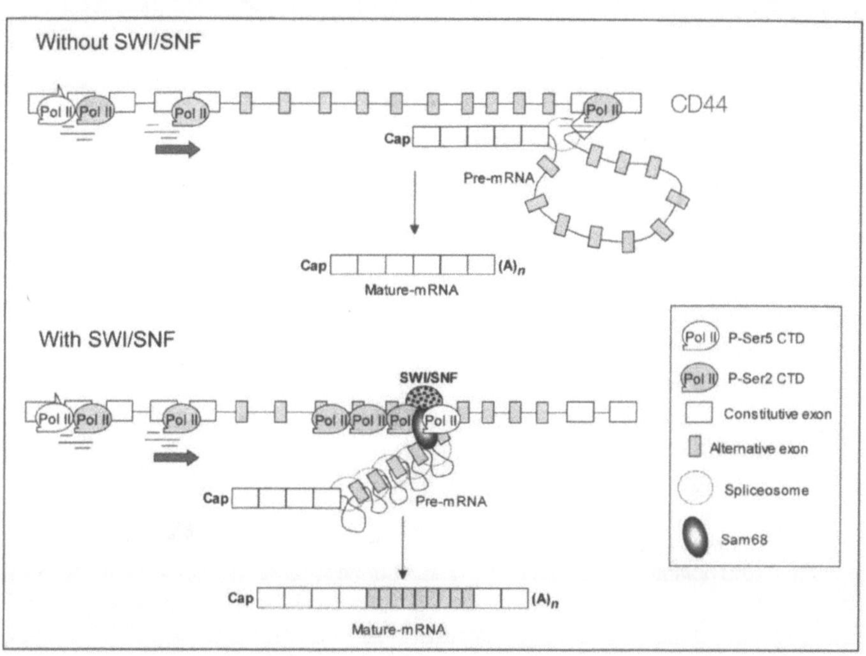

Figure 6. SWI/SNF stimulates inclusion of alternative exons in the *CD44* gene by creating a "road block" to RNAPII elongation at the variable region. The pause is the consequence of multiple protein interactions involving SWI/SNF, RNAPII, the splicing regulator Sam68 and spliceosomal components. The phosphorylation pattern of RNAPII CTD associated to Brm is changed from phospho-Ser2 to phospho-Ser5. This might cause the stalling of RNAPII molecules coming behind, even if they are phosphorylated at the elongation-competent Ser2. Based on Batsché et al.[66]

chickens.[68] Although other genes with multiple regions of alternative splicing have been characterized individually, the general prevalence of this phenomenon has been only recently examined by bioinformatic approaches which indicate that a significant fraction (25%) of human genes have such an organization.[69] This organization also raises the question of whether the different alternatively spliced regions of a gene are coordinately regulated. This has been studied by transfecting human cells with minigenes carrying two alternative EDI regions in tandem, separated by 3,400 bp spanning three constitutive exons and the corresponding introns. Mutations at splice sites or regulatory elements of the proximal (with respect to the promoter) EDI exon that either stimulate or inhibit its inclusion cause parallel effects in the inclusion levels of the distal EDI. In contrast, the same mutations introduced in the distal EDI have much smaller effects on the inclusion levels of the proximal exon.[69] Although the molecular mechanism for the coordinating effect remains to be elucidated, it is clear that coordination displays gene polarity. Most interestingly, coordination persists but polarity disappears when the rate of transcriptional elongation is high (Fig. 7) but is reestablished when elongation is inhibited by DRB. Thus, the rate of transcription elongation is not only important for splice site selection at a single alternative splicing event but also for long distance effects in splicing regulation. Other examples of long distance regulation of splice site selection have been reported in the equine *β-casein* intron 1,[70] and in the human *thrombopoietin* gene.[71] However, coordination and polarity of multiple alternative splicing events do not appear to occur in every gene as splicing of the *Drosophila Dscam* gene does not appear to be governed by such rules.[74]

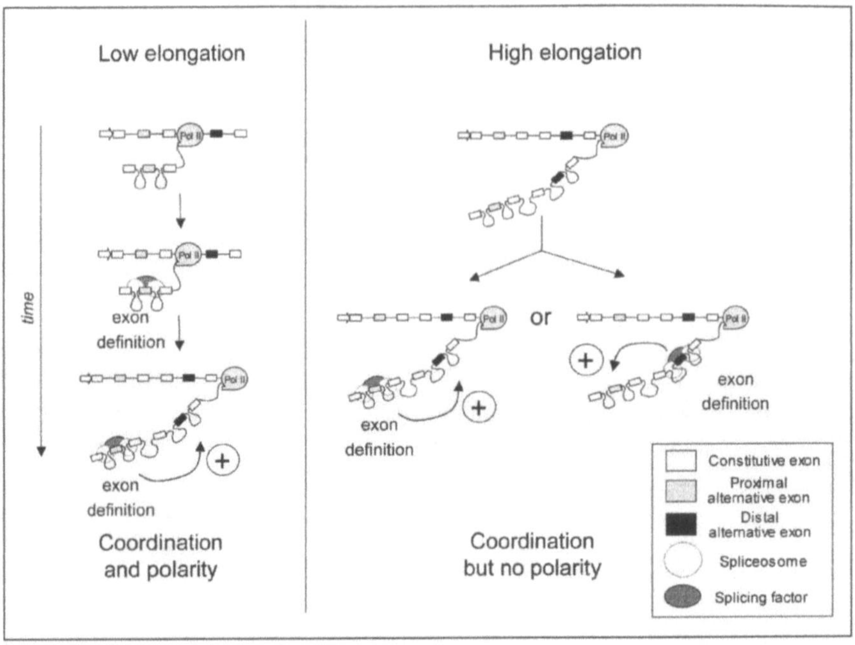

Figure 7. Model for the role of RNAPII elongation on alternative splicing polarity. Low elongation rates or internal pauses (left) allow a temporal window of opportunity for splicing complexes to assemble at the proximal alternative exon before the distal alternative exon is transcribed. As RNAPII proceeds, the exon definition complexes at the proximal alternative exon stimulate distal alternative exon inclusion in a polar way. High elongation rate or lack of internal pauses allows both proximal and distal alternative exons to be exposed simultaneously to the splicing machinery which results in the absence of polarity. Based on Fededa et al.[69]

Conclusions and Perspectives

Transcription elongation and transcription factor recruitment may contribute independently or in a concerted way to the mechanisms by which transcription controls alternative splicing. Some years ago we proposed the idea that changes in the "pausing architecture" of a gene would provoke changes in the alternative splicing pattern of its transcript. In this context, perhaps the contribution of different promoters or differential occupation of a single promoter is not crucial in the cell, but experiments of promoter swapping were important to investigate the real determinants in kinetic and recruitment coupling. Several lines of evidence point to changes in the chromatin structure in internal regions of genes as elicitors of changes in RNAPII elongation and stalling. The use of ChIP methodology has come of age to depict the "orography" of RNAPII, as well as that of proteins involved in transcription, splicingand chromatin structure along genes during cotranscriptional mRNA processing. The roles of posttranslational modifications, such as acetylation and methylation of core histones, should also be investigated. One could imagine that, in the not so distant future, a detailed map will be available of the peaks and valleys corresponding to the distributions of regulatory proteins and modifications on each gene in the genome and under different physiological or pathological conditions. Such information will likely be extremely informative for predicting the corresponding patterns of transcription and processing.

Acknowledgements

I would like to thank the present and former members of my lab for their talent, creativity and enthusiasm and for thinking in "stereo", with one ear listening to the channel of transcription and the other one to the channel of splicing: Paula Cramer, Sebastián Kadener, Guadalupe Nogués, Manuel de la Mata, Juan Pablo Fededa, Manuel Muñoz, Ignacio Schor, Ezequiel Petrillo, Mariano Alló, Soledad Pérez Santángelo, Nicolás Rascovan, Valeria Buggiano and the Srebrow group (Anabella Srebrow, Matías Blaustein, Federico Pelisch and Leandro Quadrana). I also thank my collaborators and colleagues for their continuous support and ideas: David Bentley, Michael Rosbash, Tito Baralle, Javier Cáceres, Andrés Muro, Mikhail Gelfand, Rob Chapman, Karla Neugebauer and Claudio Alonso. This work was supported by grants from the Fundación Antorchas, the Agencia Nacional de Promoción de Ciencia y Tecnología of Argentina, the Consejo Nacional de Investigaciones Científicas y Técnicas (CONICET), the European Union Network of Excellence on Alternative Splicing (EURASNET) and the University of Buenos Aires. ARK is a Howard Hughes Medical Institute international research scholar and a career investigator of the CONICET.

References

1. Bentley D. The mRNA assembly line: transcription and processing machines in the same factory. Curr Opin Cell Biol 2002; 14(3):336-342.
2. Bentley DL. Rules of engagement: cotranscriptional recruitment of pre-mRNA processing factors. Curr Opin Cell Biol 2005; 17(3):251-256.
3. Maniatis T, Reed R. An extensive network of coupling among gene expression machines. Nature 2002; 416(6880):499-506.
4. Kornblihtt AR. Promoter usage and alternative splicing. Curr Opin Cell Biol 2005; 17(3):262-268.
5. Zorio DA, Bentley DL. The link between mRNA processing and transcription: communication works both ways. Exp Cell Res 2004; 296(1):91-97.
6. Neugebauer KM. On the importance of being cotranscriptional. J Cell Sci 2002; 115(Pt 20):3865-3871.
7. Proudfoot NJ, Furger A, Dye MJ. Integrating mRNA processing with transcription. Cell 2002; 108(4):501-512.
8. Beyer AL, Osheim YN. Splice site selection, rate of splicing and alternative splicing on nascent transcripts. Genes Dev 1988; 2(6):754-765.
9. Tennyson CN, Klamut HJ, Worton RG. The human dystrophin gene requires 16 hours to be transcribed and is cotranscriptionally spliced. Nat Genet 1995; 9(2):184-190.
10. Bauren G, Wieslander L. Splicing of Balbiani ring 1 gene pre-mRNA occurs simultaneously with transcription. Cell 1994; 76(1):183-192.
11. Kotovic KM, Lockshon D, Boric L et al. Cotranscriptional recruitment of the U1 snRNP to intron-containing genes in yeast. Mol Cell Biol 2003; 23(16):5768-5779.
12. Lacadie SA, Rosbash M. Cotranscriptional spliceosome assembly dynamics and the role of U1 snRNA:5'ss base pairing in yeast. Mol Cell 2005; 19(1):65-75.
13. Gornemann J, Kotovic KM, Hujer K et al. Cotranscriptional spliceosome assembly occurs in a stepwise fashion and requires the -. Mol Cell 2005; 19(1):53-63.
14. Listerman I, Sapra AK, Neugebauer KM. Cotranscriptional coupling of splicing factor recruitment and precursor messenger RNA splicing in mammalian cells. Nat Struct Mol Biol 2006; 13(9):815-822.
15. Hicks MJ, Yang CR, Kotlajich MV et al. Linking splicing to RNAPII transcription stabilizes pre-mRNAs and influences splicing patterns. PLoS Biol 2006; 4(6):e.147
16. Das R, Dufu K, Romney B et al. Functional coupling of RNAPII transcription to spliceosome assembly. Genes Dev 2006; 20(9):1100-1109.
17. Smale ST, Tjian R. Transcription of herpes simplex virus tk sequences under the control of wild-type and mutant human RNA polymerase I promoters. Mol Cell Biol 1985; 5(2):352-362.
18. Sisodia SS, Sollner-Webb B, Cleveland DW. Specificity of RNA maturation pathways: RNAs transcribed by RNA polymerase III are not substrates for splicing or polyadenylation. Mol Cell Biol 1987; 7(10):3602-3612.
19. McCracken S, Rosonina E, Fong N et al. Role of RNA polymerase II carboxy-terminal domain in coordinating transcription with RNA processing. Cold Spring Harb Symp Quant Biol 1998; 63:301-309.
20. Dower K, Rosbash M. T7 RNA polymerase-directed transcripts are processed in yeast and link 3' end formation to mRNA nuclear export. RNA 2002; 8(5):686-697.

21. Cramer P, Pesce CG, Baralle FE et al. Functional association between promoter structure and transcript alternative splicing. Proc Natl Acad Sci USA 1997; 94(21):11456-11460.
22. Cramer P, Caceres JF, Cazalla D et al. Coupling of transcription with alternative splicing: RNA RNA-PII promoters modulate SF2/ASF and 9G8 effects on an exonic splicing enhancer. Mol Cell 1999; 4(2):251-258.
23. Pan Q, Shai O, Misquitta C et al. Revealing global regulatory features of mammalian alternative splicing using a quantitative microarray platform. Mol Cell 2004; 16(6):929-941.
24. Auboeuf D, Honig A, Berget SM et al. Coordinate regulation of transcription and splicing by steroid receptor coregulators. Science 2002; 298(5592):416-419.
25. Pagani F, Stuani C, Zuccato E et al. Promoter architecture modulates CFTR exon 9 skipping. J Biol Chem 2003; 278(3):1511-1517.
26. Robson-Dixon ND, Garcia-Blanco MA. MAZ elements alter transcription elongation and silencing of the fibroblast growth factor receptor 2 exon IIIb. J Biol Chem 2004; 279(28):29075-29084.
27. Nogues G, Kadener S, Cramer P et al. Transcriptional activators differ in their abilities to control alternative splicing. J Biol Chem 2002; 277(45):43110-43114.
28. Auboeuf D, Dowhan DH, Li X et al. CoAA, a nuclear receptor coactivator protein at the interface of transcriptional coactivation and RNA splicing. Mol Cell Biol 2004; 24(1):442-453.
29. Misteli T, Spector DL. RNA polymerase II targets pre-mRNA splicing factors to transcription sites in vivo. Mol Cell 1999; 3(6):697-705.
30. Du L, Warren SL. A functional interaction between the carboxy-terminal domain of RNA polymerase II and pre-mRNA splicing. J Cell Biol 1997; 136(1):5-18.
31. Sims RJ, Belotserkovskaya R, Reinberg D. Elongation by RNA polymerase II: the short and long of it. Genes Dev 2004; 18(20):2437-2468.
32. Saunders A, Core LJ, Lis JT. Breaking barriers to transcription elongation. Nat Rev Mol Cell Biol 2006; 7(8):557-567.
33. McCracken S, Fong N, Yankulov K et al. The C-terminal domain of RNA polymerase II couples mRNA processing to transcription. Nature 1997; 385(6614):357-361.
34. Zeng C, Berget SM. Participation of the C-terminal domain of RNA polymerase II in exon definition during pre-mRNA splicing. Mol Cell Biol 2000; 20(21):8290-8301.
35. Hirose Y, Tacke R, Manley JL. Phosphorylated RNA polymerase II stimulates pre-mRNA splicing. Genes Dev 1999; 13(10):1234-1239.
36. Dye MJ, Gromak N, Proudfoot NJ. Exon tethering in transcription by RNA polymerase II. Mol Cell 2006; 21(6):849-859.
37. Xu YX, Hirose Y, Zhou XZ et al. Pin1 modulates the structure and function of human RNA polymerase II. Genes Dev 2003; 17(22):2765-2776.
38. Bird G, Zorio DA, Bentley DL. RNA Polymerase II Carboxy-Terminal Domain Phosphorylation Is Required for Cotranscriptional Pre-mRNA Splicing and 3'-End Formation. Mol Cell Biol 2004; 24(20):8963-8969.
39. Millhouse S, Manley JL. The C-terminal domain of RNA polymerase II functions as a phosphorylation-dependent splicing activator in a heterologous protein. Mol Cell Biol 2005; 25(2):533-544.
40. de la Mata, M, Kornblihtt AR. RNAPII CTD mediates SRp20 regulation of alternative splicing. Nat Struct Mol Biol 2006; 11:973-980.
41. Laurencikiene J, Kallman AM, Fong N et al. RNA editing and alternative splicing: the importance of cotranscriptional coordination. EMBO Rep 2006; 7(3):303-307.
42. Lai MC, Teh BH, Tarn WY. A human papillomavirus E2 transcriptional activator. The interactions with cellular splicing factors and potential function in pre-mRNA processing. J Biol Chem 1999; 274(17):11832-11841.
43. Monsalve M, Wu Z, Adelmant G et al. Direct coupling of transcription and mRNA processing through the thermogenic coactivator PGC-1. Mol Cell 2000; 6(2):307-316.
44. Guillouf C, Gallais I, Moreau-Gachelin F. Spi-1/PU.1 oncoprotein affects splicing decisions in a promoter binding-dependent manner. J Biol Chem 2006; 281(28):19145-19155.
45. Davies RC, Calvio C, Bratt E et al. WT1 interacts with the splicing factor U2AF65 in an isoform-dependent manner and can be incorporated into spliceosomes. Genes Dev 1998; 12(20):3217-3225.
46. Nayler O, Stratling W, Bourquin JP et al. SAF-B protein couples transcription and pre-mRNA splicing to SAR/MAR elements. Nucleic Acids Res 1998; 26(15):3542-3549.
47. Goldstrohm AC, Albrecht TR, Sune C et al. The transcription elongation factor CA150 interacts with RNA polymerase II and the pre-mRNA splicing factor SF1. Mol Cell Biol 2001; 21(22):7617-7628.
48. Lin KT, Lu RM, Tarn WY. The WW domain-containing proteins interact with the early spliceosome and participate in pre-mRNA splicing in vivo. Mol Cell Biol 2004; 24(20):9176-9185.

49. Yuryev A, Patturajan M, Litingtung Y et al. The C-terminal domain of the largest subunit of RNA polymerase II interacts with a novel set of serine/arginine-rich proteins. Proc Natl Acad Sci USA 1996; 93(14):6975-6980.
50. Auboeuf D, Dowhan DH, Kang YK et al. Differential recruitment of nuclear receptor coactivators may determine alternative RNA splice site choice in target genes. Proc Natl Acad Sci USA 2004; 101(8):2270-2274.
51. Rosonina E, Bakowski MA, McCracken S et al. Transcriptional activators control splicing and 3'-end cleavage levels. J Biol Chem 2003; 278(44):43034-43040.
52. Rosonina E, Blencowe BJ. Analysis of the requirement for RNA polymerase II CTD heptapeptide repeats in pre-mRNA splicing and 3'-end cleavage. RNA 2004; 10(4):581-589.
53. Sato S, Tomomori-Sato C, Parmely TJ et al. A set of consensus mammalian mediator subunits identified by multidimensional protein identification technology. Mol Cell 2004; 14(5):685-691.
54. Neugebauer KM, Roth MB. Distribution of pre-mRNA splicing factors at sites of RNA polymerase II transcription. Genes Dev 1997; 11(9):1148-1159.
55. Mabon SA, Misteli T. Differential recruitment of pre-mRNA splicing factors to alternatively spliced transcripts in vivo. PLoS Biol 2005; 3(11):e374.
56. Eperon LP, Graham IR, Griffiths AD et al. Effects of RNA secondary structure on alternative splicing of pre-mRNA: is folding limited to a region behind the transcribing RNA polymerase? Cell 1988; 54(3):393-401.
57. Roberts GC, Gooding C, Mak HY et al. Cotranscriptional commitment to alternative splice site selection. Nucleic Acids Res 1998; 26(24):5568-5572.
58. Kadener S, Cramer P, Nogues G et al. Antagonistic effects of T-Ag and VP16 reveal a role for RNA RNAPII elongation on alternative splicing. EMBO J 2001; 20(20):5759-5768.
59. Travers A. Chromatin modification by DNA tracking. Proc Natl Acad Sci USA 1999; 96(24):13634-13637.
60. Lorincz MC, Dickerson DR, Schmitt M et al. Intragenic DNA methylation alters chromatin structure and elongation efficiency in mammalian cells. Nat Struct Mol Biol 2004; 11(11):1068-1075.
61. Kadener S, Fededa JP, Rosbash M et al. Regulation of alternative splicing by a transcriptional enhancer through RNA RNAPII elongation. Proc Natl Acad Sci USA 2002; 99(12):8185-8190.
62. Nogues G, Munoz MJ, Kornblihtt AR. Influence of polymerase II processivity on alternative splicing depends on splice site strength. J Biol Chem 2003; 278(52):52166-52171.
63. Lacadie SA, Tardiff DF, Kadener S et al. In vivo commitment to yeast cotranscriptional splicing is sensitive to transcription elongation mutants. Genes Dev 2006; 20(15):2055-2066.
64. de la Mata M, Alonso CR, Kadener S et al. A slow RNA polymerase II affects alternative splicing in vivo. Mol Cell 2003; 12(2):525-532.
65. Howe KJ, Kane CM, Ares M Jr. Perturbation of transcription elongation influences the fidelity of internal exon inclusion in Saccharomyces cerevisiae. RNA 2003; 9(8):993-1006.
66. Batsche E, Yaniv M, Muchardt C. The human SWI/SNF subunit Brm is a regulator of alternative splicing. Nat Struct Mol Biol 2006; 13(1):22-29.
67. Sharp PA. Split genes and RNA splicing. Cell 1994; 77(6):805-815.
68. Kornblihtt AR, Pesce CG, Alonso CR et al. The fibronectin gene as a model for splicing and transcription studies. FASEB J 1996; 10(2):248-257.
69. Fededa JP, Petrillo E, Gelfand MS et al. A polar mechanism coordinates different regions of alternative splicing within a single gene. Mol Cell 2005; 19(3):393-404.
70. Lenasi T, Peterlin BM, Dovc P. Distal regulation of alternative splicing by splicing enhancer in equine beta-casein intron 1. RNA 2006; 12(3):498-507.
71. Romano M, Marcucci R, Baralle FE. Splicing of constitutive upstream introns is essential for the recognition of intra-exonic suboptimal splice sites in the thrombopoietin gene. Nucleic Acids Res 2001; 29(4):886-894.
72. Rosonina E, Ip JY, Calarco JA et al. Role for PSF in mediating transcriptional activator-dependent stimulation of pre-mRNA processing in vivo. Mol Cell Biol 2005; 25(15):6734-6746.
73. Greenleaf AL, Weeks JR, Voelker RA et al. Genetic and biochemical characterization of mutants of an RNA polymerase II locus in D. melanogaster. Cell 1980; 21:785-792.
74. Neves G, Zucker J, Daly M et al. Stochastic yet biased expression of multiple Dscam splice variants by individual cells. Nat Genet 2004; 36(3):240-246.

Chapter 12

The Coupling of Alternative Splicing and Nonsense-Mediated mRNA Decay

Liana F. Lareau, Angela N. Brooks, David A.W. Soergel, Qi Meng and Steven E. Brenner*

Abstract

Most human genes exhibit alternative splicing, but not all alternatively spliced transcripts produce functional proteins. Computational and experimental results indicate that a substantial fraction of alternative splicing events in humans result in mRNA isoforms that harbor a premature termination codon (PTC). These transcripts are predicted to be degraded by the nonsense-mediated mRNA decay (NMD) pathway. One explanation for the abundance of PTC-containing isoforms is that they represent splicing errors that are identified and degraded by the NMD pathway. Another potential explanation for this startling observation is that cells may link alternative splicing and NMD to regulate the abundance of mRNA transcripts. This mechanism, which we call "Regulated Unproductive Splicing and Translation" (RUST), has been experimentally shown to regulate expression of a wide variety of genes in many organisms from yeast to human. It is frequently employed for autoregulation of proteins that affect the splicing process itself. Thus, alternative splicing and NMD act together to play an important role in regulating gene expression.

Introduction

One major result of the large-scale sequencing projects of the last decade has been an appreciation of the extent of alternative splicing of mammalian genes. Estimates vary, but most reports agree that over half of human genes are alternatively spliced.[1,2] What is the biological function of this extensive alternative splicing? Many propose that it is a major mechanism underlying proteome expansion,[3] but alternative splicing can also modulate the function or activity of a gene, for instance by adding or removing exons encoding protein domains or by altering the stability of the transcript or resulting protein.[4-6]

In the last few years, it has become clear that many alternative splice forms previously thought to encode truncated proteins are actually targets of NMD (Fig. 1). In mammals, a termination codon located more than about 50 nucleotides upstream of the final exon junction is generally recognized as premature and elicits NMD.[7-9] Understanding of this rule allowed for the identification of numerous transcripts that are predicted to be degraded rather than translated into protein. The prevalence of these predicted NMD-targeted transcripts calls for a reconsideration of the roles of alternative splicing and NMD. Since the mechanism of recognition by the NMD pathway is best understood in mammals and there are relatively few predicted or verified targets of NMD in other organisms, we will focus this review primarily on mammalian targets of NMD.

*Corresponding Author: Steven E. Brenner—Department of Molecular and Cell Biology, University of California, Berkeley. Email: brenner@compbio.berkeley.edu

Alternative Splicing in the Postgenomic Era, edited by Benjamin J. Blencowe and Brenton R. Graveley.

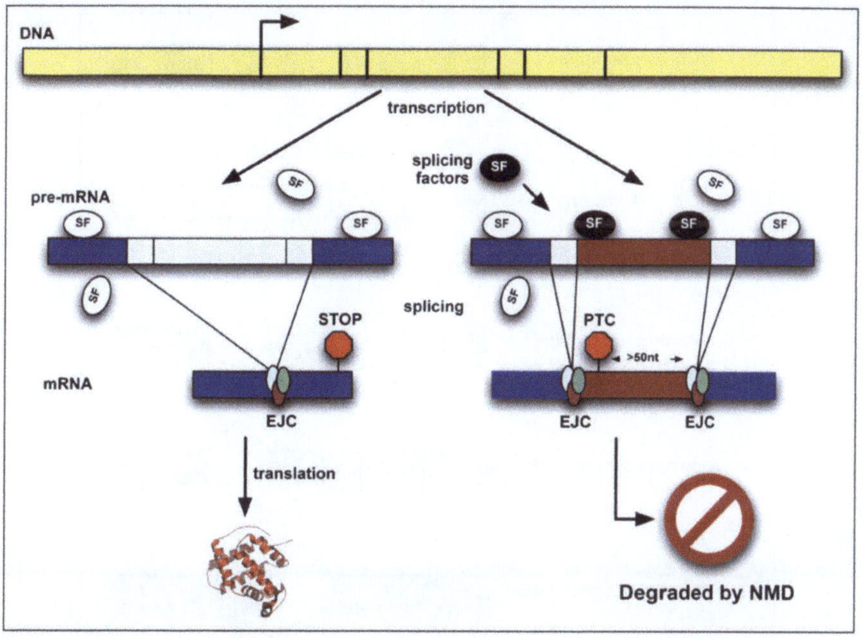

Figure 1. Some alternatively spliced transcripts are degraded by Nonsense Mediated mRNA Decay (NMD). The spliceosome deposits an Exon Junction Complex (EJC) on the mRNA ~20-24nt upstream of the splice junction, thereby marking the former location of the excised intron.[9] On the first, pioneering round of translation,[25] any in-frame stop codon found more than 50 nt upstream of the splice junction triggers NMD; such a codon is called a PTC.[8,9] Alternative splicing can lead to the inclusion of a PTC on an alternatively spliced region, or may give rise to a downstream PTC due to a frameshift. Thus, alternative splicing can give rise to unproductive transcripts. Splicing factors (labelled "SF") can alter the ratio of productive transcripts to transcripts that contain a PTC, targeting them for degradation. In this example, the dark splicing factor shown induces the inclusion of an alternative exon with a PTC, thereby decreasing the abundance of the productive isoform and downregulating protein expression. Components of the splicing machinery such as U2AF35 and PTB can similarly regulate isoform proportions. Reprinted from: Soergel D et al. In: Maquat LE, ed. Nonsense-Mediated mRNA Decay. Goergetown: Landes Bioscience 2006:175-196.[109]

Transcripts containing a PTC can arise through various patterns of alternative splicing (Fig. 2). For example, an exon inclusion event can introduce an in-frame PTC, thus targeting the transcript for NMD. Most alternative splicing events that induce a frameshift are predicted to give rise to a downstream PTC. Alternative splicing in noncoding regions can also give rise to NMD targets. For example, the splicing of a 3' UTR intron can create an exon-exon junction more than 50 nucleotides downstream of the original stop codon, which will consequently be recognized as premature.[10]

NMD was originally considered to be a quality control mechanism, protecting the cell from the potentially toxic effects of nonsense codons introduced by errors in replication, transcription, or splicing.[11,12] We now know that there are many targets of NMD,[13,14] including transcripts with uORFs, products of alternative splicing, byproducts of V(D)J recombination and transcripts arising from transposons and retroviruses.[15] Indeed, it now seems that a major effect of NMD is to downregulate specific transcripts, in addition to clearing the cell of aberrant transcripts.

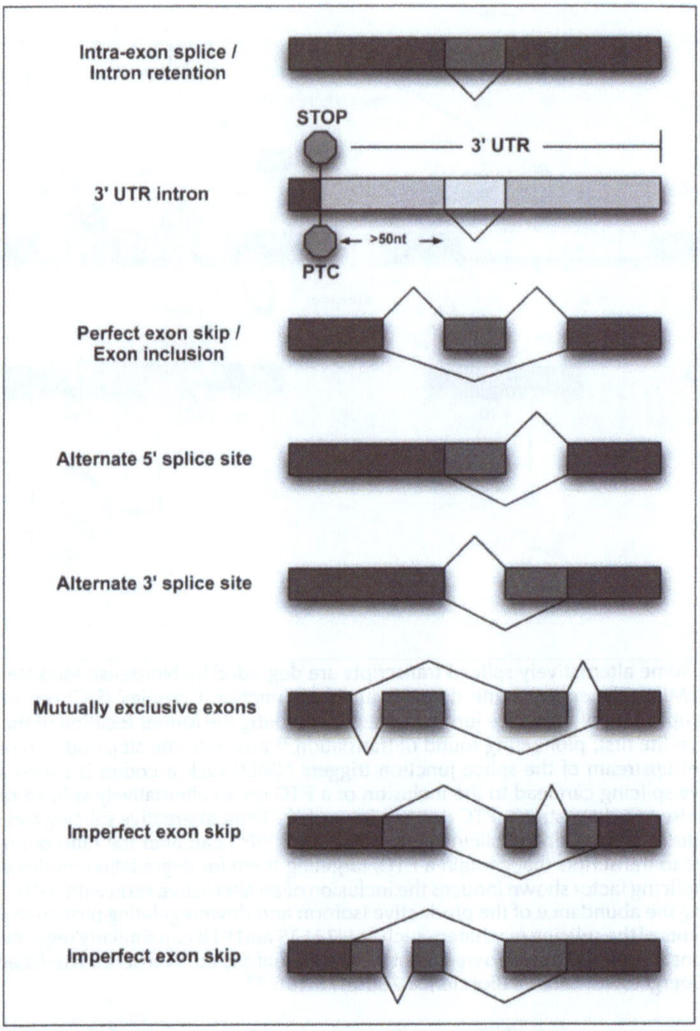

Figure 2. Patterns of alternative splicing. Alternative selection of 5' and 3' splice sites can lead to various patterns of included exons. Any exon that is included in an alternative form may harbor a PTC. Also, whenever an exon whose length is not a multiple of 3 is included or removed, the concomitant frame shift may result in a downstream PTC. Finally, splicing out an intron in the 3' untranslated region (UTR) can cause the normal stop codon to be considered premature. Reprinted from: Soergel D et al. In: Maquat LE, ed. Nonsense-Mediated mRNA Decay. Goergetown: Landes Bioscience 2006:175-196.[109]

Many Alternative Splice Forms Are Targets of NMD

While it was long known that alternative splicing may produce isoforms that are degraded by NMD, this was not appreciated as a pervasive phenomenon until genome-wide studies indicated that a substantial fraction of human genes are spliced to produce isoforms that may be targeted for NMD.

The first study to predict widespread NMD of alternative splice forms used human mRNA and EST sequences from public databases to infer alternative splice forms and identify PTCs.[16] Lewis et al considered 16,780 human mRNA sequences from the reviewed category of RefSeq, a set of well-characterized, experimentally confirmed transcript sequences.[17] Alignment of the RefSeq mRNAs to their genomic loci showed that 617 of these curated mRNA sequences, or 3.7%, contained PTCs. However, the alternative splice forms inferred by aligning EST sequences from dbEST[18] to the RefSeq-defined genomic loci substantially increased the estimated fraction of genes with PTC$^+$ isoforms (Fig. 3). Based on the EST data, over 3000 of the RefSeq genes had alternative splice forms and of these alternatively spliced genes, 45% were predicted to encode at least one splice isoform that is a target of NMD.[16] Therefore, the study found that at least 12% of human genes have a PTC$^+$ isoform.

These results have been confirmed and strengthened by more recent studies. An analysis of the isoforms contained in SWISS-PROT[19] showed that even this reliable, curated database contained presumed translation products of mRNA sequences that are likely to be degraded by NMD. Alignment of the mRNA sequence of each protein isoform reported in SWISS-PROT to the human genome identified reliable exon-intron structures for 2483 isoforms from 1363 genes. The 50-nucleotide rule predicted that 144 isoforms (5.8% of 2483) from 107 genes (7.9% of 1363) contain a PTC and are likely targets of NMD.[20]

An elegant study by Baek and Green extended the analysis of PTC$^+$ alternative splicing to consider conservation of splice forms between human and mouse.[21] This approach helps distinguish aberrant splicing events from rare but functional variants. Starting from a large set of cDNA and EST sequences, Baek and Green identified about 1500 pairs of exon inclusion/exclusion splice forms found in both human and mouse. A quarter of the conserved alternative forms contain a conserved PTC,[21] which is consistent with subsequent findings,[23] suggesting that these isoforms play a functional role and that the PTC is important to their function.

Several microarray experiments have provided direct evidence to support these computational results.[15,22] In one example of these experiments, Mendell and coworkers depleted HeLa cells of Upf1, an essential component of the NMD pathway, and used microarrays to compare mRNA abundances in these cells to mRNA abundances in mock-treated cells.[15] They found that 4.9% of the ~4000 transcripts tested showed significantly higher abundances in cells deficient in NMD, suggesting that NMD normally downregulates those transcripts. Evidence that their observations were largely due to the direct action of NMD, rather than being a downstream regulatory consequence, was provided by showing that several of the putative NMD-targeted transcripts they identified decayed faster in normal cells than in cells depleted of Upf1. They also provided evidence that the effect they observed was due to NMD by showing that the PTC$^+$ transcript abundances responded similarly to depletion of Upf2, another protein that is essential for NMD. Finally, Mendell et al also observed that 4.3% of the transcripts decreased in abundance in NMD-deficient cells. The stability of those transcripts was not altered by NMD deficiency, showing that the change in their abundance was an indirect effect. Because this microarray experiment detected changes in total transcript levels across all isoforms of a particular gene, it may not have detected changes in transcript levels of a specific PTC$^+$ isoform. Therefore, many true NMD targets would not have been identified.

To specifically detect changes in specific isoform abundances due to NMD inhibition, Pan et al used an alternative splicing microarray platform.[23] By distinguishing relative levels of PTC$^+$ versus PTC$^-$ isoforms, they found that approximately 10% of the PTC-containing isoforms increased in abundance by at least 15 percentage points upon inhibition of NMD. Although Pan et al were able to detect relatively few targets of NMD, they reported that a majority of the PTC$^+$ isoforms are present at relatively low abundances, even when NMD is inhibited. They concluded that many of these may represent nonfunctional transcripts or transcripts that are not under strong selection pressure. This conclusion is consistent with the observation that the majority of PTC-containing splice variants identified in sequenced transcripts are not conserved between human and mouse.[21,23]

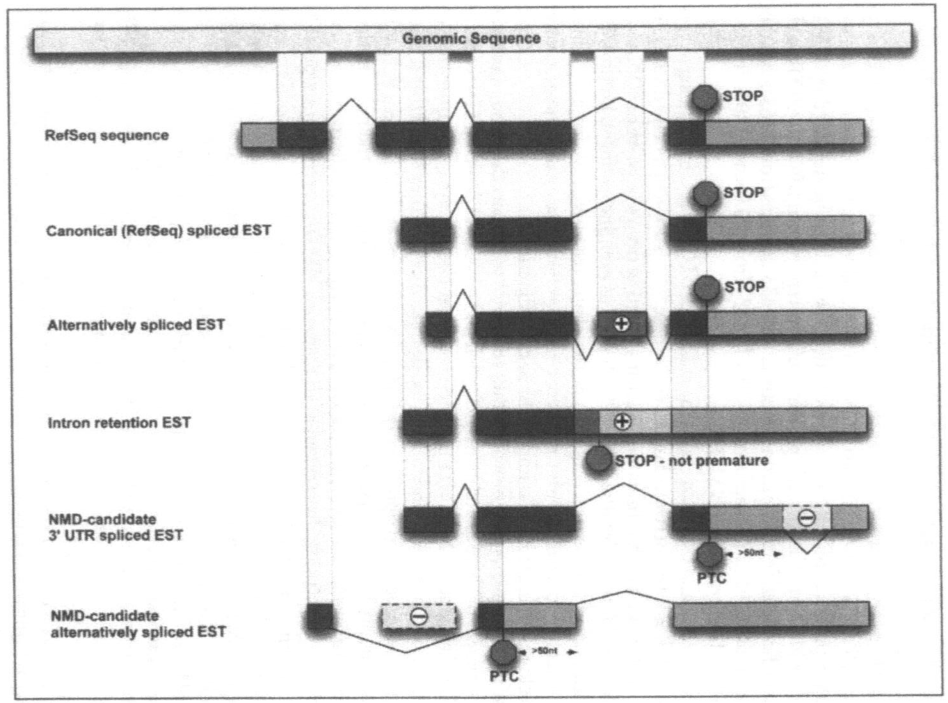

Figure 3. Inference of alternative splice forms and PTCs from RefSeq and EST data. Lewis et al aligned coding regions of RefSeq mRNAs to the genomic sequence to determine canonical splicing patterns.[16] EST alignments to the genomic sequence confirmed the canonical splices and indicated alternative splices. Canonical (RefSeq) splices are indicated above the exons, whereas alternative splices are indicated below the exons. When an alternative splice introduced a stop codon >50 nucleotides upstream of the final exon-exon junction of an inferred mRNA isoform, the stop codon was classified as a PTC and the corresponding mRNA isoform was labeled an NMD candidate. Reprinted from: Soergel D et al. In: Maquat LE, ed. Nonsense-Mediated mRNA Decay. Goergetown: Landes Bioscience 2006:175-196.[109]

There is limited information about the prevalence of alternative splicing coupled to NMD in *Drosophila*. Rehwinkel et al used a gene expression array and found that 3.4% of genes had a significant increase in overall transcript abundances when NMD was inhibited; intriguingly, the NMD protein SMG5 was among these.[24] More recently, an alternative splicing array platform capable of distinguishing distinct isoforms found an order of magnitude more isoforms that are targets of NMD in fly (manuscript in preparation). However, even this finding represents a modest level of coupling relative to the amount predicted based on the analysis of EST and cDNA transcripts from human and mouse tissues.

The striking number of predicted PTC$^+$ alternative splice forms demands more detailed explanation. Are some of these isoforms translated at levels sufficient to impact physiology? Are they an unavoidable side effect of productive alternative splicing? How many of the observed PTC$^+$ isoforms are due to transcriptional or splicing noise? To what extent do PTC$^+$ isoforms represent the coupling of alternative splicing and NMD in order to regulate gene expression? We shall consider each of these potential explanations in turn.

Do the Observed PTC$^+$ mRNA Isoforms Evade NMD to Produce Functional Protein?

The existence of numerous PTC$^+$ isoforms was first inferred from EST data.[16] One may wonder why EST evidence exists at all for isoforms that are expected to be degraded by NMD. As observed in numerous examples (Table 1), NMD substantially reduces the abundance of PTC$^+$ transcripts, but does not eliminate them entirely. One explanation for the presence of these ESTs is that NMD surveillance may not be completely effective. Furthermore, PTC$^+$ isoforms are not degraded instantly upon being spliced; rather, their degradation occurs only after a pioneer round of translation,[25] which might occur near the nuclear pore during or soon after export of the message from the nucleus (reviewed in ref. 26). Thus, we expect there to be some steady-state abundance of PTC$^+$ isoforms that have not yet been degraded, especially inside the nucleus. A series of elegant experiments and computational modeling in yeast suggest that the dominant reason for the presence of PTC$^+$ mRNAs in the cell is the temporal lag between splicing and degradation, rather than incomplete surveillance.[27] Evidently, the resulting abundance of PTC$^+$ isoforms is in many cases high enough for ESTs derived from those isoforms to be observed and deposited in dbEST. Indeed, many of the alternative splice junctions that generate a PTC are supported by multiple ESTs.

Nevertheless, less stable isoforms will be underrepresented in EST libraries. Using sequence features such as splice site strength, Baek and Green modeled the predicted inclusion rates of alternative exons.[21] They showed that PTC$^+$ isoforms are probably produced at a higher rate than they are observed in EST data and that many are degraded before they can be sequenced. Thus, the EST data underestimate the fraction of a given gene's mRNA that is PTC$^+$ and also underestimate the number of genes with PTC$^+$ alternative splicing. For this reason and also because the quality filters used in the above studies excluded many genes and isoforms, EST-based reports offer a lower bound on the number of PTC$^+$ isoforms; the true prevalence of alternative splicing and of PTC$^+$ isoforms may be substantially higher.

Some PTC$^+$ transcripts may evade NMD, increasing their likelihood of being observed and deposited in sequence databases. This evasion can happen in two ways—by the incomplete action of NMD to degrade the PTC$^+$ transcript, or by a specific mechanism that allows the transcript to evade NMD to ensure protein production. There are a few known examples in which a transcript which should be degraded according to the 50-nucleotide rule is in fact stable and is translated to produce protein. These include polycistronic transcripts on which translation is reinitiated downstream of a PTC[28-30]; *apolipoprotein B*, which is protected from NMD by an RNA editing complex[31]; some transcripts with a PTC near the initiation codon[32]; *cytokine thrombopoietin* (*TPO*) mRNA with several uORFs[33]; and an aberrant β-*globin* transcript which is protected from NMD by an unknown mechanism.[34] Although NMD does not prevent protein production entirely in such cases, it may nonetheless limit expression from PTC$^+$ transcripts substantially, as was shown for an alternative transcript of *FAH*[35] and for *ARD-1*.[28]

Table 1. Experimentally confirmed examples of unproductive splicing in natural wild-type isoforms

Name	Organism	AS → PTC	Regulated AS	PTC → Low Abundance	NMD	Notes	References
a) Unproductive splicing							
Calpain-10	Human	•		•	•		20, 45, 46
HPRT	Human	•		•	•	Unproductive transcripts are likely noise	84
POLB	Human	•		•	•	It is not clear whether AS is regulated or if splice variants are noise	84
TCR-beta	Human	•		•	•	V(D)J cleanup. NMD strength boosted by sequence elements	85, 86
ABCC4	Human, Monkey, Mouse	•		•	•	High conservation of PTC-producing exons from mouse to human suggests that they are under ESE control, or that translation is reinitiated downstream of the PTC	78
FGFR2	Rat	•		•	•	Side effect. Productive forms are tissue-specific	37, 87
cbp-2ps	Worm	•		•	•	Pseudogene of *cbp-1*, a gene that encodes a cyclic AMP response element binding protein (CREB) binding protein	88
F45D11.2, F45D11.3, F45D11.4	Worm	•		•	•	Pseudogenes from triplication of gene/pseudogene pairs from the F45D11 family of genes	88
rpl-7aps	Worm	•		•	•	Pseudogene of *rpl-7a*, which encodes ribosomal protein L7a	88
ubq-3ps, ubq-4ps	Worm	•		•	•	Pseudogenes of *ubq-1*	88
b) Regulated unproductive splicing							
FAH	Human	•	•	•	•	PTC^{+} isoform is an NMD target and produces a short protein	35

continued on next page

Table 1. Continued

Name	Organism	AS → PTC	Regulated AS	PTC → Low Abundance	NMD	Notes	References
U2AF35	Human	•	•	•	•	Mutually exclusive exons; PTC+ isoforms are an apparent side effect	89-93
MID-1	Human, Mouse, Fugu	•	•	•	•	Tissue specific alternative splicing	52
ClC-1	Human, Mouse	•	•	•	•	Misregulation of PTC-containing isoform leads to myotonic dystrophy	57, 58
SSAT	Mouse	•	•	•	•	Relative abundance of PTC+ isoform is regulated by substrates that SSAT acts upon (i.e., depletion of polyamine substrates causes an increase in relative levels of the NMD targeted isoform)	53
Nicastrin	Rat	•	•	•	•	PTC+ isoform expressed in neurons	94
MER2	Yeast	•	•	•	•	Splicing is regulated by MER1, which is produced only in meiotic cells. As a result, *MER2* transcripts are productively spliced only during meiosis. In mitotic cells, a PTC+ form is produced and degraded	63, 95
ITSN-1	Human, Mouse	•	•	•		Productive and unproductive isoforms are tissue-specific	96
NDUFS4	Human	•		•	•		97
LARD	Human	•	•			PTC+ isoforms are abundant	98
ARD-1, NIPP-1	Human	•		•		Translation reinitiation. ARD1 is downregulated by NMD, but is nonetheless expressed. Also, ARD1 and NIPP1 may influence splicing via PP1	28

continued on next page

Table 1. Continued

Name	Organism	AS → PTC	Regulated AS	PTC → Low Abundance	NMD	Notes	References
c) Autoregulatory unproductive splicing not affecting splicing factors							
RPL3	Human	•	•	•	•		99
RPL12	Worm, Human	•	•	•	•		64, 99
RPL30	Yeast	•	•	•	•		62
AtGRP7	Arabidopsis	•	•	•			73
d) Autoregulatory unproductive splicing affecting splicing factors							
PTB	Human	•	•	•	•		67, 69, 70
SC35	Human	•	•	•	•		10
TIAR / TIA-1	Human	•	•	•	•		100
CLKs	Human, Mouse, Ciona	•	•	•	•		20, 71, 72, 101, 102
AUF1	Human	•	•	•			103, 104
nPTB	Human	•	•	•		Splicing is regulated by PTB. Direct evidence of additional autoregulation has not been shown	69
SRp20, SRp30b	Worm	•		•	•		105
TRA2-beta	Human	•	•			PTC$^+$ forms are abundant, but are not translated, perhaps due to sequestration	106

Dots indicate direct experimental confirmation; lack of a dot means only that the experiment has not been performed to our knowledge. For instance, while it seems certain that *ITSN-1* PTC$^+$ isoforms are degraded by NMD, this has not been directly observed using Upf1 knockdown or another NMD assay. The few cases where an experiment was performed but yielded a negative result are noted. AS, alternative splicing

Nonetheless, documented exceptions to the 50-nt rule are rare and there are many more known cases in which the 50-nt rule is obeyed.

Even for PTC+ transcripts that do not evade NMD, the possibility remains that the single truncated protein product of the pioneer round of translation is functionally significant, since some regulatory proteins can have an effect even at a very low copy number.[36] Also, to the extent that NMD is not completely effective at detecting and degrading typical PTC+ transcripts, the transcripts that escape this process may be translated to produce truncated proteins. However, these proteins will frequently lack critical domains, rendering them inactive or even harmful. In any case, it is hard to imagine that functional roles of truncated proteins could explain the high prevalence of genes with PTC+ isoforms, especially given the wide functional diversity of these genes, and no data exist to support such a view.

While there may be exceptions, it seems unlikely that many PTC+ isoforms produce functional protein, either during the pioneer round of translation, due to incomplete surveillance, or by evading NMD altogether.

Are the Observed PTC+ mRNA Isoforms a Side Effect of Productive Alternative Splicing?

In the particular situation of mutually-exclusive exon usage, NMD may be a mechanism for removing transcripts that erroneously include both exons or neither exon. For isoforms of *FGFR2*, including both exons or neither exon introduces a frameshift and PTC into the mRNA, targeting it for degradation.[37] In this circumstance, an isoform including exon IIIb while skipping exon IIIc is productive; similarly, the isoform including exon IIIc but excluding exon IIIb is productive. However, the spliceosome may also pair the same splice sites differently such that both exons are included, or such that neither are included. Both of these latter possibilities introduce a PTC (Fig. 4).

Each splice site involved in the removal of exons IIIb and IIIc is required for the production of at least one productive isoform; the unproductive isoforms arise simply from alternate pairings of these otherwise productive splice sites. Given that the spliceosome is prone to such alternate pairings, there may be evolutionary pressure to ensure that the undesired isoforms include a PTC. This results in an inevitable side effect of the mechanism for productive alternative splicing. NMD can be used as a filter to remove these "side effect" isoforms, which may comprise a substantial fraction of the transcripts produced (up to 50% in the case of *FGFR2*).[37]

We examined the alternative isoforms inferred from human dbEST data (see above) and found that PTC+ isoforms could be explained as a side effect for 34% of the genes that produce them. That is, 66% of the genes with a PTC+ isoform have a splice site that is specific to PTC+ isoforms and that is responsible for introducing the PTC (Soergel 2005, unpublished data). If these unproductive isoforms were on the whole detrimental to the cell, then we would expect evolutionary pressure to have selected against PTC+ specific splice sites, but in fact many of them are strikingly conserved, as we discuss below. Thus, while the contribution of "side effect" isoforms may be significant, they alone cannot explain the high prevalence of PTC+ isoforms.

Do the Observed PTC+ mRNA Isoforms Represent Missplicing or Cellular Noise?

NMD was originally described as a means of clearing erroneous transcripts from the cell.[11,12] In keeping with this role, some alternative splice forms that are degraded by NMD could represent splicing errors. Such errors could arise from mutations disrupting splice sites or regulatory sequences, including mutations in intronic regions that are invisible after intron removal. Also, the splicing machinery itself could recognize incorrect splice sites. The spliceosome distinguishes true splice sites from nearby cryptic sites with impressive fidelity, but splice site recognition is a complex process and errors occur at some rate. Although there are at present no clear data on the extent of missplicing, EST libraries contain millions of transcript sequences and even extremely rare events, such as those arising from missplicing, may be represented.

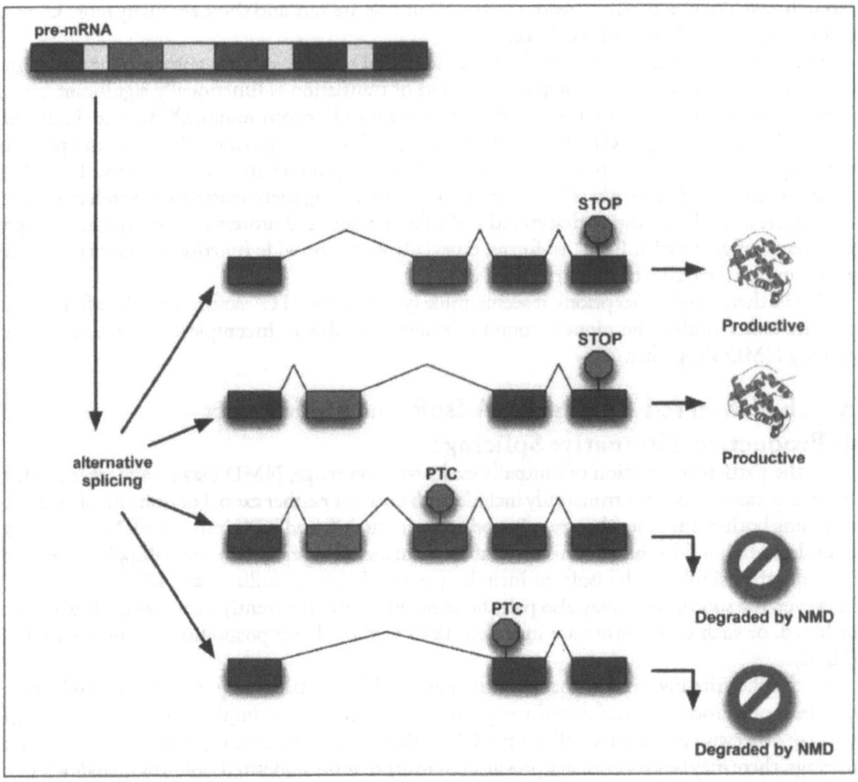

Figure 4. NMD can be employed to remove "side effect" isoforms in the case of mutually exclusive exons. Alternative splicing may generate two productive isoforms including one or the other of a pair of mutually exclusive exons. By choosing different pairings from the same set of 5' and 3' splice sites, the spliceosome may also generate isoforms including both exons, or neither exon. Frameshifts can give rise to PTCs on these undesired isoforms so that they will be degraded by NMD. Reprinted from: Soergel D et al. In: Maquat LE, ed. Nonsense-Mediated mRNA Decay. Goergetown: Landes Bioscience 2006:175-196.[109]

In EST-based computational analyses, splicing errors can be identified to some extent by filtering out splicing events that are seen only in a few ESTs, but this method cannot distinguish errors from legitimate rare splice forms. With multiple mammalian genomes available, recent work has focused on evolutionary conservation to suggest negative selection and, perhaps, functional roles for conserved alternative forms,[21,38] (also reviewed in refs. 39,40). Minor isoforms, those that occur only a fraction of the time, are less often conserved than major isoforms[41] and may sometimes represent recent mutations or splicing errors. The minor isoforms that are conserved, including PTC-containing isoforms, are more likely to be functional than minor isoforms that are seen only in one species, although species-specific isoforms may also be functional.[42]

As described above, Baek and Green identified PTC^{+} isoforms that were conserved between human and mouse to filter out aberrant splicing. They note that the inclusion of the same "accidental" alternative exon is unlikely to happen by chance in both species, but that occasional accidental skipping of the same exon could more readily happen by chance in both human and mouse. To reduce the influence of these conserved but aberrant splicing events on their data set, Baek and Green designed a statistical method to discriminate between splicing errors and functional alternative splicing. Using this method, they inferred that 80% of the conserved PTC-producing splice events they considered were legitimate, compared to 20% that appeared aberrant.[21] Thus, most of the conserved PTC-producing splice events they observed were not likely due to missplicing.

Pan and coworkers used an experimental approach to understand whether the prevalence of PTC+ transcripts is a result of functional gene regulation or splicing noise.[23] As previously described, they developed an alternative splicing microarray platform to detect the relative abundance of PTC+ versus PTC- isoforms for over a thousand cassette-exon type alternative splicing events in mouse. Their study showed that in 10 diverse untreated mouse tissues where NMD is active, most PTC+ isoforms represent less than 50% of the steady state pool of transcripts from a given gene. The low abundance is consistent either with a reduction in the levels of PTC+ isoforms due to the action of NMD, or with infrequent occurrence of the alternative splice events that produces the PTC+ isoforms.

To address these two possibilities in one cell type, Pan et al measured the changes in relative abundance of PTC+ isoforms upon NMD inhibition in HeLa cells, using a microarray profiling 3055 human cassette exons. A small percentage of PTC+ isoforms were upregulated after NMD inhibition, suggesting that their unproductive splicing could affect gene expression. Nonetheless, because the majority of PTC+ isoforms are present at low abundance even when NMD is inhibited, Pan et al inferred that most PTC+ isoforms may not contribute important functional roles. One cannot exclude the possibility that subtle changes in the abundance of some PTC+ isoforms have functional consequences, perhaps in different tissues. Nevertheless, this study suggests that the majority of PTC+ isoforms may simply be due to infrequent splicing events and represent potential cellular noise cleared by the NMD machinery.[43]

If many PTC-containing transcripts are a result of splicing noise, their prevalence in a wide variety of genes could reflect a functional role that allows for the evolution of new gene functions via alternative splicing.[40,44] The existence of NMD could have led to an increase in alternative splicing, because any splicing errors that introduced PTCs would be removed by NMD, reducing the harmful effects of missplicing. As a result, the pressure to recognize splice sites perfectly would be lowered. Functional alternative splice forms could arise through splicing errors and then become established by sequence changes that strengthen their splice sites or add regulatory elements.

Are PTC+ mRNA Isoforms Important for the Regulation of Gene Expression?

There are many examples of specific transcripts that are regulated by the coupling of alternative splicing and NMD (Table 1). This process provides an additional level of regulatory circuitry to help the cell achieve the proper level of expression for a given protein. The cell could change the level of productive mRNA after transcription by shunting some fraction of the already-transcribed pre-mRNA into an unproductive splice form and then to the decay pathway (Fig. 1). In the simplest case, some constant fraction of a gene's pre-mRNA is spliced into an unproductive, NMD-targeted form. In other cases, the proportion of transcripts targeted for degradation is regulated by an external input. Finally, autoregulatory loops can arise in which a protein affects the splicing pattern of its own pre-mRNA. The process of gene regulation through the coupled action of alternative splicing and NMD has been termed "Regulated Unproductive Splicing and Translation," or RUST.

Constitutive Unproductive Splicing

The simplest type of coupled alternative splicing and NMD is one in which the ratio of productive to unproductive splice forms is not significantly variable. In this case, the combined effect of alternative splicing with NMD reduces message abundance by a more or less constant factor. An apparent example of this is the *Calpain-10* gene, which encodes a ubiquitously expressed protease and is alternatively spliced to produce eight mRNA isoforms.[20,45] An analysis of these isoforms using SWISS-PROT and genomic sequences showed that four contained PTCs. An expression study by Horikawa et al showed that the four PTC^+ isoforms were "less abundant" in vivo than the other four.[45] Further experiments showed that the PTC^+ isoforms increased in abundance relative to the PTC^- isoforms when cells were treated with cycloheximide, which blocks translation and thereby inactivates the NMD pathway.[46] This result confirmed that all eight mRNA isoforms are produced but that the four PTC^+ isoforms are degraded by NMD. Other experimentally confirmed examples in the literature reflect apparent constitutive unproductive splicing (Table 1a). Such cases may not be regulation, but simply cellular noise, with the unproductive splicing providing little or no selective advantage or function. Of course, in each of these cases, there may be as yet unknown regulatory inputs that impact the splicing process and alter the isoform proportions.

Regulated Unproductive Splicing

There are many examples of regulated alternative splicing, particularly in tissue-specific alternative splicing events, e.g., references 47-49. The role of regulated alternative splicing is emerging as an important layer of gene regulation, much like gene regulation at the transcriptional and translational levels.[39] Twenty-three examples of regulated alternative splicing leading to NMD are shown in Table 1. In addition to changing the relative abundance of functional isoforms, changes in the splicing environment may increase or decrease the production of translated isoforms relative to PTC^+ isoforms that are degraded by NMD (Fig. 1).

The 5' and 3' splice sites recognized by the spliceosome have a range of "strengths" or binding affinities for the core spliceosome components. Selection of splice sites is also under the control of a host of regulatory splicing factors which bind to specific sequence signals in the pre-mRNA. These sequences may be exonic or intronic and may be associated with enhancement or suppression of splicing at nearby (and sometimes at distant) splice sites. Cis-regulatory sequences, such as exonic splicing enhancers (ESEs), are frequently found in clusters, suggesting a combinatorial regulation of splicing by complexes of splicing factors[50,51] (refer to chapter by Chasin).

A change in the abundances of splicing factors can shift the balance of splicing patterns towards the production of NMD-targeted isoforms, thereby reducing the abundance of productive transcripts and hence the rate of protein production. In this way, splicing factors can act as regulatory inputs to alter gene expression in a manner analogous to transcription factors. An example of this intriguing mode of gene regulation is *MID1*, which encodes a microtubule-associated protein involved in triggering the degradation of phosphatase 2A.[52] *MID1* is ubiquitously transcribed, but it is spliced in a tissue- and development-specific manner. Winter and coworkers observed numerous alternatively spliced transcripts which included novel alternative exons, in addition to nine previously known constitutive exons. Most of the transcripts with novel exons contained in-frame stop codons and subsequent alternative poly(A+) tails; the alternative polyadenylation meant that the stop codons were not premature, allowing for translation of a C-terminally truncated protein. A second class of alternative transcripts contained stop codons closely followed by an in-frame start codon, suggesting the possibility of translation reinitiation and production of N-terminally truncated protein. A third class of alternative transcripts contained premature stop codons that were associated neither with an alternate poly(A+) signal nor with an alternate start site. These transcripts were predicted to be subject to NMD according to the 50-nucleotide rule. Consistent with this prediction, Winter et al found that the abundance of human *MID1* transcripts including exon 1c (an alternative exon introducing a PTC) increased in the presence of the NMD inhibitor cycloheximide.[52] Finally, Winter et al used RT-PCR to observe that different *MID1* isoforms are produced in different tissues and at different developmental stages in both human and mouse. For instance, the PTC-introducing exon 1a was observed in five distinct transcripts in human fetal brain cells, two transcripts in fetal liver cells and none in fetal fibroblasts. These results strongly suggest that alternative splicing and NMD are being employed to regulate the overall abundance of productive *MID1* transcripts.

The gene encoding spermidine/spermine N^1-acetyltransferase (SSAT), an enzyme that regulates the intracellular levels of the polyamines spermidine and spermine, provides an interesting example of RUST.[53] SSAT acetylates spermidine and spermine, which are then excreted out of the cell.[53]

Polyamines were known to regulate gene expression of *SSAT* at the level of transcription and stabilization of the mRNA[53] and Hyvönen et al present evidence that polyamines regulate gene expression of *SSAT* by promoting the exclusion of an exon containing three in-frame PTCs. Upregulation of the PTC$^+$ isoform occurs when polyamine levels are low—a condition where the enzyme is not needed. Upon depletion of polyamines in mouse embryonic stem cells, Hyvönen et al observed an increase in the relative abundance of the PTC$^+$ isoform of *SSAT*, termed *SSAT-X*. Conversely, after treating cells with DENSpm, a polyamine analog, they observed a decrease in the amount of *SSAT-X* mRNA relative to the normal, PTC$^-$ *SSAT* mRNA.[53] To demonstrate that *SSAT-X* mRNAs are degraded by NMD, Hyvönen et al inhibited NMD in fetal fibroblasts by treatment with cycloheximide or an siRNA targeted to *UPF1*. Under both conditions, they observed an increase in the relative amount of *SSAT-X* mRNA, thus providing evidence that *SSAT-X* is a target of NMD. Furthermore, they observed a decrease in the relative amount of *SSAT-X* in cells treated with cycloheximide and DENSpm compared to cells treated with only cycloheximide, indicating that the addition of polyamines does not enhance NMD activity, but does affect alternative splicing.[53] Polyamines also affected the splicing of a PTC$^+$ isoform of an unrelated gene, *Clk1*, but did not affect the splicing of three other genes, indicating that polyamines may specifically regulate other transcripts as well. Although the mechanism which enables polyamines to regulate changes in alternative splicing of *SSAT* is not known, the results of the above experiments show a clear and novel example of RUST.

Defects in the regulation of unproductive splicing can lead to disease. Myotonic dystrophy (DM), an autosomal dominant disease, is the most common form of adult-onset muscular dystrophy. DM has been shown to be caused by either of two repeat expansions whose presence in an mRNA affect the function of several splicing factors[54] such as CUG-BP1 and thus induce splicing changes in several genes[55,56] (refer to chapter by Orengo and Cooper). The mechanism by which these repeat expansions affect the function of CUG-BP1 is not clear; however, the misregulation of CUG-BP1 has downstream effects that contribute to DM. Patients develop myotonia from lack of muscle-specific chloride channel 1 (ClC-1). The misexpression of CUG-BP1 in DM tissue results in the mis-splicing of the *ClC-1* pre-mRNA.[57]

The normal developmental splicing pattern for *ClC-1* has a PTC$^+$ splice form in embryos but a productive splice form in adult cells. In DM tissue, *ClC-1* splicing reverts to its embryonic, PTC-containing splicing pattern—which is greatly reduced in abundance, likely as a consequence of NMD.[58] Tissues from DM patients have increased steady-state levels of CUG-BP1 protein and the overexpression of CUG-BP1 in mouse skeletal muscle and heart tissues results in the embryonic splicing pattern of *ClC-1*.[59,60] In addition, expression of the CUG-BP1 protein is decreased in mouse skeletal muscle and heart tissues shortly after birth,[60,61] providing evidence that the CUG-BP1 protein influences the splicing of the PTC$^+$ isoform. Thus, it appears that normal *ClC-1* expression is governed by RUST and that the DM disease is caused by undermining the proper function of splicing factors with consequent disruption of RUST.

Autoregulatory Unproductive Splicing

There is abundant evidence that RUST is used for autoregulation, most prevalently of proteins that are part of the splicing or translation machinery. In some fascinating cases, proteins that are not generally involved in mRNA processing bind specifically to their own transcripts to affect their splicing and elicit NMD. One example of this is found in yeast. Yeast genes are generally unspliced, but in the few intron-containing genes, intron inclusion can introduce an in-frame stop codon and target the transcript for NMD. The yeast ribosomal protein RPL30 binds to its own pre-mRNA to prevent intron removal. This retained intron contains a PTC that triggers NMD.[62] The mRNAs of other yeast ribosomal protein genes, including *RPL28* (*CYH2*) and *RPS17B* (*RP51B*), sometimes retain their introns and become natural NMD targets, leaving open the possibility that their splicing is also regulated to elicit NMD.[63]

Some ribosomal proteins in *C. elegans* are similarly autoregulated. A screen for natural targets of NMD identified the genes for ribosomal proteins L3, L7a, L10a and L12. Each of these genes can be alternatively spliced to generate either a productive isoform or an unproductive isoform that contains a PTC and is therefore degraded by NMD. The ratio of productive to unproductive alternative splicing of *rpl-12* is affected by levels of RPL-12 protein, indicating that unproductive splicing of *rpl-12* is under feedback control.[64] More recently, NMD-target isoforms of the human ribosomal protein genes *rpL3* and *rpL12* were identified and the unproductive splicing of *rpL3* was shown to be autoregulated by rpL3 protein in a negative feedback loop.[99] RUST thus seems to play a similar role in the regulation of ribosomal proteins in species from yeast to human.

A striking number of splicing factors and elements of the splicing machinery are autoregulated through RUST (Fig. 5 and Tables 1c and 1d). One such example is the polypyrimidine tract

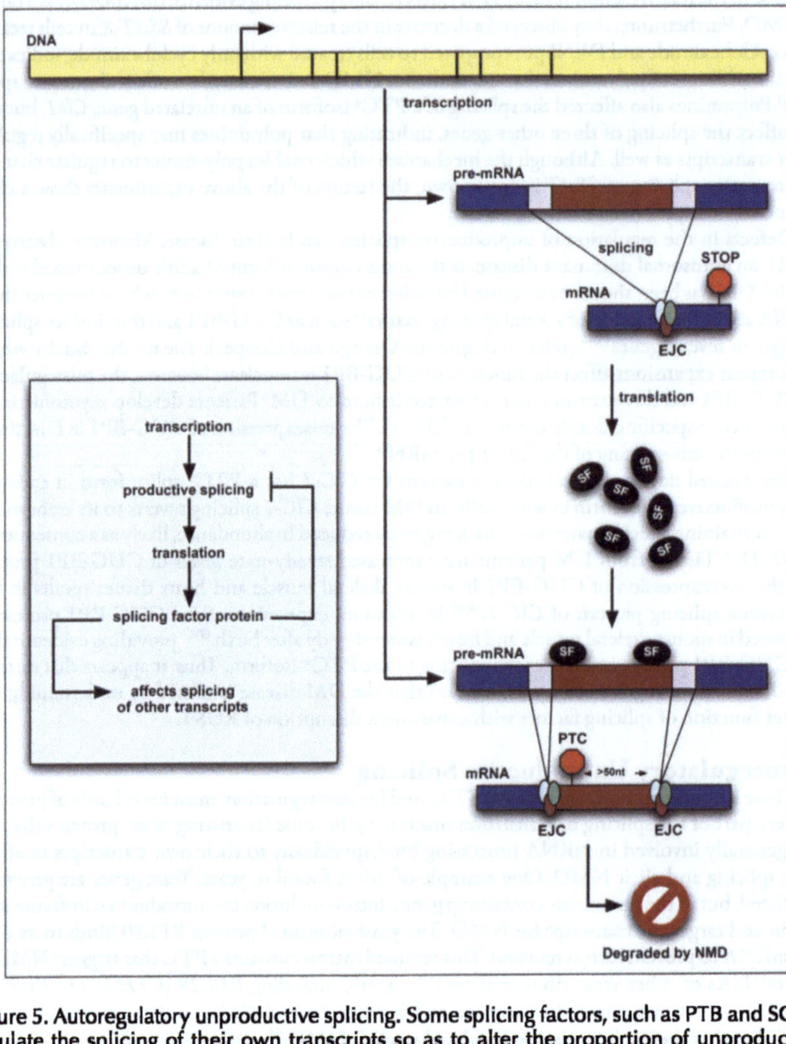

Figure 5. Autoregulatory unproductive splicing. Some splicing factors, such as PTB and SC35, regulate the splicing of their own transcripts so as to alter the proportion of unproductive isoforms.[10,67] This creates a negative feedback loop, stabilizing the concentration of the splicing factor over time. Autoregulated splicing factors are generally not specific to their own transcripts, however; they impinge on the splicing of many other pre-mRNAs as well.

binding protein (PTB), a protein whose function is to inhibit splicing by competing with U2AF for the polypyrimidine tract and perhaps through other mechanisms as well (reviewed in refs. 50,65). *PTB* is alternatively spliced to produce two major productive isoforms (one of which lacks exon 9),[66,67] one minor productive isoform lacking exons 3-9,[67,68] and two unproductive isoforms lacking exon 11. Removing exon 11 causes a frameshift leading to a downstream PTC. PTB protein has been found to promote the removal of exon 11 from its own transcripts.[67] Consequently, when PTB levels are high, PTB production is slowed by targeting *PTB* transcripts for NMD and when PTB levels are low, production is accelerated by reducing the proportion of transcripts that are degraded.[67,69,70]

A similar autoregulatory process has been reported for members of a family of splicing factors known as SR proteins. Overexpression of the SR protein SC35 upregulates the splicing of its own NMD-targeted isoform to reduce protein production.[10] Intriguingly, similar unproductive splicing is found in all human SR genes and some hnRNPs,and the alternative splicing events that create the PTC-containing isoforms are conserved in mouse orthologs.[107,108] Remarkably, some of the most conserved regions of the human and mouse genomes are associated with this unproductive splicing.

The CDC-like kinases (Clks), which regulate SR proteins, seem to be affected by RUST as well.[20] RUST appears to regulate the Clk1 protein through an indirect feedback mechanism. Clk1 has been shown to indirectly modify splicing of its own transcript, most likely through phosphorylation of SR proteins.[71] Thus, as a variation of the autoregulatory circuit described above, increased Clk1 activity may result in changes in the activity of one or more SR proteins. These SR proteins in turn affect the splicing of *Clk1* pre-mRNA to favor a PTC$^+$ transcript that is predicted to undergo NMD. This PTC$^+$ transcript is stabilized by cycloheximide, providing evidence that it is indeed normally degraded by NMD.[72] RUST regulation of SR proteins, Clks and PTB may have downstream effects on many pre-mRNAs. Thus, RUST can regulate factors that control alternative splicing of many other gene products.

Finally, splicing factors that are autoregulated by RUST may also be subject to RUST that is triggered by heterologous factors rather than autoregulation. This is seen in the alternative splicing of *PTB*, which can be affected by the splicing regulators raver1 and CELF4, forming a network of regulatory factors contributing to RUST.[67]

Conservation of Regulated Unproductive Splicing and Translation (RUST)

The coordinated use of alternative splicing and NMD is seen not only in mammals but in organisms as distant as yeast[62] and plants.[73] The mechanism of PTC recognition differs between mammals and other species, where it does not seem to depend on the location of the stop codon relative to exon junctions.[74] There have been significant advances recently in elucidating the recognition mechanism in flies and yeast,[75,76] but the rules are not clear enough to allow for computational identification of NMD targets. Nonetheless, NMD affects gene expression in a variety of different organisms.[24,77]

In several of the examples discussed above, analysis of orthologous and paralogous sequences suggests that splicing to generate PTC$^+$ alternative isoforms and thus RUST regulation, is shared across species and across protein families. In the case of *PTB*, the sequence and upstream regulatory elements of alternatively included PTC-containing exon 11 are very similar between the *Fugu rubripes* ortholog and the analyzed human gene, as well as in the human *nPTB* paralog.[67] Mouse and monkey orthologs of the human multidrug resistance associated transporter *ABCC4* share highly conserved PTC-containing exons that are orthologous to the alternatively included exons of human *ABCC4*, another apparent RUST target.[78] Particularly strong evidence of conservation of RUST is found in the Clks. Alternative splicing to exclude exon 4, introducing a frameshift and PTC, is conserved among three human paralogs (*Clk1*, *Clk2* and *Clk3*), the three orthologs of these genes in mouse and even the sole copy of the gene in the sea squirt *Ciona intestinalis*.[20] One SR protein in *Ciona* also has unproductive splice patterns matching those seen in human and mouse.[107]

The action of NMD on a gene can be retained even when the specific alternative splicing events that elicit NMD are not conserved. As discussed above, *MID1* is a human RUST target. Interestingly, while PTC$^+$ isoforms of MID1 were found in human, mouse and fugu, the responsible stop codons were introduced by alternative exons that showed no homology between these species.[52] Thus, in this case, it appears that the RUST mode of regulation was maintained while the specific sequence elements triggering it were not. This suggests that RUST has often and readily evolved to regulate specific classes of genes in organisms that already have both alternative splicing and NMD.

Why Regulated Unproductive Splicing and Translation?

A substantial portion of alternatively-spliced mRNAs seem to be targets of NMD. We have discussed possible explanations for the prevalence of unproductive splicing: do these splice forms represent biological noise or are they produced to regulate protein expression? The relatively low abundance and lack of conservation of many PTC$^+$ isoforms suggests that many of these isoforms are nonfunctional or cellular noise, but the growing body of examples nevertheless suggests that RUST plays a significant regulatory role in the cell.

Many truncated proteins encoded by alternative transcripts would be nonfunctional even if their transcripts were not removed by NMD. Is the combination of alternative splicing and NMD inherently different from alternative splicing that produces nonfunctional protein? Or does alternative splicing alone provide the important regulatory step, with NMD acting only as a convenient but inessential cleanup mechanism? Some proven cases of RUST illustrate that the coordinated action of both pathways is required for regulation. As described above, expression of the SR protein SC35 is autoregulated by RUST; its alternative splicing occurs in the 3' UTR to create an exon junction downstream of the original stop codon without changing the open reading frame.[10] The alternative splicing seems to have no role other than causing the original termination codon to be recognized as premature. Without NMD, the alternative mRNA would still encode the full-length protein, so the alternative splice event alone could not be used to regulate protein levels. It seems, then, that some genes have evolved to take advantage of the combination of alternative splicing and NMD in a role different from those filled by either process alone.

RUST seems, at first, to be a wasteful process. A gene is transcribed and spliced, only to be degraded before it can produce a protein. Yet we know that there are functional cases of RUST. The cost to the cell of transcribing apparently-extraneous RNA is clearly not prohibitive. In humans, roughly 85-95% of transcribed sequence is spliced out as introns and discarded.[79] Evidently, transcription of intron sequence is not a significant selective disadvantage and intron splicing may even provide some general selective advantage. Similarly, the cost of transcribing a pre-mRNA only to splice it into an unproductive form must be balanced by the advantages of an additional layer of regulation of gene expression or the flexibility to evolve new gene isoforms without harmful effects.

How is a process like RUST beneficial to the cell? It provides an additional level of regulation. Transcriptional regulation is the most studied means of controlling gene expression, but in some cases additional control may be beneficial. Splicing regulation occurs after the decision to transcribe a region and RUST may provide a rapid way to change the levels of productive mRNA. In extreme cases such as the dystrophin gene, transcription can take many hours[80] and the requirements of the cell might change after transcription begins but before a critical splicing decision that determines whether or not to introduce a PTC. Even when temporal regulation is not necessary, an extra layer of regulation can help fine tune or amplify transcriptional and other regulation.

RUST could either increase or decrease protein expression from steady-state levels. The splicing factor PTB illustrates this point. At steady state, 20% of the transcribed pre-mRNAs of PTB are spliced to an unproductive form.[67] In general, we expect that a RUST-regulated gene is transcribed to produce more pre-mRNA than is needed at steady state and that in normal conditions there is a base level of downregulation by unproductive splicing. This fraction of "wasted" transcripts constitutes the headroom available to the regulatory system to increase levels of productive transcript.

In a system with prevalent alternative splicing, regulation by RUST may evolve easily. For any particular gene, there are many possible alternative splicing events that could introduce a PTC and elicit NMD (Fig. 1). If the sequence of the gene changes slightly to promote one of these splicing events in certain splicing environments and the resulting downregulation of the gene by NMD is beneficial, then a basic sort of regulation has evolved. This has clearly occurred independently many times. Indeed, RUST seems to have evolved independently in every one of the SR genes.[107] Without NMD, alternative splicing can still regulate gene expression by producing nonfunctional proteins. The additional advantages of coupling splicing with NMD may be that it prevents accumulation of potentially harmful truncated proteins and that it reduces wasted translation, making unproductive splicing less costly.

Splicing factors such as PTB seem to be overrepresented among the known RUST targets. Is this a coincidence or acquisition bias, or is RUST used most often to regulate a small set of proteins that are already capable of binding pre-mRNAs? The latter may be the case for autoregulation by RUST. A protein that has an existing role in splicing may evolve autoregulation through splicing more easily than a non-RNA-binding protein. Indeed, this is a simple and elegant means of regulation for RNA binding proteins. There are only a handful of known cases in which a protein that is not a splicing factor is autoregulated by RUST and even these are predominantly ribosomal proteins that do bind RNA in other, nonsplicing contexts. However, autoregulation is by no means the only role of RUST and there is no reason for non-autoregulatory RUST to affect splicing factors preferentially. The examples listed in Table 1 indicate that RUST is involved in the regulation of a diverse set of proteins.

The potential for alternative splicing to regulate gene expression has been appreciated for many years. Bingham et al proposed that "on/off regulation at the level of splicing might be unexpectedly common," in a 1988 review featuring three cases of unproductive splicing in *Drosophila*.[81] An early paper about the splicing factor ASF discussed alternative splicing as a quantitative control of gene expression.[82] Nonsense-mediated decay adds an additional layer to the story[83]; many of the unproductive splice forms identified years ago are now known to be degraded rather than translated. Alternative splicing and NMD can be combined in an elegant way to regulate a wide variety of genes.

Acknowledgements

We would like to thank Maki Inada, Richard E. Green and Rajiv Bhatnagar for helpful discussions and comments. This work was supported by National Instituties of Health grants K22 HG00056 and R01 GM071655 and an IBM SUR grant. DAWS is supported by a predoctoral fellowship from the Howard Hughes Medical Institute. ANB is supported by the Chancellor's Fellowship for Graduate Study from the University of California, Berkeley. SEB is also supported by NIH/NIGMS R01 GM073109 and Sloan and Searle fellowships. This chapter represents an update of a version that was previously published.[109]

References

1. Boue S, Letunic I, Bork P. Alternative splicing and evolution. Bioessays 2003; 25:1031-1034.
2. Modrek B, Lee C. A genomic view of alternative splicing. Nat Genet 2002; 30:13-19.
3. Maniatis T, Tasic B. Alternative pre-mRNA splicing and proteome expansion in metazoans. Nature 2002; 418:236-243.
4. Garcia J, Gerber SH, Sugita S et al. A conformational switch in the piccolo C2A domain regulated by alternative splicing. Nat Struct Mol Biol 2004; 11:45-53.
5. Resch A, Xing Y, Modrek B et al. Assessing the impact of alternative splicing on domain interactions in the human proteome. J Proteome Res 2004; 3:76-83.
6. Xing Y, Xu Q, Lee C. Widespread production of novel soluble protein isoforms by alternative splicing removal of transmembrane anchoring domains. FEBS Lett 2003; 555:572-578.
7. Maquat LE. NMD in mammalian cells: A history. In: Maquat LE, ed. Nonsense-Mediated mRNA Decay. Georgetown: Landes Bioscience, 2006:45-58.
8. Nagy E, Maquat LE. A rule for termination-codon position within intron-containing genes: When nonsense affects RNA abundance. Trends Biochem Sci 1998; 23:198-199.

9. Lejeune F, Maquat LE. Mechanistic links between nonsense-mediated mRNA decay and pre-mRNA splicing in mammalian cells. Curr Opin Cell Biol 2005; 17:309-315.
10. Sureau A, Gattoni R, Dooghe Y et al. SC35 autoregulates its expression by promoting splicing events that destabilize its mRNAs. EMBO J 2001; 20:1785-1796.
11. Cali BM, Anderson P. mRNA surveillance mitigates genetic dominance in caenorhabditis elegans. Mol Gen Genet 1998; 260:176-184.
12. Maquat LE, Carmichael GG. Quality control of mRNA function. Cell 2001; 104:173-176.
13. He F, Jacobson A. Endogenous substrates of the yeast NMD pathway. In: Maquat LE, ed. Nonsense-Mediated mRNA Decay. Georgetown: Landes Bioscience, 2006:27-27-41.
14. Sharifi NA, Dietz HC. Physiologic substrates and functions for mammalian NMD. In: Maquat LE, ed. Nonsense-Mediated mRNA Decay. Georgetown: Landes Bioscience, 2006:97-97-109.
15. Mendell JT, Sharifi NA, Meyers JL et al. Nonsense surveillance regulates expression of diverse classes of mammalian transcripts and mutes genomic noise. Nat Genet 2004; 36:1073-1078.
16. Lewis BP, Green RE, Brenner SE. Evidence for the widespread coupling of alternative splicing and nonsense-mediated mRNA decay in humans. Proc Natl Acad Sci USA 2003; 100:189-192.
17. Pruitt KD, Maglott DR. RefSeq and LocusLink: NCBI gene-centered resources. Nucleic Acids Res 2001; 29:137-140.
18. Boguski MS, Lowe TM, Tolstoshev CM. dbEST—database for "expressed sequence tags". Nat Genet 1993; 4:332-333.
19. Boeckmann B, Bairoch A, Apweiler R et al. The SWISS-PROT protein knowledgebase and its supplement TrEMBL in 2003. Nucleic Acids Res 2003; 31:365-370.
20. Hillman RT, Green RE, Brenner SE. An unappreciated role for RNA surveillance. Genome Biol 2004; 5:R8.
21. Baek D, Green P. Nonsense-mediated decay, sequence conservation and relative isoform frequencies in evolutionarily conserved alternative splicing. Proc Natl Acad Sci USA 2005.
22. Wittmann H, Jack. hUPF2 silencing identifies physiologic substrates of mammalian nonsense-mediated mRNA decay. Molecular and cellular biology 2006; 26:1272.
23. Pan Q, Saltzman AL, Kim YK et al. Quantitative microarray profiling provides evidence against widespread coupling of alternative splicing with nonsense-mediated mRNA decay to control gene expression. Genes development 2006; 20:153.
24. Rehwinkel J, Letunic I, Raes J et al. Nonsense-mediated mRNA decay factors act in concert to regulate common mRNA targets. RNA 2005; 11:1530.
25. Ishigaki Y, Li XJ, Serin G et al. Evidence for a pioneer round of mRNA translation: mRNAs subject to nonsense-mediated decay in mammalian cells are bound by CBP80 and CBP20. Cell 2001; 106:607-617.
26. Maquat LE. Nonsense-mediated mRNA decay: Splicing, translation and mRNP dynamics. Nat Rev Mol Cell Biol 2004; 5:89-99.
27. Cao D, Parker R. Computational modeling and experimental analysis of nonsense-mediated decay in yeast. Cell 2003; 113:533-545.
28. Chang AC, Sohlberg B, Trinkle-Mulcahy L et al. Alternative splicing regulates the production of ARD-1 endoribonuclease and NIPP-1, an inhibitor of protein phosphatase-1, as isoforms encoded by the same gene. Gene 1999; 240:45-55.
29. Veldhoen N, Metcalfe S, Milner J. A novel exon within the mdm2 gene modulates translation initiation in vitro and disrupts the p53-binding domain of mdm2 protein. Oncogene 1999; 18:7026-7033.
30. Zhang J, Maquat LE. Evidence that translation reinitiation abrogates nonsense-mediated mRNA decay in mammalian cells. EMBO J 1997; 16:826-833.
31. Chester A, Somasekaram A, Tzimina M et al. The apolipoprotein B mRNA editing complex performs a multifunctional cycle and suppresses nonsense-mediated decay. EMBO J 2003; 22:3971-3982.
32. Inácio A, Silva AL, Pinto J et al. Nonsense mutations in close proximity to the initiation codon fail to trigger full nonsense-mediated mRNA decay. J Biol Chem 2004; 279:32170-32180.
33. Stockklausner C, Breit S, Neu-Yilik G et al. The uORF-containing thrombopoietin mRNA escapes nonsense-mediated decay (NMD). Nucleic acids research 2006; 34:2355.
34. Danckwardt S, Neu-Yilik G, Thermann R et al. Abnormally spliced beta-globin mRNAs: A single point mutation generates transcripts sensitive and insensitive to nonsense-mediated mRNA decay. Blood 2002; 99:1811-1816.
35. Dreumont N, Maresca A, Boisclair-Lachance JF et al. A minor alternative transcript of the fumarylacetoacetate hydrolase gene produces a protein despite being likely subjected to nonsense-mediated mRNA decay. BMC Mol Biol 2005; 6:1.
36. McAdams HH, Arkin A. It's a noisy business! genetic regulation at the nanomolar scale. Trends Genet 1999; 15:65-69.

37. Jones RB, Wang F, Luo Y et al. The nonsense-mediated decay pathway and mutually exclusive expression of alternatively spliced FGFR2IIIb and -IIIc mRNAs. J Biol Chem 2001; 276:4158-4167.
38. Yeo GW, Van Nostrand E, Holste D et al. Identification and analysis of alternative splicing events conserved in human and mouse. Proc Natil Acad Sci USA 2005; 102:2850.
39. Blencowe BJ. Alternative splicing: New insights from global analyses. Cell 2006; 126:37.
40. Lareau LF, Green RE, Bhatnagar RS et al. The evolving roles of alternative splicing. Curr Opin Struct Biol 2004; 14:273-282.
41. Modrek B, Lee CJ. Alternative splicing in the human, mouse and rat genomes is associated with an increased frequency of exon creation and/or loss. Nat Genet 2003; 34:177-180.
42. Kan Z, States D, Gish W. Selecting for functional alternative splices in ESTs. Genome Res 2002; 12:1837-1845.
43. Neu-Yilik G, Gehring NH, Hentze MW et al. Nonsense-mediated mRNA decay: From vacuum cleaner to swiss army knife. GenomeBiology.com 2004; 5:218.
44. Xing Y, Lee CJ. Negative selection pressure against premature protein truncation is reduced by alternative splicing and diploidy. Trends Genet 2004; 20:472-475.
45. Horikawa Y, Oda N, Cox NJ et al. Genetic variation in the gene encoding calpain-10 is associated with type 2 diabetes mellitus. Nat Genet 2000; 26:163-175.
46. Green RE, Lewis BP, Hillman RT et al. Widespread predicted nonsense-mediated mRNA decay of alternatively-spliced transcripts of human normal and disease genes. Bioinformatics 2003; 19(Suppl 1):118-121.
47. Ule J, Ule A, Spencer J et al. Nova regulates brain-specific splicing to shape the synapse. Nature genetics 2005; 37:844.
48. Yeo G, Holste D, Kreiman G et al. Variation in alternative splicing across human tissues. GenomeBiology.com 2004; 5:R74.
49. Sugnet CW, Srinivasan K, Clark TA et al. Unusual intron conservation near tissue-regulated exons found by splicing microarrays. PLoS Computational Biology 2006; 2:e4.
50. Black DL. Mechanisms of alternative pre-messenger RNA splicing. Annu Rev Biochem 2003; 72:291-336.
51. Wagner EJ, Baraniak AP, Sessions OM et al. Characterization of the intronic splicing silencers flanking FGFR2 exon IIIb. J Biol Chem 2005.
52. Winter J, Lehmann T, Krauss S et al. Regulation of the MID1 protein function is fine-tuned by a complex pattern of alternative splicing. Hum Genet 2004; 114:541-552.
53. Hyvönen MT, Uimari A, Keinänen TA et al. Polyamine-regulated unproductive splicing and translation of spermidine/spermine N1-acetyltransferase. RNA 2006; 12:1569.
54. Philips AV, Timchenko LT, Cooper TA. Disruption of splicing regulated by a CUG-binding protein in myotonic dystrophy. Science 1998; 280:737-741.
55. Brook JD, McCurrach ME, Harley HG et al. Molecular basis of myotonic dystrophy: Expansion of a trinucleotide (CTG) repeat at the 3' end of a transcript encoding a protein kinase family member. Cell 1992; 68:799-808.
56. Liquori CL, Ricker K, Moseley ML et al. Myotonic dystrophy type 2 caused by a CCTG expansion in intron 1 of ZNF9. Science 2001; 293:864-867.
57. Charlet BN, Savkur RS, Singh G et al. Loss of the muscle-specific chloride channel in type 1 myotonic dystrophy due to misregulated alternative splicing. Mol Cell 2002; 10:45-53.
58. Mankodi A, Takahashi MP, Jiang H et al. Expanded CUG repeats trigger aberrant splicing of ClC-1 chloride channel pre-mRNA and hyperexcitability of skeletal muscle in myotonic dystrophy. Mol Cell 2002; 10:35-44.
59. Ho TH, Bundam D, Armstrong DL. Transgenic mice expressing CUG-BP1 reproduce splicing mis-regulation observed in myotonic dystrophy. Human molecular genetics 2005; 14:1539.
60. Ranum LPW, Cooper TA. RNA-mediated neuromuscular disorders. Annual review of neuroscience 2006; 29:259.
61. Ladd AN, Stenberg MG, Swanson MS et al. Dynamic balance between activation and repression regulates pre-mRNA alternative splicing during heart development. Developmental dynamics 2005; 233:783.
62. Vilardell J, Chartrand P, Singer RH et al. The odyssey of a regulated transcript. RNA 2000; 6:1773-1780.
63. He F, Peltz SW, Donahue JL et al. Stabilization and ribosome association of unspliced pre-mRNAs in a yeast upf1- mutant. Proc Natl Acad Sci USA 1993; 90:7034-7038.
64. Mitrovich QM, Anderson P. Unproductively spliced ribosomal protein mRNAs are natural targets of mRNA surveillance in C. elegans. Genes Dev 2000; 14:2173-2184.
65. Valcárcel J, Gebauer F. Post-transcriptional regulation: The dawn of PTB. Curr Biol 1997; 7:R705-8.
66. Ghetti A, Piñol-Roma S, Michael WM et al. hnRNP I, the polypyrimidine tract-binding protein: Distinct nuclear localization and association with hnRNAs. Nucleic Acids Res 1992; 20:3671-3678.

67. Wollerton MC, Gooding C, Wagner EJ et al. Autoregulation of polypyrimidine tract binding protein by alternative splicing leading to nonsense-mediated decay. Mol Cell 2004; 13:91-100.
68. Hamilton BJ, Genin A, Cron RQ et al. Delineation of a novel pathway that regulates CD154 (CD40 ligand) expression. Mol Cell Biol 2003; 23:510-525.
69. Rahman L, Bliskovski V, Reinhold W et al. Alternative splicing of brain-specific PTB defines a tissue-specific isoform pattern that predicts distinct functional roles. Genomics 2002; 80:245-249.
70. Spellman R, Rideau A, Matlin A et al. Regulation of alternative splicing by PTB and associated factors. Biochem Soc Trans 2005; 33:457-460.
71. Duncan PI, Stojdl DF, Marius RM et al. In vivo regulation of alternative pre-mRNA splicing by the Clk1 protein kinase. Mol Cell Biol 1997; 17:5996-6001.
72. Menegay HJ, Myers MP, Moeslein FM et al. Biochemical characterization and localization of the dual specificity kinase CLK1. J Cell Sci 2000; 113(Pt 18):3241-3253.
73. Staiger D, Zecca L, Wieczorek Kirk DA et al. The circadian clock regulated RNA-binding protein AtGRP7 autoregulates its expression by influencing alternative splicing of its own pre-mRNA. Plant J 2003; 33:361-371.
74. Conti E, Izaurralde E. Nonsense-mediated mRNA decay: Molecular insights and mechanistic variations across species. Curr Opin Cell Biol 2005; 17:316-325.
75. Amrani N, Ganesan R, Kervestin S et al. A faux 3'-UTR promotes aberrant termination and triggers nonsense-mediated mRNA decay. Nature 2004; 432:112-118.
76. Gatfield D, Unterholzner L, Ciccarelli FD et al. Nonsense-mediated mRNA decay in drosophila: At the intersection of the yeast and mammalian pathways. EMBO J 2003; 22:3960-3970.
77. Arciga-Reyes L, Wootton L, Kieffer M et al. UPF1 is required for nonsense-mediated mRNA decay (NMD) and RNAi in arabidopsis. The plant journal 2006; 47:480.
78. Lamba JK, Adachi M, Sun D et al. Nonsense mediated decay downregulates conserved alternatively spliced ABCC4 transcripts bearing nonsense codons. Hum Mol Genet 2003; 12:99-109.
79. Lander ES, Linton LM, Birren B et al. Initial sequencing and analysis of the human genome. Nature 2001; 409:860-921.
80. Tennyson CN, Klamut HJ, Worton RG. The human dystrophin gene requires 16 hours to be transcribed and is cotranscriptionally spliced. Nat Genet 1995; 9:184-190.
81. Bingham PM, Chou TB, Mims I et al. On/off regulation of gene expression at the level of splicing. Trends Genet 1988; 4:134-138.
82. Ge H, Zuo P, Manley JL. Primary structure of the human splicing factor ASF reveals similarities with drosophila regulators. Cell 1991; 66:373-382.
83. Hilleren P, Parker R. Mechanisms of mRNA surveillance in eukaryotes. Annu Rev Genet 1999; 33:229-260.
84. Skandalis A, Uribe E. A survey of splice variants of the human hypoxanthine phosphoribosyl transferase and DNA polymerase beta genes: Products of alternative or aberrant splicing? Nucleic Acids Res 2004; 32:6557-6564.
85. Carter MS, Li S, Wilkinson MF. A splicing-dependent regulatory mechanism that detects translation signals. EMBO J 1996; 15:5965-5975.
86. Wang J, Vock VM, Li S et al. A quality control pathway that down-regulates aberrant T-cell receptor (TCR) transcripts by a mechanism requiring UPF2 and translation. J Biol Chem 2002; 277:18489-18493.
87. Hovhannisyan RH, Carstens RP. A novel intronic cis element, ISE/ISS-3, regulates rat fibroblast growth factor receptor 2 splicing through activation of an upstream exon and repression of a downstream exon containing a noncanonical branch point sequence. Mol Cell Biol 2005; 25:250-263.
88. Mitrovich QM, Anderson P. mRNA surveillance of expressed pseudogenes in C. elegans. Current biology 2005; 15:963.
89. Blanchette M, Labourier E, Green RE et al. Genome-wide analysis reveals an unexpected function for the drosophila splicing factor U2AF(50) in the nuclear export of intronless mRNAs. Mol Cell 2004; 14:775-786.
90. Cazalla D, Newton K, Cáceres JF. A novel SR-related protein is required for the second step of pre-mRNA splicing. Mol Cell Biol 2005; 25:2969-2980.
91. Henscheid KL, Shin DS, Cary SC et al. The splicing factor U2AF65 is functionally conserved in the thermotolerant deep-sea worm Alvinella pompejana. Biochim Biophys Acta 2005.
92. Lallena MJ, Chalmers KJ, Llamazares S et al. Splicing regulation at the second catalytic step by sex-lethal involves 3' splice site recognition by SPF45. Cell 2002; 109:285-296.
93. Pacheco TR, Gomes AQ, Barbosa-Morais NL et al. Diversity of vertebrate splicing factor U2AF35: Identification of alternatively spliced U2AF1 mRNAS. J Biol Chem 2004; 279:27039-27049.
94. Confaloni A, Crestini A, Albani D et al. Rat nicastrin gene: CDNA isolation, mRNA variants and expression pattern analysis. Brain Res Mol Brain Res 2005; 136:12-22.

95. Engebrecht JA, Voelkel-Meiman K, Roeder GS. Meiosis-specific RNA splicing in yeast. Cell 1991; 66:1257-1268.
96. Tsyba L, Skrypkina I, Rynditch A et al. Alternative splicing of mammalian intersectin 1: Domain associations and tissue specificities. Genomics 2004; 84:106-113.
97. Petruzzella V, Panelli D, Torraco A et al. Mutations in the NDUFS4 gene of mitochondrial complex I alter stability of the splice variants. FEBS Lett 2005.
98. Screaton GR, Xu XN, Olsen AL et al. LARD: A new lymphoid-specific death domain containing receptor regulated by alternative pre-mRNA splicing. Proc Natl Acad Sci USA 1997; 94:4615-4619.
99. Cuccurese M, Russo G, Russo A et al. Alternative splicing and nonsense-mediated mRNA decay regulate mammalian ribosomal gene expression. Nucleic Acids Res 2005; 33:5965.
100. Le Guiner C, Gesnel MC, Breathnach R. TIA-1 or TIAR is required for DT40 cell viability. J Biol Chem 2003; 278:10465-10476.
101. Colwill K, Pawson T, Andrews B et al. The Clk/Sty protein kinase phosphorylates SR splicing factors and regulates their intranuclear distribution. EMBO J 1996; 15:265-275.
102. Duncan PI, Stojdl DF, Marius RM et al. The Clk2 and Clk3 dual-specificity protein kinases regulate the intranuclear distribution of SR proteins and influence pre-mRNA splicing. Exp Cell Res 1998; 241:300-308.
103. Wilson GM, Sun Y, Sellers J et al. Regulation of AUF1 expression via conserved alternatively spliced elements in the 3' untranslated region. Mol Cell Biol 1999; 19:4056-4064.
104. Wilson GM, Brewer G. The search for trans-acting factors controlling messenger RNA decay. Prog Nucleic Acid Res Mol Biol 1999; 62:257-291.
105. Morrison M, Harris KS, Roth MB. Smg mutants affect the expression of alternatively spliced SR protein mRNAs in caenorhabditis elegans. Proc Natl Acad Sci USA 1997; 94:9782-9785.
106. Stoilov P, Daoud R, Nayler O et al. Human tra2-beta1 autoregulates its protein concentration by influencing alternative splicing of its pre-mRNA. Hum Mol Genet 2004; 13:509-524.
107. Lareau LF, Inada M, Green RE et al. Unproductive splicing of SR genes associated with highly conserved and ultraconserved DNA elements. Nature 2007; 446:926-929.
108. Ni JZ, Grate L, Donahue JP et al. Ultraconserved elements are associated with homeostatic control of splicing regulators by alternative splicing and nonsense-mediated decay. Genes Dev 2007; 21:708-718.
109. Soergel DAW, Lareau LF, Brenner SE. Regulation of gene expression by the coupling of alternative splicing and nonsense-mediated mRNA decay. In: Maquat LE, ed. Nonsense-Mediated mRNA Decay. Goergetown: Landes Bioscience 2006:175-196.

Chapter 13

Alternative Splicing in Disease

James P. Orengo and Thomas A. Cooper*

Abstract

Alternative splicing is a major source of diversity in the human proteome. The regulation of alternative splicing modulates the composition of this diversity to fulfill the physiological requirements of a cell. When control of alternative splicing is disrupted, the result can be a failure to meet cellular and tissue requirements resulting in dysfunction and disease. There are several well-characterized examples in which disruption of alternative splicing is a cause of disease. Investigations into how the mis-regulation of alternative splicing causes disease complements investigations of normal regulatory processes and enhances our understanding of regulatory mechanisms in general. Ultimately, an understanding of how alternative splicing is altered in disease will facilitate strategies directed at reversing or circumventing mis-regulated splicing events.

Introduction

Mutations that cause disease by affecting splicing either disrupt cis-acting splicing signals or the trans-acting components that are required for recognition or modulation of splice site use. Mutations that disrupt cis-acting elements or trans-acting factors have significantly different implications. Cis-acting splicing mutations alter expression of only the mutated allele while mutations that affect splicing factors required for constitutive or regulated splicing have the potential to affect large numbers of genes. Cis-acting mutations most often result in aberrant splicing, which can be defined as the expression of nonnatural mRNAs that are not expressed in normal circumstances. Two thirds of the time, aberrant splicing alters the reading frame of mRNAs such that they are often degraded by nonsense-mediated decay. Therefore, the mechanism by which most cis-acting splicing mutations cause disease is by loss of function of the mutated allele.

Mutations in trans-acting factors that regulate alternative splicing result in expression of natural splice variants, however, these mRNAs are expressed at inappropriate levels or contexts. Such aberrant regulation of splicing causes disease because the inappropriately expressed isoforms cannot fulfill the functional requirements of the cell, or they have a function that is detrimental to normal cell physiology.

Early estimates were that 15% of point mutations that cause disease did so by disrupting splicing.[1] This prediction was made prior to knowledge that exons contain a vast array of cis elements that are required for appropriate exon recognition. More recent estimates based on extensive surveys of mutations within individual genes now suggest that the primary effect of up to 50% of disease-causing mutations is to disrupt splicing.[2] The extent to which splicing mutations cause disease is still under debate as some propose that most disease causing mutations disrupt splicing.[3]

There have been several recent reviews highlighting aberrant splicing.[2,4] Here we describe several examples in which alterations in alternative splicing and its regulation result in disease.

*Corresponding Author: Thomas A. Cooper—Departments of Pathology and Molecular and Cellular Biology, Baylor College of Medicine, Houston, TX 77030. Email: tcooper@bcm.edu

Alternative Splicing in the Postgenomic Era, edited by Benjamin J. Blencowe and Brenton R. Graveley.

Tauopathies: Mutations in the *MAPT* Gene

A large number of disease-causing mutations within the *MAPT* gene encoding tau cause disease by disrupting the balanced expression of mRNA isoforms that contain or lack exon 10. The known causes of *MAPT*-associated mis-splicing disorders described below are due to mutations in cis-acting elements that determine exon 10 inclusion, however, there is evidence that alterations in the functions of trans-acting factors that regulate splicing of exon 10 also play a role in tauopathies.

Tau is a microtubule binding protein that enhances microtubule assembly and controls microtubule dynamics. Tau-mediated regulation of microtubules is particularly important for neuronal processes such as neurite outgrowth and axonal transport. Tau pathology has several components. First, pathological tau accumulates in inclusions or neurofibrillary tangles in a number of diverse neurodegenerative diseases such as Alzheimer's disease, progressive supranuclear palsy (PSP), corticobasal degeneration, argyrophilic grain disease and frontotemporal lobar degeneration.[5] Second, tau is highly phosphorylated at up to 38 sites. Phosphorylation is developmentally regulated with increased phosphorylation in embryonic compared to adult isoforms and pathological tau located in inclusions is the hyper-phosphorylated form.[6] Third, the essential roles of microtubules in neuron function make these cells particularly susceptible to aberrant tau function. Therefore, the mechanisms by which altered tau expression can result in disease include loss of a tau-associated MT function, gain of an aberrant function due to hyperphosphorylation and the accumulation and resulting toxicity of neurofibrillary tangles.

It is still a matter of debate whether the neurofibrillary tangles found in Alzheimer's disease, for example, are a cause or a consequence of pathology. However, the demonstration that mutations in *MAPT* cause another neurodegenerative disorder, frontotemporal dementia with parkinsonism linked with chromosome 17 (FTDP-17), established a clear cause-effect relationship between mutations in the *MAPT* gene and neurodegenerative disease.[7,8] In FTDP-17 there are two pathways to pathogenesis. Approximately half of the mutations that cause disease are missense mutations that directly alter tau function and disrupt microtubule binding capability or increase the propensity for self aggregation. In the other pathway, which is the subject of this section, mutations in alternative exon 10 or adjacent intronic regions disrupt the normal ratio of isoforms generated by alternatively splicing. The total level of tau protein isoforms remain unchanged, only the ratio is altered.

MAPT exon 10 encodes a 31 amino acid segment that encodes the fourth of four microtubule-binding domains. The CNS normally expresses a 1:1 ratio of mRNAs including and excluding exon 10 which encode the 4R and 3R protein isoforms, respectively. FTDP-17 mutations primarily cause over expression of 4R isoforms at a 2-3 to 1 ratio.[8] It is unknown how changes in the 4R/3R ratio result in neurodegeneration. Potential mechanisms include reduced affinity of tau for microtubules resulting in microtubule dysfunction and the aggregation of excess free tau favored by altered ratios of splice variants.[5,9]

Alternative splicing of exon 10 is regulated during CNS development such that the 3R isoforms are expressed in embryonic brain and the 4R isoforms gradually appear after birth. Of the three *MAPT* alternative splicing events only the 3R to 3R + 4R developmental transition is conserved in human, mouse, rat and guinea pig.[10] Exon 10 and its flanking introns contain a myriad of cis-acting elements, reflecting the precise developmental regulation required to maintain the appropriate 4R/3R balance in the adult CNS. In fact, *MAPT* exon 10 represents an excellent example of the features that establish a specific level of basal exon inclusion. Weak 3' and 5' splice sites are balanced by the presence of purine and A/C-rich exonic splicing enhancers (ESEs) and exonic splicing silencers (ESSs). A stem loop structure is proposed to sequester the 5' splice site which promotes exon skipping.[8] All fifteen FTDP-17 mutations that affect exon 10 splicing are located from 10 nucleotides upstream to 29 nucleotides downstream of the exon.[6] Mutations causing FTDP-17 either strengthen the 3' or 5' splice sites, strengthen an ESE, weaken an ESS, or alter the stability of the 5' splice site stem loop. All of these mutations change the 4R/3R ratio toward the 4R isoform.[11,12]

Several proteins (such as p54, Tra2β, SF2/ASF, SRp30c, SRp55 and RBM4) have been shown to affect *MAPT* exon 10 splicing and binding sites for several of these proteins are disrupted by disease-causing mutations.[13-17] Intriguingly, recent results suggest that alterations in the activities of trans-acting factors can also cause mis-expression of 4R/3R ratios. For example, alternative splicing of clk2, which encodes a protein kinase that phosphorylates SR proteins, including Tra2β, is altered by overexpression of CLK2 and Tra2β. Differences in alternative splicing of *MAPT*, *Tra2β* and clk2 in CNS regions most affected in Alzheimer's disease compared to age matched controls could reflect a change in the activities of the splicing regulators and effects on their targets.[18] Barring the possibility of general splicing changes secondary to CNS damage, these results suggest the potential for cross regulation between a *MAPT* splicing regulator and its kinase that could be altered in pathological conditions.

MAPT exon 10 is actively used as a paradigm to develop therapeutic methods to correct aberrant splicing. Two general approaches have been pursued to correct the balance of isoforms without affecting *MAPT* transcription. One approach, pioneered by R. Kole, has been to use modified antisense oligonucleotides complementary to the splice sites flanking the alternative exon.[19] Oligonucleotides complementary to the 3' or 5' splice sites inhibited inclusion of exons containing FTDP-17 mutations in exogenously expressed minigenes of *MAPT* exon 10. The oligonucleotides were 2'-O-methyl modified to increase their half-life and to prevent RNase H-mediated RNA digestion of the DNA-RNA hybrid. Antisense oligonucleotides were shown to induce exon skipping of endogenous *MAPT* exon 10 in rat pheochromocytoma PC12 cells and the effects were specific to exon 10 because splicing of alternative exons 2 and 3 were not affected. Western blot analysis demonstrated a shift to 3R protein isoforms without a change in total tau protein levels. Importantly, inducing the shift to 3R isoforms had a significant effect on microtubule function based on cell morphology and microtubule distribution, compared to cells treated to control oligonucleotides.[20]

Spliceosome-mediated RNA trans-splicing (SMaRT) is a different approach to correct aberrant splicing in which the desired wild type mRNA segment can replace a pathological mRNA segment by trans-splicing.[21] In this strategy, an RNA is exogenously expressed that contains a sequence complementary to an intronic region upstream of the mutation in the pre-mRNA, followed by a strong 3' splice site and an exon containing the wildtype sequence. The endogenous mutant pre-mRNA anneals to the exogenous RNA such that the two RNAs are spliced together producing an mRNA encoding the wild type protein. This is a new application for SMaRT, which previously was used only to target constitutively spliced exons and opens the possibility of using trans-splicing to decrease the pathological 4R/3R ratio in FTDP-17.[22] The general challenge of applying these approaches in patients is in delivering the therapeutic agents efficiently and specifically to the appropriate tissues. A particular challenge for correcting aberrant splicing of *MAPT* exon 10 is to restore and maintain the correct 4R/3R ratio.

Myotonic Dystrophy

Myotonic dystrophy (DM) is an autosomal dominant, multi-systemic disorder with an incidence of approximately 1 in 8000. Individuals affected with DM develop debilitating progressive muscular degeneration, myotonia, cardiac arrhythmias, dilated cardiomyopathy, sleep and neuropsychiatric disorders, dementia, a mild form of type 2 diabetes, defective smooth muscle function leading to digestive motility abnormalities, infertility, testicular atrophy, frontal balding and cataracts.[23] There are two forms of myotonic dystrophy, DM1 and DM2, which exhibit strikingly similar symptoms but are caused by microsatellite expansions in separate and unrelated genes. DM1, which represents the majority of DM cases, is caused by an expanded CTG repeat in the 3' untranslated region (UTR) of the *DMPK* gene.[24-26] The second form, DM2, is caused by an expanded CCTG repeat in intron 1 of the *ZNF9* gene.[27] The expanded alleles causing DM1 or DM2 express transcripts containing the expanded CUG and CCUG repeats, respectively. These transcripts remain trapped in the nucleus were they form foci (Fig. 1). There is strong evidence to indicate that the expanded repeat transcripts and not disruption of *DMPK* or *ZNF9* gene

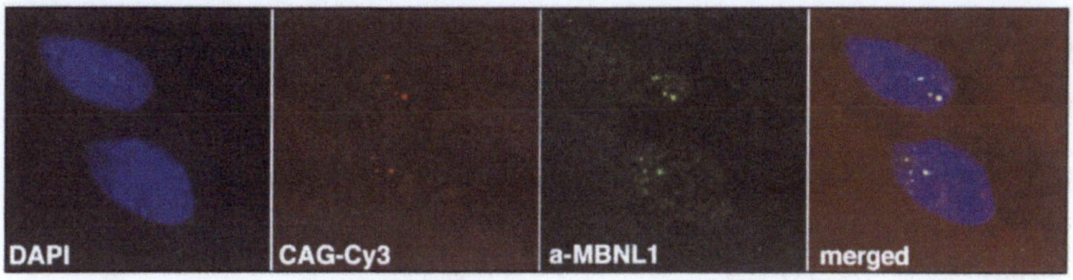

Figure 1. CUG repeat-containing nuclear foci in DM1 fibroblasts. Cultured DM1 fibroblasts stained for CUG repeat containing RNA using fluorescently labeled protein-nucleic acid (PNA) probes and polyclonal antibody against MBNL1.

expression, is responsible for pathogenesis in DM.[28] First, the fact that the expression of unrelated genes containing expanded repeats cause similar symptoms in DM1 and DM2 suggests a common molecular mechanism that is independent of the function of the genes containing the expansion. Second, a transgenic mouse model (HSA^{LR}) expressing a human skeletal alpha actin mRNA containing 250 CTG repeats, specifically in skeletal muscle, recapitulates many features of DM1 including histological abnormalities, foci formation and myotonia.[29] Third, *DMPK* knock-out mice do not exhibit a significant DM1 phenotype.[30,31] Fourth, none of the large number of DM cases that have been characterized thus far are due to mutations that result in *DMPK* or *ZNF9* loss of function; the only known alterations in these genes are the expanded CTG or CCTG repeats. Therefore, the repeat expansions rather than a loss of function of the mutated alleles appear to be the cause of disease. In addition, the consequences of expression of a toxic transcript from a single expanded allele is consistent with the dominant inheritance of the disease.

A major feature of DM is disrupted regulation of alternative splicing. In contrast to diseases caused by mutations that disrupt splicing in cis, the expanded repeat RNA alters the activities of splicing regulatory factors causing a transdominant misregulation of alternative splicing. Twenty-one genes have been found to undergo misregulated alternative splicing in DM heart, skeletal muscle, or CNS (Table 1). All of these alternative splicing events are developmentally regulated in the absence of disease, however in DM, the embryonic splicing pattern is aberrantly expressed in adult tissue.

While the functional consequences of the majority of misregulated splicing events in DM are unknown, two splicing events have cause-effect relationships with DM clinical features. First, in individuals with DM, the embryonic mRNA isoforms encoding the muscle-specific chloride channel (CLCN1) are aberrantly expressed. These isoforms contain premature termination codons, which lead to decreased *CLCN1* mRNA levels secondary to nonsense mediated mRNA decay.[32,33] The resulting loss of chloride conductance causes the myotonia (inability to relax muscle contractions) in individuals with DM.[34] Second, the embryonic isoform of the insulin receptor (*IR*) that is mis-expressed in DM skeletal muscle has decreased signaling efficiency consistent with the insulin resistance frequently observed in individuals with DM.[35,36] It is likely that other misregulated splicing events contribute to different aspects of DM pathogenesis. Of particular interest are splicing events that contribute to the major determinants of morbidity and mortality in affected individuals such as muscle wasting, cardiac dysfunction and CNS abnormalities.

The expression of the CUG- and CCUG-repeat containing RNAs in DM disrupts the regulatory activities of at least two families of splicing factors, MBNL (Muscleblind-like) and CELF (CUG-BP and ETR-3-like factors), which in turn results in many of the splicing abnormalities observed in DM. MBNL and CELF proteins antagonistically regulate the splicing of at least three of the pre-mRNAs that are affected in DM tissues. Significantly, both are also involved in regulating the normal developmental splicing transitions of the same pre-mRNAs.[37-39]

Muscleblind was initially identified as a gene required in *Drosophila* for muscle and photoreceptor cell development.[40,41] *Drosophila* MBNL and the three mammalian homologues (MBNL1, MBNL2 and MBNL3)[42,43] all contain four Cys_3His-type zinc finger domains and have been

Table 1. Alternative splicing changes in DM

Gene	Exon/Intron Pattern in DM	Tissue	Reference
ALP	Exon 5a inclusion	Skeletal muscle Heart	39
	Exon 5b exclusion		89
APP	Exon 7 exclusion	Brain	47
CAPN3	Exon 16 exclusion	Skeletal muscle	39
CLCN1	Intron 2 retention	Skeletal muscle	33
	Exon 7a and 8a inclusion		32
FHOS	Exon 11a exclusion	Skeletal muscle	39
GFAT1	Exon 10 exclusion	Skeletal muscle	39
IR	Exon 11 exclusion	Skeletal muscle	35
KCNAB1	Exon 2 exclusion	Heart	89
MBNL1	Exon 7 inclusion	Skeletal muscle	39
MBNL2	Exon 7 inclusion	Skeletal muscle	39
MTMR1	Exon 2.1 and 2.3 exclusion	Skeletal muscle Heart	90
			38
NMDAR1	Exon 5 inclusion	Brain	47
NRAP	Exon 12 exclusion	Skeletal muscle	39
RYR1	Exon 70 exclusion	Skeletal muscle	91
SERCA1	Exon 22 exclusion	Skeletal muscle	91
			39
SERCA2	Intron 19 retention	Skeletal muscle	91
TAU	Exon 2 and 3 exclusion	Brain	47
	Exon 10 exclusion		92
z-Titin	Exon Zr4, Zr5 inclusion	Skeletal muscle	39
m-Titin	M-line exon 5 inclusion	Skeletal muscle Heart	39
			89
TNNT2	Exon 5 inclusion	Heart	62
TNNT3	Exon F (fetal) inclusion	Skeletal muscle	51
ZASP	Exon 11 inclusion	Skeletal muscle Heart	39
			89

shown to bind CUG-repeat RNAs of greater than 20 repeats.[43] CUG repeats of this length (or longer) fold into an interrupted, A-form-like double-stranded RNA.[44,45] Additionally, the MBNL proteins colocalize with CUG and CCUG repeat nuclear foci in DM cells[42,46] which depletes the nucleoplasmic pool of MBNL.[47,48] In support of the hypothesis that splicing abnormalities in DM are due to a loss of MBNL function, siRNA-mediated depletion of MBNL leads to altered splicing patterns resembling those seen in DM for several target pre-mRNAs.[49,50] In addition, mice lacking the predominant isoforms of MBNL1 (*MBNL1*$^{\Delta 3/\Delta 3}$) display a phenotype similar to DM including cataracts, myotonia, histological changes in skeletal muscle and extensive misregulation of alternative splicing.[39,51] Finally, expression of MBNL1 in skeletal muscle of the HSA^{LR} DM1 mouse model reversed myotonia and several characteristic splicing changes.[52]

Although there is a substantial amount of evidence for the role of MBNL in the splicing abnormalities of DM, a few key observations indicate that loss of MBNL function may not be the only component of DM pathogenesis. First, fluorescence recovery after photobleaching (FRAP) studies show comparable binding and nuclear dynamics of MBNL on foci of expanded CUG or CAG RNA, while splicing abnormalities are specific to cells expressing CUG repeats.[53] Second, both the *HSA*^{LR} and *MBNL1*$^{\Delta 3/\Delta 3}$ mouse models exhibit histological abnormalities in skeletal muscle but not overt muscle degeneration and atrophy, the main cause of death in humans. Third, while the expression of

MBNL1 in *HSA*^{LR} mice reversed splicing abnormalities, it did not reverse the histological abnormalities of *HSA*^{LR} mice suggesting that there is another component to DM pathogenesis.[52]

It is clear that the CELF family of proteins also contribute to DM pathogenesis. This family includes six genes which encode proteins that regulate many posttranscriptional RNA processing steps including alternative splicing, mRNA stability, translation and RNA editing.[54-59] CUG-binding protein 1 (CUG-BP1) was initially identified based on its ability to bind single stranded $(CUG)_8$ repeats and was the first CELF family protein to be identified.[60] In contrast to MBNL, CUG-BP1 does not bind to expanded CUG repeats and does not colocalize to expanded repeat RNA foci in vivo.[61] However, CUG-BP1 protein levels are abnormally elevated in DM tissues and overexpression of CUG-BP1 in cultured cells or in transgenic mice alters the splicing of several target pre-mRNAs in a way that mirrors the patterns seen in DM.[33,35,38,50,62,63] Moreover, CUG-BP1 was shown to be upregulated in a DM mouse model in which a doxycycline-inducible transgene expresses only 5 CUG repeats in the context of the *DMPK* 3'UTR. These mice display cardiac conduction defects, myotonia, histological abnormalities and splicing changes similar to DM patients, which are reversible upon doxycycline removal.[64] Intriguingly, there are no RNA foci in tissues from these mice and presumably they are not affected by sequestration of nuclear MBNL.[64] These results suggest that CUG-BP1 and other members of the CELF family are likely to play a critical role in DM pathogenesis.

Overall, there is evidence that both the loss of MBNL and gain of CELF activities affect pathogenesis in individuals with DM as the disease can be partially phenocopied in different mouse models by either reducing MBNL or increasing CELF activity.[38,51] It is likely that the mechanism of DM pathogenesis also involves alterations in other post transcriptional processing events regulated by the CELF and MBNL proteins[58,59] as well as through other, yet to be identified factors. A great deal of mechanistic information has been learned since the discovery that micro-satellite expansions within noncoding regions can cause disease; however, the connections between the MBNL and CELF protein families and the major pathogenic features of the disease remain to be firmly established.

Facioscapulohumeral Muscular Dystrophy (FSHD)

FSHD is the third most common form of inherited progressive muscular dystrophy exhibiting dominant inheritance and affecting predominantly facial and shoulder girdle muscle by the second decade of life. The mutation causing ~95% of cases was identified in 1992 but the mechanism of pathogenesis has remained a mystery.[65] The majority of FSHD cases are caused by deletion of tandem 3.3 kb repeats, called D4Z4, located on the subtelomeric region of chromosome 4q35. The D4Z4 region is proposed to contain elements that repress transcription and the loss of these elements has been postulated to result in overexpression of one or more of three closely located genes, *FRG1*, *FRG2* and *ANT1*.[66] While still under debate, this model nicely explains dominant inheritance as well as the observations that larger deletions removing more of the tandem repeats correlates with higher expression levels of these genes and increased disease severity. A recent report found that mice overexpressing *FRG1*, but not *FRG2* or *ANT1*, induced a muscular dystrophy that resembles FSHD.[67] FRG1 was previously identified as a component of purified spliceosomes[68] and is found in nuclear regions associated with splicing factors[69] suggesting a role in some aspect of splicing. Interestingly, FRG1 overexpressing mice exhibited defects in the alternative splicing of two genes that are also affected in myotonic dystrophy: *Tnnt3* and *Mtmr1*. Splicing abnormalities in these two genes were also identified in muscle cultures from FSHD patients,[67] however, results of splicing analysis using tissue samples from individuals with FSHD remain to be described. Overexpression of FRG1 in C2C12 mouse muscle cell cultures induced splicing changes in these genes, supporting the contention that the splicing changes were a direct effect of protein overexpression rather than a secondary effect due to muscle degeneration.[67] In addition, alternative splicing of *Tnnt3* and *Mtmr1* was unaffected in muscle cultures from individual with Duchenne's and merosin deficient muscular dystrophies.[67] The relationship between splicing changes induced by FRG1 overexpression and specific features of the disease phenotype remain to be determined. For example, are splicing changes in muscle responsible for muscular dystrophy and, if so, which genes are most critical? In addition, the splicing patterns in FSHD samples do not revert to the embryonic splicing patterns as observed in DM. Therefore, if the FRG1 mice reproduce the mechanism as well as the phenocopy of FSHD, this is likely to represent a disease mechanism that is distinct from DM.

Cancer and Mis-Regulated Splicing

A literature search of splicing alterations associated with cancer pulls out thousands of citations indicating just how widespread such changes are. It is useful to consider three aspects in which altered splicing is relevant to cancer. First is the question of the cause-effect relationships between altered splicing and cellular transformation or metastatic spread. Which splicing changes associated with cancer contribute to the cause or progression of the disease? Second, splicing signatures, in combination with histological and additional molecular markers, can contribute high-resolution information to diagnostic (what is it?) and prognostic (what is the typical outcome?) indications.[70] The value of this approach requires the identification of a splicing signature that consistently correlates with tumor type or grade, prognostic indicators or level of responsiveness to specific treatment regimes. Finally, cell-specific alternative splicing can be utilized in gene therapy approaches for cell-specific therapeutic delivery.

Known cause-effect relationships between specific splicing events and cancer are of particular interest. In general, as for the other disease-associated splicing events described above, splicing abnormalities that are causative of cancer can be attributed to cis-acting mutations within a single gene, or due to a change in a gene encoding a trans-acting splicing factor, which predisposes those cells for progression to a malignant phenotype. Better understood are cis-acting mutations within cancer relevant genes such as tumor suppressors, oncogenes, genes associated with cell adhesion or migration and apoptosis. Such mutations result in altered gene expression and have been identified either as somatic mutations associated with tumorigenesis or germ line mutations associated with syndromes that include a predisposition for tumor formation. Genes within this category have recently been described.[71,72] In addition, the potential roles of alternative splicing in regulating apoptosis has also been reviewed.[73,74]

There is growing evidence for abnormalities in trans-acting factors as a cause of splicing mis-regulation and for these abnormalities contributing to tumorigenicity or metastasis. Splicing factors most commonly associated with cancer belong to the SR protein family (refer to chapter by Lin and Fu). One of the first connections between SR protein expression, alternative splicing and tumorigenicity was established using an elegant mouse model of mammary gland tumorigenesis in which preneoplastic cells were serially transplanted into mammary fat pads and characterized for progression to adenocarcinoma and metastatic spread.[75] Analysis of SR protein expression demonstrated a correlation between neoplasia and induction of SR proteins detectable by the monoclonal antibody mAb104, particularly those in the 30 kDa range. Also found were changes in the alternative splicing of *CD44*, which encodes a cell adhesion molecule and some of these changes have been correlated with metastasis. Conclusions from these studies were that tumors ultimately express a complex mixture of SR proteins compared to preneoplasias and that the major changes in SR protein expression and *CD44* splicing occurred in the transition between preneoplasias and adenocarcinoma rather than between adenocarcinoma and metastasis. Even so, the connection between tumorigenesis and SR protein expression in this model was not straightforward because the expression of SR proteins in preneoplasias was not predictive of tumor development or invasive potential.[75] Thus, a cause-effect relationship between SR protein expression and changes in *CD44* splicing was not established.

Tra2-β1, an SR-related protein, has recently been shown to regulate splicing of *CD44* and has been linked to invasive breast cancer.[76] Overexpression of Tra2-β1 promoted inclusion of *CD44* exons v4 and v5 and this regulation required a C/A-rich ESE within exon 4. Immunohistochemical analysis revealed increased nuclear expression of Tra2-β1 in breast cancer tumors compared to surrounding normal tissue and this correlated with increased inclusion of the *CD44* variable exons. These results confirm the earlier correlations between *CD44* splicing changes and invasive cancer and establish that expression of yet another SR protein family member is increased in these tumors and alters the splicing of *CD44*.

Strong correlations have also been established between expression of the SR protein SF2/ASF, alternative splicing of a target proto-oncogene, ***recepteur d'origine nantais*** (***Ron***) and cellular behavior that is consistent with metastatic potential.[77] Ron is a member of the MET-proto-oncogene family, a tyrosine kinase receptor that binds the ligand macrophage-stimulating protein (MSP).[78] Ron is expressed in epithelial cells and binding of MSP results in Ron autophosphorylation which promotes protein:protein interactions that result in cell proliferation, anti-apoptotic responses, cell dissociation, increased cell mobility and a propensity to invade extracellular space.[79] Ron can induce an epithelial-mesenchymal transition (EMT) which is characteristic of normal embryonic development as well as metastatic progression of tumors.[78]

ΔRon is a constitutively active Ron isoform generated by skipping of exon 11 that is upregulated in 75% of breast cancers and which increases metastatic properties of tumor cells.[77] Using KATOIII gastric carcinoma cells which predominantly express ΔRon, as well as transformed cell lines that predominantly express full length Ron, Biamonti and colleagues demonstrated that regulation of ΔRon expression by the SR protein ASF/SF2 directly affected cell migratory behavior.[77] First, using a minigene containing a portion of *Ron* containing exon 11 and the flanking exons the authors identified an ESE within exon 12 that binds to SF2/ASF and which is required for exon 11 skipping, specifically in KATOIII cells. Interestingly, this ESE functions specifically with SF2/ASF as other SR proteins, such as SRp20 and SRp40, did not stimulate exon skipping. The connection between SF2/ASF expression, induction of the ΔRon isoform and cellular changes indicative of EMT and metastasis was convincingly demonstrated in both SF2/ASF-overexpression and RNAi knockdown experiments (Fig. 2). For example, SF2/ASF overexpression in 293 cells induced ΔRon expression as well as several features of EMT including increased cell motility. SF2/ASF is expressed in higher levels in KATOIII cells compared to 293, consistent with the observed differences in *Ron* splicing. Importantly, siRNA knockdown of SF2/ASF in KATOIII cells caused a switch to predominantly Ron rather than ΔRon and with a corresponding decrease in cell motility. Similarly, siRNA knockdowns of Ron in 293 cells overexpressing SF2/ASF also decreased cell

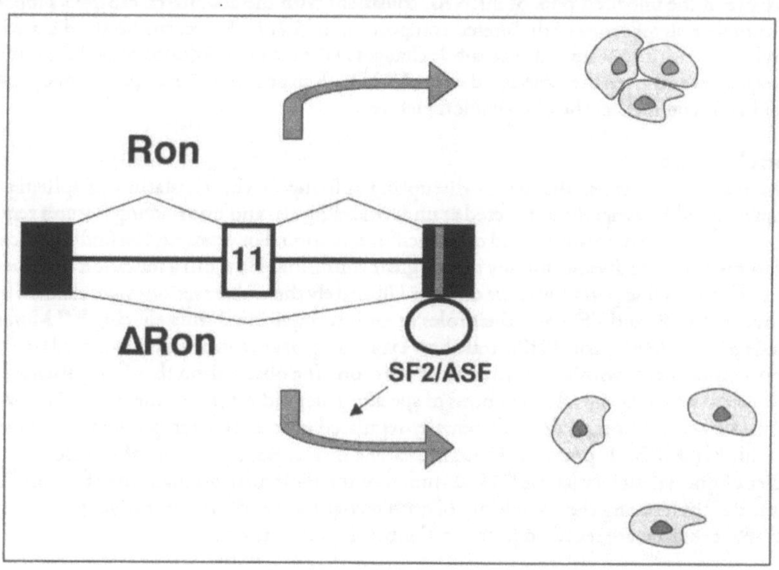

Figure 2. SF2/ASF regulation of *Ron* alternative splicing affects cell migratory behavior.

motility indicating that increased motility in cells expressing SF2/ASF requires and is very likely due to, a switch in splicing to the ΔRon isoform. These results suggest that SF2/ASF and factors that affect SF2/ASF function (e.g., kinases, phosphatases) represent potential therapeutic targets and that cell migratory behavior in culture could serve as a potential assay to screen compounds for such therapeutic potential.

Prader-Willi Syndrome (PWS)

Prader-Willi syndrome is a congenital syndrome characterized by mental retardation, short stature, obesity and behavioral abnormalities.[80] PWS is caused by loss of the paternal copy of a region of 15q11-13 that contains multiple genes including 47 copies of the *HBII-52* snoRNA gene. SnoRNAs typically function as guide RNAs (i.e., they base pair with target RNAs) to direct modifications such as 2'-O-methylation and pseudouridylation in ribosomal RNA, small nuclear RNA and tRNA.[81] No potential target sites were found for HBII-52 in the usual snoRNA targets, but it was noticed that *HBII-52* is complementary to an alternatively spliced exon within the *5-HT$_{2C}$R* gene, which encodes the serotonin receptor. *HBII-52* binds between two alternative 5' splice sites of *5-HT$_{2C}$R* exon V and transient overexpression of *HBII-52* snoRNA resulted in enhanced use of the downstream *5-HT$_{2C}$R 5'* splice site.[82] This results in the expression of a functional 5-HT$_{2C}$R receptor while use of the upstream 5' splice site results in expression of a truncated protein. The snoRNA binding site contains an ESS and it is though that binding of *HBII-52* to the *5-HT$_{2C}$R* pre-mRNA blocks the activity of the ESS, thereby activating the downstream 5' splice site.

The story is complicated by the fact that five adenosines within and upstream of the region in *5-HT$_{2C}$R* that is complementary to *HBII-52* undergo RNA editing events that result in amino acid changes that decrease serotonin responsiveness 10 to 100 fold.[83] RNA editing is carried out by adenosine deaminases that recognize double-stranded RNA structures within the unspliced pre-mRNA. Thus, one possibility was that the *HBII-52* snoRNAs could direct the editing of the *5-HT$_{2C}$R* pre-mRNA. However, *5-HT$_{2C}$R* editing was found not to be affected by the *HBII-52* snoRNA.[82,84] While the ratios of total *5-HT$_{2C}$R* splice variants was not changed in PWS hippocampus samples (which lack *HBII-52* snoRNA), there was decreased usage of the downstream 5' splice site in the unedited pool of mRNAs, consistent with the absence of *HBII-52* snoRNA.[82] The complete significance of the altered composition of *5-HT$_{2C}$R* isoforms to the disease is not entirely clear, but it appears that even subtle changes in the ratios of isoforms could have profound consequences and given the established roles of 5-HT$_{2C}$R in behavior,[85] changes in subpopulations of 5-HT$_{2C}$R could affect the PWS clinical picture.

Conclusions

Mutations that cause disease by disrupting splicing or the regulation of splicing have complemented investigations directed at understanding cis- and trans-acting features required for both basal exon recognition and cell-specific regulation. For example, the finding that silent mutations can cause disease initially caused great confusion. How can a mutation that does not affect protein coding potential cause disease? Ultimately these observations were clarified by the discovery of ESEs and ESSs and their roles in exon recognition during splicing.[86-88] Mutations in and around *MAPT* exon 10 illustrate how exon recognition can be finely balanced by a dense array of cis-elements within the environs of the exon. The observation that DM patients fail to appropriately express the adult isoforms of specific genes indicates that the disease is caused by a subtle defect in a normal developmentally-regulated gene expression program controlled by the CELF and MBNL proteins. There is clearly a bidirectional flow of information between studies of normal molecular events and studies of the molecular mechanisms of disease. These results are illuminating the complexity of exon recognition and the alternative splicing regulatory networks that are required for normal physiological processes.

Acknowledgements

The work in the authors' lab is supported by NIH and the MDA.

References

1. Krawczak M, Reiss J, Cooper DN. The mutational spectrum of single base-pair substitutions in messenger RNA splice junctions of human genes—causes and consequences. Hum Genet 1992; 90:41-54.
2. Cartegni L, Chew SL, Krainer AR. Listening to silence and understanding nonsense: exonic mutations that affect splicing. Nat Rev Genet 2002; 3:285-298.
3. Lopez-Bigas N, Audit B, Ouzounis C et al. Are splicing mutations the most frequent cause of hereditary disease? FEBS Lett 2005; 579:1900-3.
4. Buratti E, Baralle M, Baralle FE. Defective splicing, disease and therapy: searching for master checkpoints in exon definition. Nucleic Acids Res 2006; 34:3494-510.
5. Lee VM, Goedert M, Trojanowski JQ. Neurodegenerative tauopathies. Annu Rev Neurosci 2001; 24:1121-59.
6. Pittman AM, Fung HC, de Silva R. Untangling the tau gene association with neurodegenerative disorders. Hum Mol Genet 15 Suppl 2006; 2:R188-95.
7. Spillantini MG et al. Mutation in the tau gene in familial multiple system tauopathy with presenile dementia. Proc Natl Acad Sci USA 1998; 95:7737-7741.
8. Hutton M et al. Association of missense and 5'-splice-site mutations in tau with the inherited dementia FTDP-17. Nature 1998; 393:702-705.
9. Makrides V et al. Microtubule-dependent oligomerization of tau. Implications for physiological tau function and tauopathies. J Biol Chem 2003; 278:33298-304.
10. Takuma H, Arawaka S, Mori H. Isoforms changes of tau protein during development in various species. Brain Res Dev Brain Res 2003; 142:121-7.
11. Malkani R et al. A MAPT mutation in a regulatory element upstream of exon 10 causes frontotemporal dementia. Neurobiol Dis 2006; 22:401-3.
12. D'Souza I, Schellenberg GD. Regulation of tau isoform expression and dementia. Biochim Biophys Acta 2005; 1739:104-115.
13. Wu JY, Kar A, Kuo D et al. SRp54 (SFRS11), a Regulator for tau Exon 10 Alternative Splicing Identified by an Expression Cloning Strategy. Mol Cell Biol 2006; 26:6739-47.
14. Kar A, Havlioglu N, Tarn WY et al. RBM4 interacts with an intronic element and stimulates tau exon 10 inclusion. J Biol Chem 2006; 281:24479-88.
15. Jiang Z et al. Mutations in tau gene exon 10 associated with FTDP-17 alter the activity of an exonic splicing enhancer to interact with Tra2 beta. J Biol Chem 2003; 278:18997-9007.
16. D'Souza I, Schellenberg GD. Arginine/serine-rich protein interaction domain-dependent modulation of a tau exon 10 splicing enhancer: altered interactions and mechanisms for functionally antagonistic FTDP-17 mutations Delta280K AND N279K. J Biol Chem 2006; 281:2460-9.
17. Wang Y et al. Tau exons 2 and 10, which are misregulated in neurodegenerative diseases, are partly regulated by silencers which bind a SRp30c. SRp55 complex that either recruits or antagonizes htra2beta1. J Biol Chem 2005; 280:14230-9.
18. Glatz DC et al. The alternative splicing of tau exon 10 and its regulatory proteins CLK2 and TRA2-BETA1 changes in sporadic Alzheimer's disease. J Neurochem 2006; 96:635-44.
19. Sazani P, Kole R. Therapeutic potential of antisense oligonucleotides as modulators of alternative splicing. J Clin Invest 2003; 112:481-6.
20. Kalbfuss B, Mabon SA, Misteli T. Correction of alternative splicing of tau in frontotemporal dementia and parkinsonism linked to chromosome 17. J Biol Chem 2001; 276:42986-49293.
21. Puttaraju M, Jamison SF, Mansfield SG et al. Spliceosome-mediated RNA trans-splicing as a tool for gene therapy. Nat Biotechnol 1999; 17:246-52.
22. Chao H et al. Phenotype correction of hemophilia A mice by spliceosome-mediated RNA trans-splicing. Nat Med 2003; 9:015-9.
23. Harper PS. Myotonic Dystrophy In: Warlow CP, Van Gijn J, eds. London: W.B. Saunders, 2001.
24. Mahadevan M et al. Myotonic Dystrophy mutation—An unstable CTG repeat in the 3' untranslated region of the gene. Science 1992; 255:1253-1255.
25. Fu YH et al. An Unstable Triplet Repeat in a Gene Related to Myotonic Muscular Dystrophy. Science 1992; 255:1256-1258.
26. Brook JD et al. Molecular basis of myotonic dystrophy: expansion of a trinucleotide (CTG) repeat at the 3' end of a transcript encoding a protein kinase family member. Cell 1992; 68:799-808.
27. Liquori CL et al. Myotonic dystrophy type 2 caused by a CCTG expansion in intron 1 of ZNF9. Science 2001; 293:864-867.
28. Ranum LP, Cooper TA. RNA-Mediated Neuromuscular Disorders. Annu Rev Neurosci 2006; 29:259-277.
29. Mankodi A et al. Myotonic dystrophy in transgenic mice expressing an expanded CUG repeat. Science 2000; 289:1769-1773.
30. Berul CI, Maguire CT, Gehrmann J et al. Progressive atrioventricular conduction block in a mouse myotonic dystrophy model. J Interv Card Electrophysiol 2000; 4:351-358.

31. Reddy S et al. Mice lacking the myotonic dystrophy protein kinase develop a late onset progressive myopathy. Nature Genet 1996; 13:325-335.
32. Mankodi A et al. Expanded CUG repeats trigger aberrant splicing of ClC-1 chloride channel pre-mRNA and hyperexcitability of skeletal muscle in myotonic dystrophy. Mol Cell 2002; 10:35-44.
33. Charlet-BN et al. Loss of the muscle-specific chloride channel in type 1 myotonic dystrophy due to misregulated alternative splicing. Mol Cell 2002; 10:45-53.
34. Berg J, Jiang H, Thornton CA et al. Truncated ClC-1 mRNA in myotonic dystrophy exerts a dominant-negative effect on the Cl current. Neurology 2004; 63:2371-5.
35. Savkur RS, Philips AV, Cooper TA. Aberrant regulation of insulin receptor alternative splicing is associated with insulin resistance in myotonic dystrophy. Nat Gen 2001; 29:40-47.
36. Savkur RS et al. Insulin receptor splicing alteration in myotonic dystrophy type 2. Am J Hum Genet 2004; 74:1309-1313.
37. Ladd AN, Stenberg MG, Swanson MS et al. Dynamic balance between activation and repression regulates pre-mRNA alternative splicing during heart development. Dev Dyn 2005; 233:783-793.
38. Ho TH, Bundman D, Armstrong DL et al. Transgenic mice expressing CUG-BP1 reproduce splicing mis-regulation observed in myotonic dystrophy. Hum Mol Genet 2005; 14:1539-1547.
39. Lin X et al. Failure of MBNL1-dependent postnatal splicing transitions in myotonic dystrophy. Hum Mol Genet Advanced online publication 2006.
40. Artero R et al. The muscleblind gene participates in the organization of Z-bands and epidermal attachments of Drosophila muscles and is regulated by Dmef2. Dev Biol 1998; 195:131-143.
41. Begemann G et al. Muscleblind, a gene required for photoreceptor differentiation in Drosophila, encodes novel nuclear Cys3His-type zinc-finger-containing proteins. Development 1997; 124:4321-4331.
42. Fardaei M et al. Three proteins, MBNL, MBLL and MBXL, colocalize in vivo with nuclear foci of expanded-repeat transcripts in DM1 and DM2 cells. Hum Mol Genet 2002; 11:805-814.
43. Miller JW et al. Recruitment of human muscleblind proteins to (CUG) n expansions associated with myotonic dystrophy. EMBO J 2000; 19:4439-4448.
44. Mooers BH, Logue JS, Berglund JA. The structural basis of myotonic dystrophy from the crystal structure of CUG repeats. Proc Natl Acad Sci USA 2005; 102:16626-31.
45. Napierala M, Krzyosiak WJ. CUG repeats present in myotonin kinase RNA form metastable "slippery" hairpins. J Biol Chem 1997; 272:31079-31085.
46. Mankodi A et al. Ribonuclear inclusions in skeletal muscle in myotonic dystrophy types 1 and 2. Ann Neurol 2003; 54:760-768.
47. Jiang H, Mankodi A, Swanson MS et al. Myotonic dystrophy type 1 is associated with nuclear foci of mutant RNA, sequestration of muscleblind proteins and deregulated alternative splicing in neurons. Hum Mol Genet 2004; 13:3079-3088.
48. Cardani R, Mancinelli E, Rotondo G et al. Muscleblind-like protein 1 nuclear sequestration is a molecular pathology marker of DM1 and DM2. Eur J Histochem 2006; 50:177-182.
49. Ho TH et al. Muscleblind proteins regulate alternative splicing. EMBO J 2004; 23:3103-3112.
50. Dansithong W, Paul S, Comai L et al. MBNL1 is the primary determinant of focus formation and aberrant insulin receptor splicing in DM1. J Biol Chem 2005; 280:5773-5780.
51. Kanadia RN et al. A muscleblind knockout model for myotonic dystrophy. Science 2003; 302:1978-1980.
52. Kanadia RN et al. Reversal of RNA missplicing and myotonia after muscleblind overexpression in a mouse poly(CUG) model for myotonic dystrophy. Proc Natl Acad Sci USA 2006; 24:24.
53. Ho TH et al. Colocalization of muscleblind with RNA foci is separable from mis-regulation of alternative splicing in myotonic dystrophy. J Cell Sci 2005; 118:2923-2933.
54. Ladd AN, Charlet-BN, Cooper TA. The CELF family of RNA binding proteins is implicated in cell-specific and developmentally regulated alternative splicing. Mol Cell Biol 2001; 21:1285-1296 (2001).
55. Ladd AN, Nguyen HN, Malhotra K et al. CELF6, a member of the CELF family of RNA binding proteins, regulates MSE-dependent alternative splicing. J Biol Chem 2004; 279:17756-17764.
56. Anant S et al. Novel role for RNA-binding protein CUGBP2 in mammalian RNA editing. CUGBP2 modulates C to U editing of apolipoprotein B mRNA by interacting with apobec-1 and ACF, the apobec-1 complementation factor. J Biol Chem 2001; 276:47338-47351.
57. Mukhopadhyay D, Houchen CW, Kennedy S et al. Coupled mRNA stabilization and translational silencing of cyclooxygenase-2 by a novel RNA binding protein, CUGBP2. Mol Cell 2003; 11:113-126.
58. Timchenko NA, Welm AL, Lu X et al. CUG repeat binding protein (CUGBP1) interacts with the 5' region of C/EBPbeta mRNA and regulates translation of C/EBPbeta isoforms. Nucleic Acids Res 1999; 27:4517-4525.
59. Adereth Y, Dammai V, Kose N et al. RNA-dependent integrin alpha3 protein localization regulated by the Muscleblind-like protein MLP1. Nat Cell Biol 2005; 7:1240-7.
60. Timchenko LT et al. Identification of a (CUG)n triplet repeat RNA-binding protein and its expression in myotonic dystrophy. Nucl Acids Res 1996; 24:4407-4414.

61. Michalowski S et al. Visualization of double-stranded RNAs from the myotonic dystrophy protein kinase gene and interactions with CUG-binding protein. Nucleic Acids Res 1999; 27:3534-3542.
62. Philips AV, Timchenko LT, Cooper TA. Disruption of splicing regulated by a CUG-binding protein in myotonic dystrophy. Science 1998; 280:737-741.
63. Timchenko NA et al. RNA CUG repeats sequester CUGBP1 and alter protein levels and activity of CUGBP1. J Biol Chem 2001; 276:7820-7826.
64. Mahadevan MS et al. Reversible model of RNA toxicity and cardiac conduction defects in myotonic dystrophy. Nat Genet 2006; 38:1066-1070.
65. Tawil R, Van Der Maarel SM. Facioscapulohumeral muscular dystrophy. Muscle Nerve 2006; 34:1-15.
66. Gabellini D, Green MR, Tupler R. Inappropriate gene activation in FSHD: a repressor complex binds a chromosomal repeat deleted in dystrophic muscle. Cell 2002; 110:339-48.
67. Gabellini D et al. Facioscapulohumeral muscular dystrophy in mice overexpressing FRG1. Nature 2006; 439:973-7.
68. Rappsilber J, Ryder U, Lamond AI et al. Large-scale proteomic analysis of the human spliceosome. Genome Res 2002; 12:1231-45.
69. van Koningsbruggen S et al. FRG1P is localised in the nucleolus, Cajal bodies and speckles. J Med Genet 2004; 41:e46.
70. Zhang C et al. Profiling alternatively spliced mRNA isoforms for prostate cancer classification. BMC Bioinformatics 2006; 7:202.
71. Srebrow A, Kornblihtt AR. The connection between splicing and cancer. J Cell Sci 2006; 119:2635-41.
72. Venables JP. Aberrant and alternative splicing in cancer. Cancer Res 2004; 64:7647-54.
73. Schwerk C, Schulze-Osthoff K. Regulation of apoptosis by alternative pre-mRNA splicing. Mol Cell 2005; 19:1-13.
74. Wu JY, Tang H, Havlioglu N. Alternative pre-mRNA splicing and regulation of programmed cell death. Prog Mol Subcell Biol 2003; 31:153-185.
75. Stickeler E, Kittrell F, Medina D et al. Stage-specific changes in SR splicing factors and alternative splicing in mammary tumorigenesis. Oncogene 1999; 18:3574-3582.
76. Watermann DO et al. Splicing factor Tra2-beta1 is specifically induced in breast cancer and regulates alternative splicing of the CD44 gene. Cancer Res 2006; 66:4774-80.
77. Ghigna C et al. Cell motility is controlled by SF2/ASF through alternative splicing of the Ron protooncogene. Mol Cell 2005; 20:881-90.
78. Wang MH, Wang D, Chen YQ. Oncogenic and invasive potentials of human macrophage-stimulating protein receptor, the RON receptor tyrosine kinase. Carcinogenesis 2003; 24:1291-300.
79. Trusolino L, Comoglio PM. Scatter-factor and semaphorin receptors: cell signalling for invasive growth. Nat Rev Cancer 2002; 2:289-300.
80. Goldstone AP. Prader-Willi syndrome: advances in genetics, pathophysiology and treatment. Trends Endocrinol Metab 2004; 15:12-20.
81. Kiss T. Small nucleolar RNAs: an abundant group of noncoding RNAs with diverse cellular functions. Cell 2002; 109:145-8.
82. Kishore S, Stamm S. The snoRNA HBII-52 regulates alternative splicing of the serotonin receptor 2C. Science 2006; 311:230-2.
83. Wang Q et al. Altered G protein-coupling functions of RNA editing isoform and splicing variant serotonin2C receptors. J Neurochem 2000; 74:1290-300.
84. Niswender CM, Sanders-Bush E, Emeson RB. Identification and characterization of RNA editing events within the 5-HT2C receptor. Ann NY Acad Sci 1998; 861:38-48.
85. Tohda M, Nomura M, Nomura Y. Molecular pathopharmacology of 5-HT2C receptors and the RNA editing in the brain. J Pharmacol Sci 2006; 100:427-32.
86. Cooper TA, Ordahl CP. Nucleotide substitutions within the cardiac troponin T alternative exon disrupt pre-mRNA alternative splicing. Nuc Acids Res 1989; 17:7905-7921.
87. Hampson RK, Follette LL, Rottman FM. Alternative processing of bovine growth hormone mRNA is influenced by downstrean exon sequence. Mol Cell Biol 1989; 9:1604-1610.
88. Mardon HJ, Sebastio G, Baralle FE. A role for exon sequence in alternative splicing of the human fibronectin gene. Nucl Acids Res 1987; 15:7725-7733.
89. Mankodi A, Lin X, Blaxall BC et al. Nuclear RNA foci in the heart in myotonic dystrophy. Circ Res 2005; 97:1152-5.
90. Buj-Bello A et al. Muscle-specific alternative splicing of myotubularin-related 1 gene is impaired in DM1 muscle cells. Hum Mol Genet 2002; 11:2297-2307.
91. Kimura T et al. Altered mRNA splicing of the skeletal muscle ryanodine receptor and sarcoplasmic/endoplasmic reticulum Ca2+-ATPase in myotonic dystrophy type 1. Hum Mol Genet 2005; 14:2189-2200.
92. Sergeant N et al. Dysregulation of human brain microtubule-associated tau mRNA maturation in myotonic dystrophy type 1. Hum Mol Genet 2001; 10:2143-2155.

INDEX

P

R

S

T

U

MIX
Papier aus verantwortungsvollen Quellen
Paper from responsible sources
FSC® C105338

If you have any concerns about our products,
you can contact us on
ProductSafety@springernature.com

In case Publisher is established outside the EU,
the EU authorized representative is:
Springer Nature Customer Service Center GmbH
Europaplatz 3, 69115 Heidelberg, Germany

Printed by Libri Plureos GmbH
in Hamburg, Germany